Mechatronic Systems

Devices, Design, Control, Operation and Monitoring

Mechanical Engineering Series
Frank Kreith & Roop Mahajan - Series Editors

Published Titles

Mechatronic Systems

Devices, Design, Control, Operation and Monitoring

Edited by

Clarence W. de Silva

CRC Press
Taylor & Francis Group
Boca Raton   London   New York

CRC Press is an imprint of the
Taylor & Francis Group, an **informa** business

CRC Press
Taylor & Francis Group
6000 Broken Sound Parkway NW, Suite 300
Boca Raton, FL 33487-2742

Library of Congress Cataloging-in-Publication Data

Mechatronic systems : devices, design, control, operation and monitoring / Clarence W. de Silva, ed.
    p. cm. -- (Mechanical engineering series)
  Includes bibliographical references and index.
  ISBN 978-0-8493-0775-1 (alk. paper)
  1. Mechatronics. I. Silva, Clarence W. de. II. Title. III. Series.

TJ163.12.M4113 2008
621--dc22
                                                       2007017632

Visit the Taylor & Francis Web site at
http://www.taylorandfrancis.com

and the CRC Press Web site at
http://www.crcpress.com

Dedication

To my friends at the National University of Singapore (alphabetically):
Associate Professor Marcello Ang; Professor Ben M. Chen; Professor Tong Heng Lee;
Professor Jim A.N. Poo; and Associate Professor Kok Kiong Tan.

For valuable support and professional collaboration.

Table of Contents

Section III   Control Technologies

Section IV   Mechatronic Design and Optimization

Section V Monitoring and Diagnosis

Foreword

The multidisciplinary field of mechatronics brings together mechanical engineering, electrical and electronic engineering, control engineering, and computer science in a synergistic manner. In recent years, this field has advanced rapidly and gained maturity, through the development of an increasing number of degree programs, extensive research activity, product and system developments, and an increasingly broad range of industrial applications. Above all, as is the case at my own university, both undergraduate- and graduate-level degree programs are gaining acceptance and are in great demand worldwide. With this background, I am honored to have been invited to write this foreword for the book *Mechatronic Systems—Devices, Design, Control, Operation, and Monitoring*, edited by Professor Clarence de Silva. This volume brings together distinguished scholars and researchers from several disciplines, institutions, and countries, with the intent of advancing technical knowledge in the theory, design, development, and application in the field of mechatronics.

The origins of this book may be traced back to a special partnership between the National University of Singapore (NUS) and the University of British Columbia (UBC) through the NUS-UBC Applied Science Research Centre. The centre was established in August 2004 as a research partnership between the Faculty of Applied Science at UBC and our colleagues at NUS. The centre (see www.researchcentre. apsc.ubc.ca) was formed with the primary goals of initiating, encouraging, facilitating, and fostering research collaborations between NUS and UBC in the areas of engineering and applied science. In October 2005, the Centre was the primary sponsor of the *International Symposium on Collaborative Research in Applied Science* that was held at UBC, with Professor de Silva as the symposium chair. The symposium was well attended, generated much enthusiasm, and sparked lively debate. Most papers of the symposium were in the area of mechatronics. A selected group of authors among the symposium's participants and other prominent researchers in mechatronics were invited to contribute chapters to the present book.

As dean of the Faculty of Applied Science at UBC, I am proud of my university's association with the publication of this valuable book. I have no doubt that it will lead to further insights, new research, enhanced collaborations, and increased practical applications in the field of mechatronics, and will thereby prove to be of significant benefit to a broad range of researchers, institutions, and industries, and thus ultimately to society at large.

Professor Michael Isaacson, Ph.D., P.Eng.
Dean, Faculty of Applied Science
The University of British Columbia

Preface

With individual chapters authored by distinguished professionals in their respective topics, this book provides for engineers, designers, researchers, educators, and students, a convenient and up-to-date reference with information on mechatronic devices and systems, including technologies and methodologies for their analysis, design, control, monitoring, and diagnosis. The book consists of 29 chapters, grouped into 5 sections: mechatronic devices, communication technologies, control technologies, mechatronic design and optimization, and monitoring and diagnosis. Cross-referencing is used throughout to indicate other places in the book where further information on a particular topic is provided.

In the book, equal emphasis is given to theory and practical application. The chapters are grouped into those covering mechatronic devices and applications, linking and communication within and outside a mechatronic systems, control of mechatronic systems, design of mechatronic devices and systems, and monitoring and fault diagnosis of mechatronic systems. Analytical formulations, numerical methods, design approaches, control techniques, and commercial tools are presented and illustrated using examples and case studies. Practical implementations, field applications, and experimentation are described and demonstrated.

Mechatronics concerns synergistic and concurrent use of mechanics, electronics, computer engineering, and intelligent control systems in modeling, analyzing, designing, developing, and implementing smart electromechanical products. As the modern machinery and electromechanical devices are typically being controlled using analog and digital electronics and computers, the technologies of mechanical engineering in such a system can no longer be isolated from those of electronic and computer engineering. For example, in a robot systems or a micro-machine, mechanical components are integrated with analog and digital electronic components to provide single functional units or products. Similarly, devices with embedded and integrated sensing, actuation, signal processing, and control have many practical advantages. In the framework of mechatronics, a unified approach is taken to integrate different types of components and functions, both mechanical and electrical, in modeling, analysis, design, and implementation, with the objective of harmonious operation that meets a desired set of performance specifications.

In the mechatronic approach, a multidomain (mixed) system consisting of subsystems that have primarily mechanical (including fluid and thermal) or primarily electrical (including electronic) character is treated using integrated engineering concepts. In particular, electromechanical analogies, consistent energy transfer (e.g., kinetic, potential, thermal, fluid, electrostatic, and electromagnetic energies) through energy ports and integrated design methodologies may be incorporated, resulting in benefits with regard to performance, efficiency, reliability, and cost.

Mechatronics has emerged as a bona fide field of practice, research, and development, and simultaneously as an academic discipline in engineering. The present book is geared toward the focus on integrated education and practice as related to electromechanical and multidomain systems. In view of the analytical methods, practical considerations, design issues, and experimental techniques that are presented throughout the book, it serves as a useful reference tool and an extensive information source for engineers in industry and laboratories, researchers, and students in the field of mechatronics.

Clarence W. de Silva
Vancouver

Acknowledgments

I wish to express my gratitude to the authors of the chapters for their valuable and highly professional contributions. The assistance of my research associate and lab manager, Ying Wang, in managing the manuscript production was quite valuable. I am very grateful to Michael Slaughter, executive editor–engineering, CRC Press, for his enthusiasm and support throughout the project. The editorial and production staff at CRC Press and its affiliates, particularly Jim McGovern, Robin Lafazan, and John Lavender have done an excellent job in getting the book out in print. It is with much sadness that I note here the tragic loss of Liz Spangenberger of CRC Press, who had eagerly helped me with the publication of many of my books. Finally, I wish to lovingly acknowledge the patience and understanding of my family.

The Editor

Clarence W. de Silva, P.Eng., Fellow ASME and Fellow IEEE, is Professor of Mechanical Engineering, University of British Columbia, Vancouver, Canada, and has occupied the NSERC Research Chair in Industrial Automation since 1988. Prior to that he served as a faculty member at Carnegie Mellon University (1978–1987) and as a Fulbright Visiting Professor at the University of Cambridge (1987–1988). De Silva has earned Ph.D. degrees from the Massachusetts Institute of Technology (1978) and Cambridge University, England (1998). De Silva has also occupied the Mobil Endowed Chair Professorship in the Department of Electrical and Computer Engineering at the National University of Singapore (2000). He has served as a consultant to several companies including IBM and Westinghouse in the United States, and has led the development of six industrial machines. He is recipient of the Henry M. Paynter Outstanding Investigator Award and Yasundo Takahashi Education Award of the Dynamic Systems and Control Division of the American Society of Mechanical Engineers (ASME); Killam Research Prize; Outstanding Engineering Educator Award of IEEE Canada; Lifetime Achievement Award of the World Automation Congress; IEEE Third Millennium Medal; Meritorious Achievement Award of the Association of Professional Engineers of British Columbia; and the Outstanding Contribution Award of the Systems, Man, and Cybernetics Society of the Institute of Electrical and Electronics Engineers (IEEE). He has authored 16 technical books including *Sensors and Actuators—Control System Instrumentation* (Taylor & Francis/CRC Press, 2007); *Vibration: Fundamentals and Practice*, Second Edition (Taylor & Francis/CRC Press, 2006); *Mechatronics—An Integrated Approach* (Taylor & Francis/CRC Press, 2005); *Soft Computing and Intelligent Systems Design—Theory, Tools, and Applications* (with F. Karry, Addison-Wesley, 2004); *Intelligent Control: Fuzzy Logic Applications* (CRC Press, 1995); *Control Sensors and Actuators* (Prentice Hall, 1989); over 170 journal papers; and a similar number of conference papers and book chapters. He has served as editor of twelve books and on the editorial boards of twelve international journals. He is the Editor-in-Chief, of *International Journal of Control and Intelligent Systems*; and was Editor-in-Chief, *International Journal of Knowledge-Based Intelligent Engineering Systems*; Senior Technical Editor, *Measurements and Control*; and Regional Editor, North America, *Engineering Applications of Artificial Intelligence—the International Journal of Intelligent Real-Time Automation*. He is a Lilly Fellow, NASA-ASEE Fellow, Senior Fulbright Fellow at Cambridge University, Fellow of the Advanced Systems Institute of British Columbia, Killam Fellow, and Fellow of the Canadian Academy of Engineering.

Contributors

S. Adibi
University of Waterloo
Waterloo, Ontario
Canada

M.R. Alrasheed
University of British Columbia
Vancouver, British Columbia
Canada

Y. Altintas
University of British Columbia
Vancouver, British Columbia
Canada

M.H. Ang, Jr.
National University of Singapore
Singapore

S. Behbahani
University of British Columbia
Vancouver, British Columbia
Canada

A.E. Brockwell
Carnegie Mellon University
Pittsburgh, Pennsylvania

H.S.O. Chan
National University of Singapore
Singapore

Z. Chen
Ningbo University
Ningbo, Zhejiang
China

C.M. Chew
National University of Singapore
Singapore

M. Chiao
University of British Columbia
Vancouver, British Columbia
Canada

C.W. de Silva
University of British Columbia
Vancouver, British Columbia
Canada

R. Du
The Chinese University
    of Hong Kong
Hong Kong

K. Erkorkmaz
University of Waterloo
Waterloo, Ontario
Canada

T. Fan
The University of British Columbia
Vancouver, British Columbia
Canada

Z. Feng
Ningbo University
Ningbo, Zhejiang
China

Y. Fu
The Chinese University
    of Hong Kong
Hong Kong

M.S. Gadala
University of British Columbia
Vancouver, British Columbia
Canada

Z. Guo
Beijing University of Aeronautics
    and Astronautics
Beijing, China

G.S. Hong
National University of Singapore
Singapore

S. Huang
National University of Singapore
Singapore

N.W. Koh
National University of Singapore
Singapore

T.H. Lee
National University of Singapore
Singapore

V.C.M. Leung
University of British Columbia
Vancouver, British Columbia
Canada

H. Li
University of Guelph
Guelph, Ontario
Canada

Y. Li
Beijing University of Aeronautics
    and Astronautics
Beijing, China

S.Y. Lim
Singapore Institute of
    Manufacturing Technology
Singapore

M. Mallakzadeh
University of British Columbia
Vancouver, British Columbia
Canada

J. Mao
Beijing University of Aeronautics
and Astronautics
Beijing, China

H. Marzi
St. Francis Xavier University
Antigonish, Nova Scotia
Canada

T. Nanayakkara
University of Moratuwa
Sri Lanka

K.S. Neo
National University of Singapore
Singapore

H.Y.T. Ngan
University of Hong Kong
Hong Kong

H. Onozuka
National University
of Singapore
Singapore

G.K.H. Pang
University of Hong Kong
Hong Kong

F. Pirmoradi
University of British Columbia
Vancouver, British Columbia
Canada

L. Piyathilaka
Rinzen Laboratories
Sri Lanka

A.N. Poo
National University
of Singapore
Singapore

A.S. Putra
National University of Singapore
Singapore

M. Rahman
National University of Singapore
Singapore

F. Sassani
University of British Columbia
Vancouver, British Columbia
Canada

T. Siu
University of British Columbia
Vancouver, British Columbia
Canada

A. Subasinghe
Rinzen Laboratories
Sri Lanka

J. Sun
National University of Singapore
Singapore

K.K. Tan
National University of Singapore
Singapore

C.S. Teo
National University of Singapore
Singapore

M. Velliste
University of Pittsburgh
Pittsburgh, Pennsylvania

W.H. Wang
National University of Singapore
Singapore

Y. Wang
University of British Columbia
Vancouver, British Columbia
Canada

Z.G. Wang
National University
of Singapore
Singapore

V.W.S. Wong
University of British Columbia
Vancouver, British Columbia
Canada

Y.S. Wong
National University
of Singapore
Singapore

H. Xia
National University
of Singapore
Singapore

L. Yang
National University
of Singapore
Singapore

S.X. Yang
University of Guelph
Guelph, Ontario
Canada

C.H. Yeung
University of British Columbia
Vancouver, British Columbia
Canada

Y. Yue
Beijing University of Aeronautics
and Astronautics
Beijing, China

1

Technology Needs for Mechatronic Systems

C.W. de Silva

Summary

Mechatronics is a multidisciplinary engineering field which involves a synergistic integration of several areas such as mechanical engineering, electrical and electronic engineering, control engineering, and computer engineering. In this chapter the field of mechatronics is introduced, the technology needs for such systems are indicated, and some important issues in the design and development of a mechatronic product or system are highlighted. Technologies of sensing, actuation, signal conditioning, interfacing, communication, and control are particularly important for mechatronic systems. Intelligent mechatronic systems require further technologies for representation and processing of knowledge and intelligence, and particularly those technologies that impart intelligent characteristics to the system. The design and development of a mechatronic system will require an integrated and concurrent approach to deal with the subsystems and subprocesses of a multidomain (mixed) system. The subsystems of a mechatronic system should not be designed or developed independently without addressing the system integration, subsystem interactions and matching, and the intended operation of the overall system. Such an integrated and concurrent approach will make a mechatronic design more optimal than a conventional design. In this chapter, some important issues in the design and development of a mechatronic product or system are highlighted.

1.1 Introduction

The subject of mechatronics concerns the synergistic application of mechanics, electronics, controls, and computer engineering in the development of electromechanical products and systems, through an integrated design approach [1]. A mechatronic system will require a multidisciplinary approach for its design, development, and implementation. In the traditional development of an electromechanical system, the mechanical components and electrical components are designed or selected separately and then integrated, possibly with other components, and hardware and software. In contrast, in the mechatronic approach, the entire electromechanical system is treated concurrently in an integrated manner by a multidisciplinary team of engineers and other professionals. Naturally, a system formed by interconnecting a set of independently designed and manufactured components will not provide the same level of performance as a mechatronic system, which employs an integrated and concurrent approach for design, development, and implementation [2]. The main reason is straightforward. The best match and compatibility between component functions can be achieved through an integrated and unified approach to design and development, and best operation is possible through an integrated implementation. Generally, a mechatronic product will be more efficient and cost effective, more precise and accurate, more reliable, more flexible and functional, and less mechanically complex, compared to a nonmechatronic product that needs a similar level of effort in its development. Performance of a nonmechatronic system can be improved through sophisticated control, but this is achieved at an additional cost of sensors, instrumentation, and control hardware and software, and with added complexity and control burden. Mechatronic products and systems include modern automobiles and aircraft, smart household appliances, medical robots, space vehicles, and office automation devices. In this chapter, some important issues in the design and development of a mechatronic product or system will be highlighted and the associated technology needs will be indicated.

1.2   Mechatronic System Technologies

A typical mechatronic system consists of a mechanical skeleton, actuators, sensors, controllers, signal conditioning/modification devices, computer/digital hardware and software, interface devices, and power sources. Different types of sensing, and information acquisition and transfer are involved among all these various types of components. For example, a servomotor, which is a motor with the capability of sensory feedback for accurate generation of complex motions, consists of mechanical, electrical, and electronic components (see Figure 1.1). The main mechanical components are the rotor and the stator. The electrical components include the circuitry for the field windings and rotor windings (not in the case of permanent-magnet rotors, as shown in Figure 1.1), and circuitry for power transmission and commutation (if needed). Electronic components include those needed for sensing, (e.g., optical encoder for displacement and speed sensing, and tachometer for speed sensing) [3].

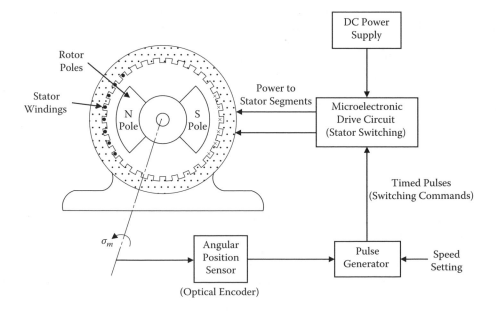

FIGURE 1.1 Brushless dc servomotor is a mechatronic device.

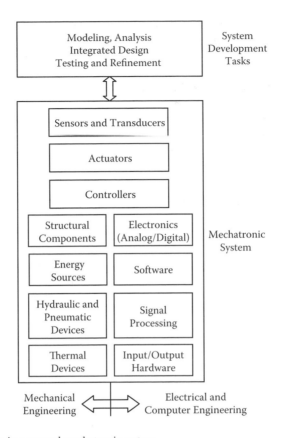

FIGURE 1.2 Technologies in a general mechatronic system.

Technology issues and needs of a general mechatronic system are indicated in Figure 1.2. It is seen that they span the traditional fields of mechanical engineering, electrical and electronic engineering, control engineering, and computer engineering. Each aspect or issue within the system may take a multidomain character. For example, as noted before, an actuator (e.g., DC servomotor) itself may represent a mechatronic device within a larger mechatronic system such as an automobile or a robot.

A mechatronic system may be treated as a control system, consisting of a plant (which is the process, machine, device, or system to be controlled), actuators, sensors, interfacing and communication structures, signal modification devices, and controllers and compensators. The function of the mechatronic system is primarily centered at the plant. Actuators, sensors, and signal modification devices might be integral with the plant itself, or might be needed as components that are external to the plant, for proper operation of the overall mechatronic system. The controller is an essential part of a mechatronic system. It generates control signals to the actuators in order to operate (drive) the plant in a desired manner. Sensed signals might be used for system monitoring and feedforward control, in addition to feedback control. These various components may not be present as physically separate and autonomous units in a mechatronic system in general, even though they may be separately identified from a functional point of view. For example, an actuator and a sensor might be an integral part of the plant itself.

As an example, consider a robotic manipulator. The joint motors are usually considered as a part of the manipulator because, from the perspective of robot dynamics, it is virtually impossible to uncouple the actuators from the main structure of the robot. Specifically, the torque transmitted to a manipulator link will depend on the (magnetic) torque of the motor at that joint as well as on the motor speed. Furthermore, magnetic torque will depend on the back e.m.f. in the rotor, which in turn will be determined by the motor speed. The transmitted torque will determine the link motion (displacement, speed, and acceleration), which is directly related to the motor motion (say, through a gear ratio). In the presence of such dynamic coupling, it is not proper to treat the actuator as a component external to the plant. However, the sensors (e.g., tachometers and encoders) at the joints can be treated separately from the plant because their dynamic coupling with the manipulator structure is usually negligible.

For a mechatronic system, design technologies are as important as the instrumentation technologies previously mentioned. In fact, in some situation, the design of a mechatronic system may be interpreted as the process of integrating (physical/functional) components such as actuators, sensors, signal modification devices, interfacing and communicating structures, and controllers with a plant so that the plant in the overall mechatronic system will respond to inputs (or commands) in a desired manner. From this point of view, design is an essential procedure in the instrumentation of a mechatronic system. The instrumentation will include the design of a component structure (including addition and removal of components and interconnecting them into various structural forms and locations), selection of components (giving consideration to types, ratings, and capacities), interfacing various components (perhaps through signal modification devices, properly considering impedances, signal types and signal levels), adding controllers and compensators (including the selection of a control structure), implementing control algorithms, and tuning (selecting and adjusting the unknown parameters of) the overall mechatronic system. Many of these instrumentation tasks are also design tasks.

In a true mechatronic sense, the design of a multidomain multicomponent system of the nature identified in Figure 1.2 will require simultaneous consideration and integrated design of all its components. Such an integrated and "concurrent" design will call for a fresh look at the design process itself, and also a formal consideration of information and energy transfer between components within the system. It is expected that the mechatronic approach will result in higher quality of products and services, improved performance, and increased reliability, approaching some form of optimality. This will enable the development and production of electromechanical systems efficiently, rapidly, and economically. Relevant technologies for mechatronic engineering should concern all stages of design, development, integration, instrumentation, control, testing, operation, and maintenance of a mechatronic system.

When performing an integrated design of a mechatronic system, the concepts of energy/power present a unifying thread. The reasons are clear. First, in an electromechanical system, ports of power/energy exist, which link electrical dynamics and mechanical dynamics. Hence, modeling, analysis, and optimization of a mechatronic system can be carried out using a hybrid-system (or mixed-system, or multi-domain system) formulation (a model) that integrates mechanical aspects and electrical aspects of the system. Second, an optimal design will aim for minimal energy dissipation and maximum energy efficiency. There are related implications; for example, greater dissipation of energy will mean reduced overall efficiency and increased thermal problems, noise, vibration, malfunctions, and wear and tear. Again, a hybrid model that presents an accurate picture of energy/power flow within the system will provide an appropriate framework for the mechatronic design.

By definition, a mechatronic design should result in an optimal final product. In particular, a mechatronic design as a result of its unified and synergistic treatment of components and functionalities with respect to a suitable performance index (single or multiple-objective and multi-criteria), should be "better" than a traditional design where the electrical design and the mechanical design are carried out separately and sequentially. The mechatronic approach should certainly be better than a simple interconnection of components that can do the intended tasks.

1.3 Intelligent Mechatronic Devices

A mechatronic system generally has some degree of "intelligence" built into it. An intelligent mechatronic system (IMS) is a system that can exhibit one or more intelligent characteristics of a human. As much as neurons themselves in a brain are not intelligent but certain behaviors that are effected by those neurons are, the basic physical elements of a mechatronic system are not necessarily intelligent but the system can be programmed to behave in an intelligent manner [4]. An intelligent mechatronic device embodies

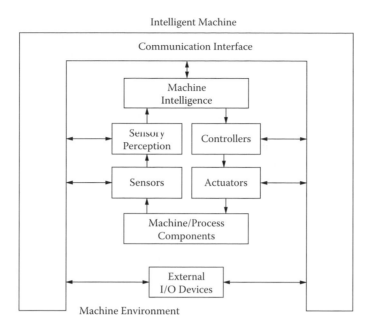

FIGURE 1.3 An intelligent mechatronic device.

machine intelligence. An IMS, however, may take a broader meaning than an intelligent computer. The term may be used to represent any electromechanical process, plant, system, device, or machinery that possesses machine intelligence. Sensors, actuators, and controllers will be integral components of such a system and will work cooperatively in making the behavior of the system intelligent. Sensing while understanding, or "feeling" what is sensed, is known as sensory perception, and this is very important for intelligent behavior. Humans use vision, smell, hearing, and touch (tactile sensing) in the context of their intelligent behavior. Intelligent mechatronic systems too should possess some degree of sensory perception. The "mind" of an IMS is represented by machine intelligence. For proper functioning of an IMS, it should have effective communication links between various components. An IMS may consist of an electromechanical structure for carrying out the intended functions of the system. Computers that can be programmed to perform "intelligent" tasks such as playing chess or understanding a natural language are known to employ artificial intelligence (AI) techniques for those purposes, and may be classified as intelligent computers. When integrated with a dynamic electromechanical structure such as robotic hands and visual, sonic, chemical, and tactile interfaces, they may be considered as intelligent mechatronic systems. Taking these various requirements into consideration, a general-purpose structure of an intelligent mechatronic device is given in Figure 1.3.

In broad terms, an IMS may be viewed to contain a knowledge system and a structural system. The knowledge system effects and manages intelligent behavior of the system, loosely analogous to the brain, and consists of various knowledge sources and reasoning strategies. The structural system consists of physical hardware and devices that are necessary to perform the system objectives yet do not necessarily need a knowledge system for their individual functions. Sensors, actuators, controllers (nonintelligent), communication interfaces, mechanical devices, and other physical components fall into this category. The broad division of the structure of an IMS, as mentioned previously, is primarily functional rather than physical. In particular, the knowledge system may be distributed throughout the system, and individual components by themselves may be interpreted as being "intelligent" as well (for example, intelligent sensors, intelligent controllers, intelligent multiagent systems). It needs to be emphasized that an actual implementation of an IMS will be domain specific, and much more detail than what is alluded to in Figure 1.3 may have to be incorporated into the system structure. Even from the viewpoint of system efficiency, domain-specific and special-purpose implementations are preferred over general-purpose mechatronic systems. Advances in digital electronics, technologies of semiconductor processing, and micro-electromechanical systems (MEMS) have set the stage for the integration of intelligence into sensors, actuators, and controllers. The physical segregation between these devices may well be lost in due time as it becomes possible to perform diversified functionalities such as sensing, actuation, conditioning (filtering, amplification, processing, modification, etc.), transmission of signals, and intelligent control, all within the same physical device. Due to the absence of adequate analytical models, sensing assumes an increased importance in the operation and control of intelligent mechatronic systems. The associated technologies are important in the field of mechatronics.

Smart mechatronic devices will exhibit an increased presence and significance in a wide variety of applications. The trend in the applications has been towards mechatronic technologies where intelligence is embedded at the component level, particularly in sensors and actuators, and distributed throughout the system. Application areas such as industrial automation, service sector, and mass transportation have a significant potential for using intelligent mechatronics, and incorporating advanced sensor technology and intelligent control. Tasks involved may include handling, cleaning, machining, joining, assembly, inspection, repair, packaging, product dispensing, automated transit, ride quality control, and vehicle entraining. In industrial plants, for example, many tasks are still not automated, and use human labor. It is important that intelligent mechatronic systems perform their tasks with minimal intervention of humans, maintain consistency and repeatability of operation, and cope with disturbances and unexpected variations in the machine, its operating environment, and performance objectives. In essence, these systems should be autonomous and should have the capability to accommodate rapid reconfiguration and adaptation. For example, a production machine should be able to quickly cope with variations ranging from design changes for an existing product to the introduction of an entirely new product line. The required flexibility and autonomous operation translate into a need for a higher degree of intelligence in the supporting devices. This will require proper integration of such devices as sensors, actuators, and controllers, which themselves may have to be "intelligent" and, furthermore, appropriately distributed throughout the system. Design, development, production, and operation of intelligent mechatronic systems, which integrate technologies of sensing, actuation, signal conditioning, interfacing, communication, and intelligent control, have been possible today through ongoing research and development in the field of intelligent mechatronic systems.

1.4   Modeling and Design

A design may use excessive safety factors and worst-case specifications (e.g., for mechanical loads and electrical loads). This will not provide an optimal design or may not lead to the most efficient performance. Design for optimal performance may not necessarily lead to the most economical (least costly) design, however. When arriving at a truly optimal design, an objective function that takes into account all important factors or criteria (performance, quality, cost, speed, ease of operation, safety, environmental impact, etc.) has to be optimized. A complete design process should generate the necessary details of a system for its construction or assembly. Of course, at the beginning of the design process, the desired system does not exist. In this context, a model of the anticipated system can be very useful. In view of the complexity of a design process, particularly when striving for an optimal design, it is useful to incorporate system modeling as a tool for design iteration [5].

Modeling and design can go hand in hand in an iterative manner. In the beginning, by having some information about the system (e.g., intended functions, performance specifications, past experience, and knowledge of related systems) and using the design objectives, it will be possible to develop a model of sufficient (low to moderate) detail and complexity. By analyzing and carrying out computer simulations of the model, it will be possible to generate useful information that will guide the design process (e.g., generation of a preliminary design). In this manner, design decisions can be made and the model can be refined using the available (improved) design. This iterative link between modeling and design is schematically shown in Figure 1.4.

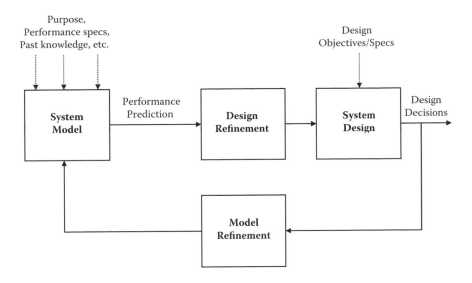

FIGURE 1.4 Link between modeling and design.

1.5 Mechatronic Design Concept

A mechatronic system will consist of many different types of interconnected components and elements. As a result, there will be energy conversion from one form to another, particularly between electrical energy and mechanical energy. This enables one to use energy as the unifying concept in the analysis and design of a mechatronic system. Let us explore this idea further.

In an electromechanical system, there exists an interaction (or coupling) between electrical dynamics and mechanical dynamics [1]. Specifically, electrical dynamics affect the mechanical dynamics and vice versa. Traditionally, a "sequential" approach has been adopted in the design of mixed systems such as electromechanical systems. For example, the mechanical and structural components are designed first, electrical and electronic components are selected or developed and interconnected next, and a computer is selected and interfaced with the system subsequently, and so on. The dynamic coupling between various components of a system dictates, however, that an accurate design of the system should consider the entire system as a whole in a concurrent manner rather than designing the electrical/electronic aspects and the mechanical aspects separately and sequentially. When independently designed components are interconnected, several problems can arise:

1. When two independently designed components are interconnected, the original characteristics and operating conditions of the two will change due to loading or dynamic interactions.
2. Perfect matching of two independently designed and developed components will be practically impossible. As a result, a component can be considerably underutilized or overloaded in the interconnected system, both conditions being inefficient and undesirable.
3. Some of the external variables in the components will become internal and "hidden" due to interconnection and associated "dynamic" coupling, which can result in potential problems that cannot be explicitly monitored through sensing and cannot be directly controlled.

The need for an integrated and concurrent design for electromechanical systems can be identified as a primary motivation for the growth of the field of mechatronics.

Design objectives for a system are expressed in terms of the desired performance specifications. By definition, a "better" design is where the design objectives (design criteria and specifications) are met more closely. The "principle of synergy" in mechatronics means that an integrated and concurrent design should result in a better product than one obtained through an uncoupled or sequential design. Note that

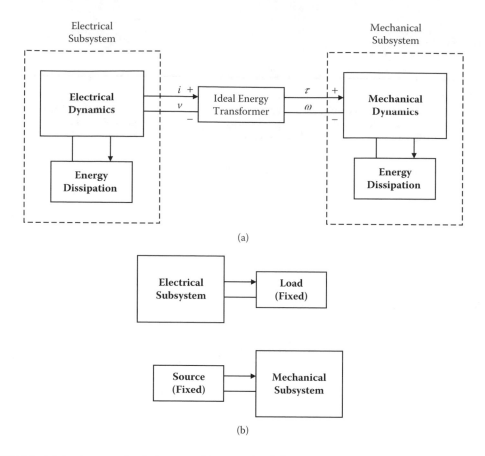

FIGURE 1.5   (a) An electromechanical system; (b) conventional design.

an uncoupled design is where each subsystem is designed separately (and sequentially) while keeping the interactions with the other subsystems constant (i.e., ignoring the dynamic interactions).

The concept of mechatronic design can be illustrated using an example of an electromechanical system, which can be treated as a coupling of an electrical subsystem and a mechanical subsystem. An appropriate model for the system is shown in Figure 1.5(a). Note that the two subsystems are coupled using a loss-free (pure) energy transformer while the losses (energy dissipation) are integral with the subsystems. In this system, assume that, under normal operating conditions, the energy flow is from the electrical subsystem to the mechanical subsystem (i.e., it behaves like a motor rather than a generator). At the electrical port connecting to the energy transformer, there exists a current i (a "through" variable) flowing in, and a voltage v (an "across" variable) with the shown polarity. The product vi is the electrical power, which is positive out of the electrical subsystem and into the transformer. Similarly, at the mechanical port coming out of the energy transformer, there exists a torque τ (a through variable) and an angular speed ω (an across variable) with the sign convention given in Figure 1.5(a). Accordingly, a positive mechanical power $\omega\tau$ flows out of the transformer and into the mechanical subsystem. The ideal transformer implies

$$vi = \omega\tau \qquad\qquad (1.1)$$

In a conventional uncoupled design of the system, the electrical subsystem is designed by treating the effects of the mechanical subsystem as a fixed load, and the mechanical subsystem is designed by treating the electrical subsystem as a fixed energy source, as indicated in Figure 1.5(b). Suppose that, in this manner, in an optimal design, the electrical subsystem achieves a "design index" of I_{ue}, and the mechanical subsystem achieves a design index of I_{um}. Note here that the design index is a measure of the degree to which the particular design satisfies the design specifications (design objectives).

When the two uncoupled designs (subsystems) are interconnected, there will be dynamic interactions. As a result, neither the electrical design objectives nor the mechanical design objectives will be satisfied at the optimal levels dictated by I_{ue} and I_{um}, respectively. Instead, they will be satisfied at the lower levels given by the design indices I_e and I_m. A truly mechatronic design will attempt to bring I_e and I_m as close as possible to I_{ue} and I_{um}, respectively. This may be achieved, for example, by minimizing the quadratic cost function

$$J = \alpha_e (I_{ue} - I_e)^2 + \alpha_m (I_{um} - I_m)^2 \tag{1.2}$$

subject to

$$\begin{bmatrix} I_e \\ I_m \end{bmatrix} = \boldsymbol{D}(\boldsymbol{p}) \tag{1.3}$$

where $\boldsymbol{D}$ denotes the transformation that represents the design process, and $\boldsymbol{p}$ denotes information, including system parameters, that is available for the design.

Even though this formulation of the mechatronic design problem appears rather simple and straight-forward, the reality is otherwise. In particular, the design process, as denoted by the transformation $\boldsymbol{D}$, can be quite complex and typically nonanalytic. Furthermore, minimization of the cost function J is by and large an iterative practical scheme and, undoubtedly, a knowledge-based and nonanalytic procedure. This complicates the process of mechatronic design. In any event, the design process will need the information represented by $\boldsymbol{p}$.

1.5.1 Mechatronic Design Quotient (MDQ)

The problem of mechatronic design may be treated as a maximization of a "mechatronic design quotient" or *MDQ* [1,6]. In particular, an alternative formulation of the optimization problem given by (1.2) and (1.3) would be the maximization of the mechatronic design quotient

$$MDQ = \frac{a_e I_e^2 + \alpha_m I_m^2}{a_e I_{ue}^2 + \alpha_m I_{um}^2} \tag{1.4}$$

subject to (1.3). Even though equation 1.4 is formulated for two categories of technologies or devices m and e, it may be generalized to three or more categories. However, the strength and applicability of the MDQ approach stem from the possibility that the design process may be hierarchically separated. Then, an MDQ may be optimized for one design layer involving two more technology groups in that layer before proceeding to the next lower design layer where each technology group is separately optimized by considering several technology/component groups within that group together with an appropriate MDQ for that lower-level design problem. In this manner, a complex design optimization may be achieved through several design optimizations at different design levels. The final design may not be precisely optimal, yet intuitively adequate for practical purposes—say, in a conceptual design.

1.6 Evolution of Mechatronics

Mechanical engineering products and systems that employ some form of electrical engineering principles and devices have been developed and used since the early part of the 20th century. These systems included the automobile, electric typewriter, aircraft, and elevator. Some of the power sources used in these systems were not necessarily electrical, but there were batteries and/or the conversion of thermal power into electricity through generators. These electromechanical systems were not mechatronic systems because they did not use an integrated approach characterizing mechatronics for their analysis, design, development, and implementation.

Rapid advances in electromechanical devices and systems were possible particularly due to developments in control engineering, which began for the most part in the early 1950s, and still more rapid advances in digital computer and communication as a result of integrated circuit (IC) and microprocessor technologies, starting from the late 1960s. With these advances, engineers and scientists felt the need for an integrated multidisciplinary approach to design and hence a mechatronic approach. Yasakawa Electric in Japan was the first to coin the term mechatronics, for which the company obtained a trademark in 1972. Subsequently, in 1982, the company released the trademark rights. Even though a need for mechatronics was felt even in those early times, no formal discipline and educational programs existed for engineers to be educated and trained in this area. Research and development activities, mainly in automated transit systems and robotics in the 1970s and 1980s, undoubtedly paved the way for the evolution of the field of mechatronics. With today's sophisticated technologies of mechanics and materials, analog and digital electronics, sensors, actuators, controllers, electromechanical design, and micro-electromechanical systems (MEMS) with embedded sensors, actuators, and microcontrollers, the field of mechatronics has attained a good degree of maturity. Now, many universities around the world offer undergraduate and graduate programs in mechatronic engineering, which have become highly effective and popular among students, instructors, employees, and employers alike.

1.7 Application Areas

The application areas of mechatronics are numerous and involve those that concern multidomain (mixed) systems and particularly electromechanical systems. These applications may involve

1. Modifications and improvements to conventional designs by using a mechatronic approach
2. Development and implementation of original and innovative mechatronic systems

In either category, the applications will employ sensing, actuation, control, signal conditioning, component interconnection and interfacing, and communication, generally using tools of mechanical, electrical and electronic, computer, and control engineering. Some important areas of application are indicated in the following text.

Transportation is a broad area where mechatronic engineering has numerous applications. In ground transportation in particular, automobiles, trains, and automated transit systems use mechatronic devices. They include airbag deployment systems, antilock braking systems (ABS), cruise control systems, active suspension systems, and various devices for monitoring, toll collection, navigation, warning, and control in intelligent vehicular highway systems (IVHS). In air transportation, modern aircraft designs with advanced materials, structures, electronics, and control benefit from the concurrent and integrated approach of mechatronics to develop improved designs of flight simulators, flight control systems, navigation systems, landing gear mechanisms, traveler comfort aids, etc.

Manufacturing and production engineering is another broad field that uses mechatronic technologies and systems. Factory robots (for welding, spray painting, assembly, inspection, and so on), automated guided vehicles (AGVs), modern computer-numerical control (CNC) machine tools, machining centers, rapid (and virtual) prototyping systems, and micromachining systems are examples of mechatronic applications. High-precision motion control is particularly important in these applications [7].

In medical and healthcare applications, robotic technologies for examination, surgery, rehabilitation, drug dispensing, and general patient care are being developed and used. Mechatronic technologies are being applied for patient transit devices, various diagnostic probes and scanners, beds, and exercise machines.

In a modern office environment, automated filing systems, multifunctional copying machines (copying, scanning, printing, FAX, and so on), food dispensers, multimedia presentation and meeting rooms, and climate control systems incorporate mechatronic technologies.

In household applications, home security systems with robots, vacuum cleaners with robots, washers, dryers, dishwashers, garage door openers, and entertainment centers use mechatronic devices and technologies.

In the computer industry [8], hard disk drives (HDD), disk retrieval, access and ejection devices, and other electromechanical components can considerably benefit from mechatronics. The impact goes further because digital computers are integrated into a vast variety of other devices and applications.

In civil engineering applications, cranes, excavators, and other machinery for building, earth removal, mixing, and so on will improve their performance by adopting a mechatronic design approach.

In space applications, mobile robots such as NASA's Mars exploration Rover, space-station robots, and space vehicles are fundamentally mechatronic systems.

It is to be noted that there is no end to the type of devices and applications that can incorporate mechatronics. In view of this, the traditional boundaries between engineering disciplines will become increasingly fuzzy, and the field of mechatronics will grow and evolve further through such merging of disciplines.

1.8 Conclusion

In this chapter, the multidisciplinary field of mechatronics was introduced, the technology needs for such systems were indicated, and some important issues in the design and development of a mechatronic product or system were highlighted. Mechatronics is a multidisciplinary engineering field that involves a synergistic integration of several areas such as mechanical engineering, electrical and electronic engineering, control engineering, and computer engineering. Technologies of sensing, actuation, signal conditioning, interfacing, communication, and control are particularly important for mechatronic systems. Intelligent mechatronic systems (IMS) require further technologies for representation and processing of knowledge and intelligence, and particularly those technologies that impart "intelligent" characteristics to the system. The design and development of a mechatronic system will require an integrated approach to deal with the subsystems and subprocesses of a multidomain (mixed) system—specifically, an electromechanical system. The subsystems of a mechatronic system should not be designed or developed independently without addressing the system integration, subsystem interactions and matching, and the intended operation of the overall system. Such an integrated and concurrent approach will make a mechatronic design more optimal than a conventional design.

References

1. De Silva, C.W., *Mechatronics—An Integrated Approach*, Taylor & Francis/CRC Press, Boca Raton, FL, 2005.
2. Shetty, D. and Kolk, R.A., *Mechatronics System Design*, PWS Publishing, Boston, MA, 1997.
3. De Silva, C.W., *Sensors and Actuators—Control System Instrumentation*, Taylor & Francis/CRC Press, Boca Raton, FL, 2007.
4. Karray, F. and de Silva, C.W., *Soft Computing and Intelligent Systems Design*, Addison Wesley, New York, 2004.
5. Necsulescu, D., *Mechatronics*, Prentice Hall, Upper Saddle River, NJ, 2002.
6. De Silva, C.W., Sensing and Information Acquisition for Intelligent Mechatronic Systems, *Science and Technology of Information Acquisition and Their Applications, Proceedings of the Symposium on Information Acquisition*, Chinese Academy of Sciences, Hefei, China, November 2003, pp. 9–18.
7. Tan, K.K., Lee, T.H., Dou, H., and Huang, S., *Precision Motion Control*, Springer-Verlag, London, 2001.
8. Chen, B.M., Lee, T.H., and Venkataramanan, V., *Hard Disk Drive Servo Systems*, Springer-Verlag, London, 2002.

I

Mechatronic Devices

2

Robotic Application of Mechatronics

H. Li

S.X. Yang

Summary

Mechatronic systems are integrated multidomain systems. They involve sensors, actuators, system modeling, locomotion, as well as parameter and state estimation. In this chapter, a mechatronic system is introduced. A fully autonomous mobile robot is built by using the behavior-based artificial intelligence (AI) approach. Several levels of competences and behaviors are implemented in this system. The improvement in system ability is accomplished by adding new modules to the system. Fuzzy control laws for steering control of the autonomous nonholonomic mobile robot are designed.

2.1 Introduction

In this chapter, the development of a special-purpose mechatronic system is presented. The specific objective is to develop an autonomous robot for transporting goods in an actual farm where the ground may be very rugged. The developed robot may also be used in construction sites to transport materials and tools. The focus in the development is on implementing efficient tools for enabling the robot to react to changes in the real world. The intended use has a direct effect not only on the physical structure of the robot but also on the control methods that are developed.

The mobile robot should be able to navigate in an outdoor terrain without colliding with any obstacles. With its ultrasonic sensors, the developed robot is able to avoid obstacles that are closer than a predefined distance. If the robot collides with an obstacle when it moves, it is able to stop and move away from the obstacles to avoid a repeated collision. Because the robot is developed to work in various unknown environments, there is no requirement for a preprogrammed path for it. Therefore, the robot should be able to see and react to changes in its environment.

On the physical side, because the robot is designed to transport a large load through a rough terrain, both the front and rear wheels are able to pivot horizontally to allow the robot to overcome large obstacles such as rocks. Large, powerful motors capable of carrying the robot and its loads over rough terrains and steep inclines are selected. It also has a platform which is sufficient to carry six large baskets of fruit or vegetable.

The implementation of the operation strategy of a complex mechatronic system can be approached by decomposing the global tasks into several simpler, well-specified behaviors that are easier to design and tune independently of each other. In the present development, behaviors are implemented in the robot at several levels within a hybrid reactive architecture. The robot is able to achieve the control objectives, and the robot trajectory is found to be smooth in spite of the interaction among different behaviors, unexpected obstacles, and noise. A behavior-based approach is used to design the control system for the robot. This approach is able to deal with multiple, changing goals in a dynamic and unpredictable environment [1]. A behavior-based system has multiple integrated competences. Layers of a control system are built to correspond to each level of competence. Individual layers are able to work on individual goals simultaneously. The suppression mechanism mediates the actions that are taken. The advantage here is that there is no need to make an early decision on which goal should be pursued. The results of pursuing all the goals to some level of conclusion can be used for determining the ultimate decision. The system is situated in its environment. It is directly connected to its problem domain through sensors and actuators/effectors. The system is able to change and affect its environment instantaneously by reacting through the effectors. The problem domain can be a dynamic environment, and the system can react within a specified time. The environment for the system is a complex real-world environment. In the present development, several levels of competence for the developed robot are defined. A level of competence is a specification of a set of desired behaviors that the mobile robot will encounter in the real world. A higher level of competence indicates a more specific and complex desired class of behaviors.

Autonomous navigation has been the subject of many studies in AI where different approaches have been attempted to solve the problem. Research has been done in both holonomic and nonholonomic mobile robots. Autonomous navigation is related to the ability of proper motion of a mobile robot in achieving a goal, without human interaction, in an environment on which no prior information is available. The mobile robot is guided using online information that is acquired during navigation. Such tasks require different abilities to execute actions that lead to goal achievement. In recent years, fuzzy strategies have been applied to this problem, resulting in satisfactory results. In the present work, a fuzzy logic control system is designed to make decisions on how to steer the front wheels of the autonomous nonholonomic mobile robot to the desired orientation. Then, reaching a desired target can be achieved simply by turning the steering wheels toward the target.

2.2   Behavior-Based Mobile Robot

The mobile robot is built in accordance with the behavior-based approach. The robot, which is shown in Figure 2.1, can navigate autonomously by reacting to its sensory inputs or be controlled by an operator using a joystick. Bumpers, ultrasonic sensors, and a vision system provide the sensory inputs. The robot can autonomously follow a target object while avoiding collision with obstacles.

2.2.1   Design of the Mechatronic System

The robot is designed to be able to carry a payload of over 100 kg through rough terrain. For this, strong structural components and powerful motors are selected in building the robot. The frame is custom made out of structural steel, and is 53 in. long, 22 in. wide, and 18 in. high. The robot has four wheels, each driven by a separate dc motor. Figure 2.1 shows the developed mobile robot. There is no real steering mechanism for the robot, and the speed of the individual motors determine its trajectory. The robot can turn right by making the front left motor move faster than the front right motor. Similarly, the robot

FIGURE 2.1 Prototype autonomous mobile robot.

can turn left by making the front right motor move faster than the front left motor. If the two front motors turn at the same speed, the robot will move straight without turning. This idea resembles differential steering except that the motors themselves pivot horizontally. This design gives the robot greater mobility than differential steering. The robot has a platform that is large enough to carry six large baskets of fruit or vegetable. The camera is mounted on the top of a 36-in. high frame, and it is able to turn 180° while the robot is moving backward. The robot has four large, powerful motors capable of carrying itself and its payload over rough terrains and steep inclines. The selected motors are 24 V, 1/2 hp dc motors. They are connected to the motor controllers as shown in Figure 2.2. High capacity batteries are used to enable the robot to operate for several hours without recharging. Two 12 V, 100 A.h batteries are connected in series to supply 24 V for all the electronics.

2.2.2   Levels and Layers

Traditionally, mobile robot builders decompose the problem into several subproblems: sensing, mapping sensor data into a world representation, path planning, task execution, and motor control. This can be regarded as a horizontal decomposition of the problem into vertical slices. For the robot developed in the present work, the problem is decomposed vertically instead of horizontally. Several levels of competence are implemented. Higher levels of competence provide additional constraints on the earlier levels. Layers of a control system corresponding to each level of competence can be built by adding a new layer to an existing set, thereby moving to a higher level of competence. The following levels of competence are incorporated:

0. Emergency stop.
1. Maintain the robot along the desired trajectory.
2. Stop automatically on colliding with an obstacle.

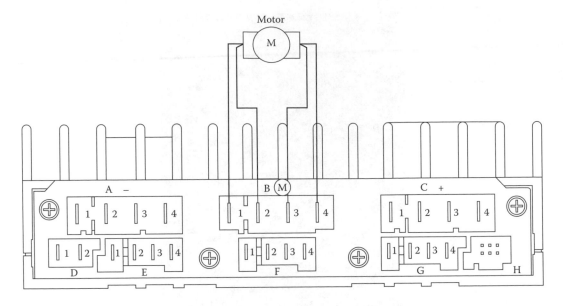

FIGURE 2.2   Connection diagram for the motor.

3. Move by the operator using the joystick.
4. Avoid contact with objects, including those during backup and on the side.
5. Reason about the world concerning object identification, heading for a target, and performing tasks related to objects.

The hardware is built in accordance with the layers just outlined and as shown in Figure 2.3. Four motors are connected to the front motor controller and the rear motor controller, which are controlled by the front motor PIC (a family of microcontrollers produced by Microchip Company) and the rear motor PIC. The motor controllers perform the tasks of moving the robot at a desired speed and along a desired

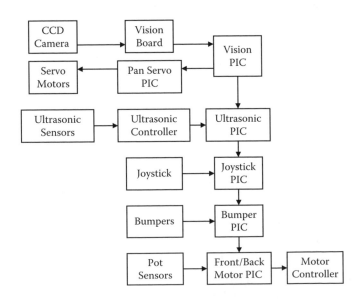

FIGURE 2.3   Hardware structure of the robot.

trajectory. The bumper sensors are used to achieve Level 2 competence. A joystick and ultrasonic sensors are connected to implement Level 3 and Level 4 competences. A color CCD camera and a vision board developed by Applied AI Systems, Inc., are employed to perform the tasks required to achieve Level 5 competence.

2.2.3 The Electronics

The electronics of the control system incorporate 8-bit PIC microcontrollers, referred to as PICs. Because the design is carried out in different levels, the hardware and the software for each level are designed, tested, and evolved individually. Figure 2.4 shows the schematic diagram of the PIC board. Each PIC has two pulse-width modulation (PWM) modules, as many as 5 analog inputs, and at least 10 digital I/Os. One USART port is also available. Because there is no analog output from the PIC, a PWM signal produced by the PIC is passed through an RC low-pass filter circuit to produce the average DC voltage of the signal.

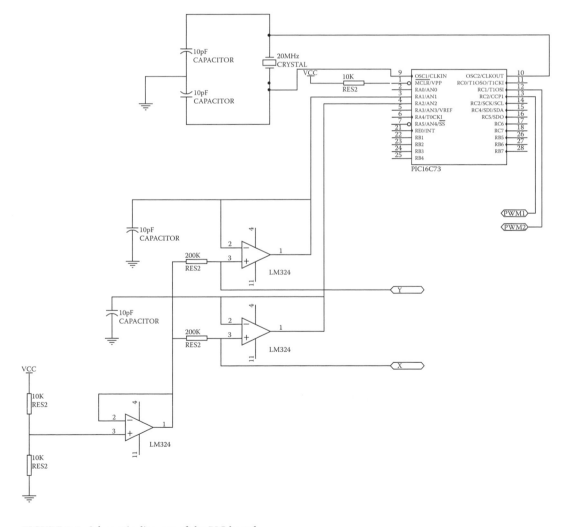

FIGURE 2.4 Schematic diagram of the PIC board.

2.3   Fuzzy Controller for Path Tracking

In this section, tracking control of the developed mobile robot is discussed. The kinematics of the nonholonomic robot model is introduced. Then a simple fuzzy controller for real-time navigation of the nonholonomic mobile robot is presented.

2.3.1   Kinematic Model of the Nonholonomic Mobile Robot

The path-tracking problem of the developed mobile robot with kinematic constraints in the two-dimensional (2-D) workspace is presented now. A nonholonomic constraint for a mobile robot is a nonintegrable equation involving the configuration parameters and their derivatives. Such a constraint does not reduce the dimension of the space of configurations attainable by the robot but reduces the dimension of the space of possible differential motions at any given configuration. Consider the mobile robot in Figure 2.5 where the rear wheels are aligned with the vehicle while the front wheels are allowed to pivot about the vertical axes. The constraints on the system arise by allowing the wheels to roll and pivot. In a sufficiently large empty space, a mobile robot can be driven to any position with any orientation; hence, the configuration space of the robot has four dimensions: two for translation, one for rotation, and one for the steering angle. Let (x, y, θ, Φ) denote the configuration of the robot, parameterized by the location of the front wheels. The kinematic model of the mobile robot may be represented as

$$\dot{x} = u_3 \cos\theta \tag{2.1}$$

$$\dot{y} = u_3 \sin\theta \tag{2.2}$$

$$\dot{\theta} = \frac{u_3}{l}\tan\phi \tag{2.3}$$

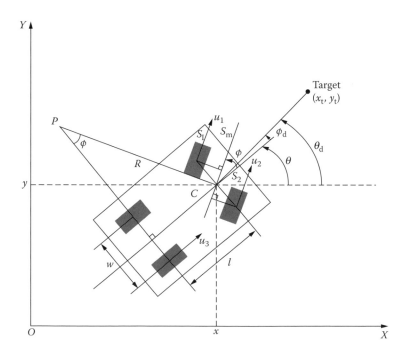

FIGURE 2.5   Model of the four-wheeled mobile robot.

where u_3 denotes the forward velocity of the vehicle; θ is the angle of the vehicle body with respect to the horizontal line; Φ is the steering angle with respect to the vehicle body; (x, y) is the location of the center point of the front wheels, and l is the distance between the front and the rear wheels.

The steering angle of the front wheels is adjusted by using a method resembling the differential steering method, as mentioned before. If both front wheels turn in tandem, the robot moves in a straight line. If one front wheel turns faster than the other, the robot follows a curved path. Accordingly, steering the robot is just a matter of varying the speeds of the front wheels. Because the turn radius of the robot is quite large compared with the radius of the wheels, referring to Figure 2.5, we have the following relationships:

$$S_1 = \left(R - \frac{w}{2}\cos\phi \right)\phi \tag{2.4}$$

$$S_m = R\phi \tag{2.5}$$

$$S_2 = \left(R + \frac{w}{2}\cos\phi \right)\phi \tag{2.6}$$

where S_1 and S_2 denote the displacement (distance traveled) of the front left and front right wheels, respectively; R is the turn radius of the center point of the front wheels; w is the distance between wheels (from center-to-center along the length between the two front wheels or two back wheels); Φ is the angle of turn in radians; and S_m is the displacement at the center point of the front wheels. Once the geometry of the differential steering system is established, it is easy to develop algorithms for controlling the steering angle Φ of the robot. Because the robot is treated as a rigid body, to develop a forward kinematic equation for the differential steering system, a frame of reference is specified. In this frame, an arbitrarily chosen point is treated as the stationary reference, and all other points in the system are treated as moving relative to the reference point. Here we consider the center point of the front wheels as the origin of the frame of reference of the robot.

By differentiating equations 2.4–2.6, we obtain the velocity of the two front wheels as

$$\dot{S}_1 = \dot{\phi}\left(R - \frac{w}{2}\cos\phi \right) + \frac{w}{2}\phi\dot{\phi}\sin\phi \tag{2.7}$$

$$\dot{S}_2 = \dot{\phi}\left(R + \frac{w}{2}\cos\phi \right) - \frac{w}{2}\phi\dot{\phi}\sin\phi \tag{2.8}$$

so that

$$u_2 - u_1 = \dot{\phi}w(\cos\phi - \phi\sin\phi) \tag{2.9}$$

where u_1 and u_2 correspond to the forward velocity of the front left wheel and the front right wheel, respectively. Then we have

$$\dot{\phi} = \frac{u_2 - u_1}{w(\cos\phi - \phi\sin\phi)} \tag{2.10}$$

2.3.2 Fuzzy Controller for the Differential Steering System

A block diagram of the fuzzy controller is given in Figure 2.6. The desired Φ_d is determined by calculating the position of the target in the image representing the environment detected by the color closed-circuit Charge Coupled Device (CCD) camera. The actual turn angle Φ is obtained from the potentiometers mounted on the front wheels of the mobile robot. The command signal Φ_d is transmitted from the vision board through

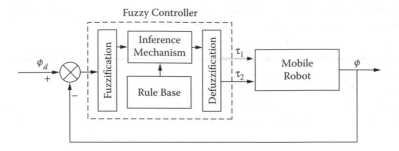

FIGURE 2.6 The fuzzy logic controller.

the serial port to the chip on which the fuzzy controller resides. The actual angle Φ is injected into the chip through the analog port of the chip. The error between the command signal and the actual position, as well as the change in error of the signal, are calculated and fed into the fuzzy controller embedded in the chip. From equation 2.10, it can be seen that the difference between the velocities of the front wheels determines the turn speed. The fuzzy controller [2] is designed to generate PWM signals corresponding to u_1 and u_2 for the front left motor and the front right motor, respectively, to make the turn angle Φ of the front wheels reach the desired angle Φ_d.

The fuzzification procedure maps the crisp input values to the linguistic fuzzy terms with membership values between 0 and 1. In the present work, five membership functions are used for both error $e_\phi = \phi - \phi_d$ and change in error $\dot{e}_\phi = \dot{\phi} - \dot{\phi}_d$.

The rule base stores the rules governing the input–output relationship of the fuzzy controller. The inference mechanism is responsible for decision making in the control system using approximate reasoning. The operation involved is "AND" because two inputs e and $\dot{e}$ are involved.

The control rules are designed based on expert knowledge and testing. Furthermore, the control rules meet the stability requirements as derived using Lyapunov's direct method. For example, if e_ϕ is "poslarge" and is increasing rapidly ($\dot{e}_\phi$ is poslarge), then the left motor should be much faster than the right motor; i.e., $u_2 - u_1$ should be "NL." Based on this knowledge, we obtain 25 fuzzy rules. Table 2.1 presents abstract knowledge of the expert related to controlling the turn angle, given the error and its derivative as inputs. The input and output linguistic variables are summarized in the table.

The defuzzification procedure maps the fuzzy output from the inference mechanism to a crisp signal. We use the center of gravity (COG) defuzzification method [2] for combining the recommendations represented by the implied fuzzy sets from all the rules. Let b_i denote the center of the membership function of the consequent of rule (i), and $\int \mu_{(i)}$ is the area under the membership function $\mu_{(i)}$. The COG method computes the crisp value μ^c to be

$$\mu^c = \frac{\sum_i b_i \int \mu_{(i)} dx}{\sum_i \int \mu_{(i)} dx} \tag{2.11}$$

TABLE 2.1   Rule Table for the Steering System

				$\dot{e}_\phi$		
$u_2 - u_1$		Pos Large	Pos Small	Zero	Neg Small	Neg Large
	Poslarge	NL	NL	NL	NS	ZO
	Possmall	NL	NL	NS	ZO	PS
e_ϕ	Zero	NL	NS	ZO	PS	PL
	Negsmall	NS	ZO	PS	PL	PL
	Neglarge	ZO	PS	PL	PL	PL

Here $\int \mu_{(i)}$ can be easily computed because symmetric triangular output membership functions are used that peak at the value 1 and have a base width of w. From geometry, it can be easily shown that the area under a triangle "truncated" at a height of h is equal to $w.(h - \frac{h^2}{2})$.

Because the output membership functions are symmetric, the center of the implied fuzzy set will be the same as the center of the consequent fuzzy set from which it is computed. If the output membership functions are not symmetric, then their centers, which are needed in the computation of the COG, will change depending on the membership value of the premise. This will require recomputation of the center at each time instant.

2.4   Visual Landmark Recognition System

In the vision layer of the robot built here, a genetic algorithm is designed and programmed into the vision board to recognize the predefined landmarks — a set of numbers written on red square boards. The recognition results are then injected into the registers in the finite-state machine that represents the vision layer to generate desired behaviors such as going forward, reversing, turning left, turning right, and so on.

2.5   Experiments and Results

In order to verify the performance of the robot system, first several behaviors are tested separately. The tracking-layer behaviors are designed to make the robot track the desired path in the farm. The bumper-layer behaviors are triggered whenever the robot collides with an obstacle such as a rock in front of it. The ultrasonic obstacle avoidance behaviors are triggered when there are obstacles around the robot.

2.5.1   Tracking on a Rugged and Steep Hill

In order to make sure that the developed mobile robot is able to transport a large load over a relatively long distance in the farm where the ground is very rugged, a load of approximately 100 kg is placed on the platform. Then the robot is made to move and track the desired path. Figure 2.7 shows that the robot is able to satisfactorily climb a 20° incline where the surface is very rough.

FIGURE 2.7   Robot navigation over a rugged terrain.

FIGURE 2.8 Typical terrain in the experimental farm.

2.5.2   Experiment in a Real Farm

The developed mobile robot is demonstrated to be able to navigate with limited human supervision. The layout of the paths in the experimental farm is illustrated in Figure 2.8.

The robot starts from the left end of the lowest horizontal path while following the operator. The operator turns left at the first intersection when the robot also turns left successfully. Then the robot reaches the second intersection with a landmark number 1 at the corner. This time it is capable of recognizing the landmark and automatically making a right turn. While the robot turns right, the obstacle avoidance behavior is triggered in order to avoid the obstacles at the corner. After that, the robot keeps tracking the operator until it reaches the landmark number 3 at the right end of the middle horizontal path. It then pans the camera to the rear and starts to move backward. Subsequently, it keeps going backward while following the operator until it reaches the landmark number 2 at the left end. The camera is redirected, and the robot starts going forward again. Next, the robot makes a left turn while following the operator. During the experiment, the ultrasonic sensors are also triggered, and the robot demonstrates the ability to avoid colliding with obstacles that are too close to it. In the end, the human operator stands in front of the bumper sensors. The robot shows the ability to back up after it touches the operator who stands in front of it.

References

1. Brooks, R.A., A robust layered control system for a mobile robot, *IEEE Journal of Robotics and Automation*, Vol. 2, No. 1, 14–23, March 1986.
2. Karray, F. and de Silva, C.W., *Soft Computing and Intelligent Systems Design*, Addison-Wesley, New York, 2004.

3

Swiss Lever Escapement Mechanism

Y. Fu

R. Du

Summary

This chapter investigates the Swiss Lever escapement mechanism in mechanical watch movement. First the structure and working principle are described. How the key function of the mechanism is realized by the special geometry is explained. Then the dynamic model is developed and an impulsive differential equation is applied to describe the motion with shock inputs that are involved. Next, simulation studies of the mechanism are presented, and conclusions are drawn.

3.1 Introduction

The mechanical watch is one of the most intricate engineering devices that mankind has ever invented. The first mechanical watch appeared in the middle of the sixteenth century. Since then, it has been studied by many people, including such geniuses as Galileo, Huygens, and Hooke. Generally speaking, a mechanical watch may be treated as a mechatronic device, and is made of five parts as shown in Figure 3.1. These include the winding mechanism, the power supply (mainspring), the gear train, the display, and the escapement mechanism.

Among the five parts, the escapement mechanism is the most important and is often referred to as the brain of the watch movement. It is called so because the energy is allowed to *escape* each time a gear tooth is moved. According to the literature, there are over one hundred different types of escapement mechanisms, but they all have the same function: to provide a stable oscillation feedback to regulate the timekeeping accuracy.

The earliest record of escapement dates back to the thirteenth century. The first escapement mechanism was named *verge* escapement. The feedback control of the verge escapement is discussed in References [1,2,3]. This simple mechanism consists of only a crown gear and a pair of rotating bodies. The pendulum clock appeared almost in the same period. In Reference [4], a computer simulation model of an escapement mechanism with a pendulum is developed using MATLAB® Simulink®. Reference [5] recounts the history of the escapement mechanism from the first design in 1657 to the late 1800s and describes how

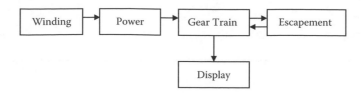

FIGURE 3.1 Main components of a mechanical watch movement mechanism.

horologists improved on the escapement mechanism step by step. More information is given in Reference [6]. In Reference [7], the geometries of several different escapement mechanisms are described.

Presently, the *Swiss lever escapement* is the most popular one. In comparison to the other existing escapement mechanisms, this device has a number of advantages. First, although it is somewhat complicated, it is almost symmetric and easy to fabricate. Second, it distributes the energy accurately and hence ensures timekeeping accuracy. Therefore, the Swiss lever mechanism is used by almost all commercial products. It is interesting to know that the basic operation principle of the Swiss lever mechanism has been well understood and described in the literature, for example, in Reference [8]. However, searching through literature, it appears that dynamic modeling of the mechanism, which will enable one to understand the working details and design, has not received much attention.

3.2   Structure and Working Principle

Figure 3.2 shows the Swiss lever escapement mechanism. It consists of five components: a balance wheel, a hairspring, two banking pins, a pallet fork, and an escape wheel. Note that one end of the hairspring is fixed, whereas the other end is attached to the balance wheel. The balance wheel oscillates periodically under the driving force from the escape wheel through the pallet fork and the restoring force of the hairspring. It is the guard pin, which is a synthetic ruby on the balance wheel, that sends and receives impulses from the pallet fork to the balance wheel. The banking pin limits the rotation of the pallet fork. The escapement wheel rotates intermittently at a specific speed according to the frequency of the system.

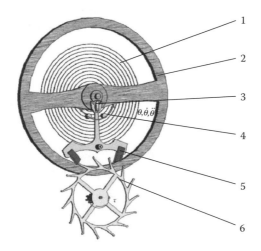

FIGURE 3.2 The Swiss lever escapement mechanism: 1—hairspring, 2—balance wheel, 3—guard pin, 4—banking pin, 5—pallet fork, 6—escape wheel.

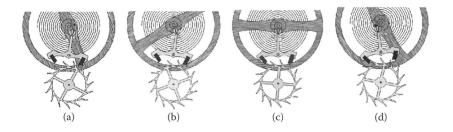

FIGURE 3.3 Four states of escapement in a half-period: (a) ascending supplementary arc, (b) unlocking, (c) impulse, (d) descending supplementary arc.

The operation of the Swiss lever escapement mechanism is well known [8]. Figure 3.3 illustrates the states of escapement in half a period. Thre are a total of five shocks during this period. In Figure 3.3(a), the balance wheel is turning clockwise with a tooth of the escape wheel locked on the right leg of the pallet fork. Hence, this state is called the supplementary arc, and the speed increases. Figure 3.3(b) shows the unlocking. The guard pin (also called the impulse pin) on the balance wheel enters and impacts the pallet fork, giving the first shock. Then it pushes the pallet fork to unlock the tooth. The escape wheel rotates clockwise through a small angle during unlocking. After that, it leaves the right leg of the pallet fork due to inertia. As a result, two recoils are generated for the escape wheel. The balance wheel and the escape wheel continue to move for a small period with a sinusoidal motion until the escape wheel contacts the right pallet under the driving input of the external torque. Thereby, the second shock occurs. Figure 3.3(c) shows this impulse. Under the driving torque of the escape wheel, one tooth of the escape wheel advances to impulse the balance wheel through the pallet fork and the impulse pin. The third shock occurs when the pallet fork pushes the impulse pin. The impulse ends when the escape wheel tooth leaves the pallet fork. The pallet fork keeps on advancing with the balance wheel and is stopped by the guard pin. In the meantime, the escape wheel accelerates and halts when a tooth reaches the right leg of the pallet fork, which gives the fourth shock. After that the impulse pin leaves the pallet fork. The escape wheel continues to lock the pallet fork until it is stopped by the banking pin. This is the fifth shock. In Figure 3.3(d), the balance wheel continues to turn clockwise until its energy is exhausted. It is also called supplementary arc, but the speed is decreasing. After finishing the half-period, the balance wheel returns in the counterclockwise direction following the same states.

Above all, the input to the system is the torque on the escape wheel. The oscillation frequency is mainly determined by mass-spring components, say, the balance wheel and the hairspring. Some energy is dissipated due to the shocks and made up during the impulse process to maintain the oscillation. The output is a discontinuous intermittent rotation of the escape wheel. There is a feedback regulation of interval variation between the intermissions.

Usually, a half-period is called one beat. The number of beats within one hour is abbreviated as bph to describe the speed of a movement. A higher number denotes more accurate movement. Most watches operate at 18000, 21600, or 28800 bph. Some precise watches can reach 36000 bph.

Example 3.1

A movement has a frequency of 21600 bph. The moment of inertia of the balance wheel is 0.000002 kg·m². Estimate the hairspring constant K.

Solution

The oscillation frequency f can be calculated as

$$f = \frac{21600}{3600 \times 2} = 3 \text{ Hz}$$

The spring constant is equal to

$$K = 4\pi^2 f^2 J_b = 0.00071 \text{ N} \cdot \text{rad}^{-1}$$

3.3 Dynamical Modeling

In order to derive the dynamical model, the following assumptions are made:

1. All components in the escapement mechanism are rigid, except the hairspring.
2. There is no friction between the axle and the bearing.
3. The center of gravity of the balance wheel is at the axis of rotation.
4. The relationship between the hairspring moment and the angle of displacement is linear.
5. The mass of the hairspring can be neglected.

In addition, it is assumed that the balance wheel, the pallet fork, and the escape wheel have the moments of inertia J_b, J_p, and J_e, respectively. The geometry of the escapement is shown in Figure 3.4. The contact radii during the impulse between the escape wheel and the pallet fork are r_1 and r_2. The contact radii during the impulse between the pallet fork and the balance wheel are r_3 and r_4. The contact radii during unlocking between the escape wheel and the pallet fork are r_5 and r_6. The friction radii during unlocking between the escape wheel and the pallet fork are r_7 and r_8. The friction radii during impulse between the escape wheel and the pallet fork are r_9 and r_{10}. l_1, l_2 are the distances from the pallet to the axle of the escape wheel and the pallet fork, respectively, when the escape wheel begins to impulse the pallet; l_3, l_4 are the distances from the impulse pin to the axle of the pallet fork and the balance wheel, respectively. The angular displacements of the balance wheel, pallet fork, and the escape wheel are θ_b, θ_p, and θ_e, respectively. The angular velocities are $\dot{\theta}_b$, $\dot{\theta}_p$, and $\dot{\theta}_e$, and the corresponding angular accelerations are $\ddot{\theta}_b$, $\ddot{\theta}_p$, and $\ddot{\theta}_e$, respectively. The initial angular displacement is θ_{b0}, and the thresholds defined by the guard pin are θ_{b1} and θ_{b6}. The torque on the escape wheel is τ. Also, it is assumed that $r_1, r_2, r_3, r_4, r_5, r_6,$ r_7, r_8, r_9, r_{10} are constants.

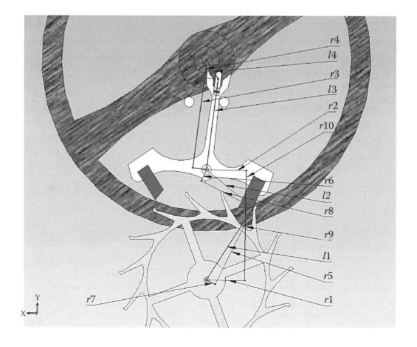

FIGURE 3.4 Definition of dimension parameters in the model.

3.3.1 Modeling of Each Step

As mentioned earlier, a total of eleven steps are involved within a cycle. Because the motion is rather symmetric, for convenience, define

$$\sigma = \begin{cases} +1, & \text{1st half - period} \\ -1, & \text{2nd half - period} \end{cases} \tag{3.1}$$

which will simplify the mathematical representation.

Step 1: The balance wheel experiences the ascending supplementary arc while the pallet fork and the escape wheel are locked. The motion is governed by the equation

$$J_b \ddot{\theta}_b + K\theta_b = 0 \tag{3.2}$$

The initial condition is $\theta_b = \theta_{b0}, \dot{\theta}_b = 0$, at $t = 0$. This step ends when the impulse pin reaches the pallet fork. The time at this instant is denoted by t_1, and the angle and velocity of the balance wheel are denoted by θ_{b1} and $\dot{\theta}_{b10}$, respectively, which will be used in the subsequent step. Note that t_1 is a function of θ_{b1} and the initial condition.

Step 2: This corresponds to the first shock when the impulse pin hits the pallet fork. The collision is inelastic and the relationship among the balance wheel, the pallet fork, and the escape wheel is as follows:

$$\frac{J_b}{r_4 r_6} \dot{\theta}_{b10} = \frac{J_b}{r_4 r_6} \dot{\theta}_{b11} - \frac{J_p}{r_3 r_6} \dot{\theta}_{p1} + \sigma \frac{J_e}{r_3 r_5} \dot{\theta}_{e1} \tag{3.3}$$

Note that

$$\frac{\dot{\theta}_{b11}}{\dot{\theta}_{p1}} = -\frac{l_3}{l_4} \tag{3.4}$$

$$\frac{\dot{\theta}_{p1}}{\dot{\theta}_{e1}} = \sigma \frac{l_1}{l_2} \tag{3.5}$$

It follows that

$$\frac{J_b}{r_4 r_6} \dot{\theta}_{b10} = \left(\frac{J_b}{r_4 r_6} + \frac{J_p l_4}{r_3 r_6 l_3} + \frac{J_e l_2 l_4}{r_3 r_5 l_1 l_3} \right) \dot{\theta}_{b11}$$

$$= -\left(\frac{J_b l_3}{r_4 r_6 l_4} + \frac{J_p}{r_3 r_6} + \frac{J_e l_2}{r_3 r_5 l_1} \right) \dot{\theta}_{p1} = \sigma \left(\frac{J_b l_1 l_3}{r_4 r_6 l_2 l_4} + \frac{J_p l_1}{r_3 r_6 l_2} + \frac{J_e}{r_3 r_5} \right) \dot{\theta}_{e1} \tag{3.6}$$

Consequently,

$$\Delta \dot{\theta}_b = -\frac{J_p r_4 r_5 l_1 l_4 + J_e r_4 r_6 l_2 l_4}{J_b r_3 r_5 l_1 l_3 + J_p r_4 r_5 l_1 l_4 + J_e r_4 r_6 l_2 l_4} \dot{\theta}_{b10} \tag{3.7}$$

$$\Delta \dot{\theta}_p = -\frac{J_b r_3 r_5 l_1 l_4}{J_b r_3 r_5 l_1 l_3 + J_p r_4 r_5 l_1 l_4 + J_e r_4 r_6 l_2 l_4} \dot{\theta}_{b10} \tag{3.8}$$

$$\Delta \dot{\theta}_e = \sigma \frac{J_b r_3 r_5 l_2 l_4}{J_b r_3 r_5 l_1 l_3 + J_p r_4 r_5 l_1 l_4 + J_e r_4 r_6 l_2 l_4} \dot{\theta}_{b10} \tag{3.9}$$

where $\Delta \dot{\theta} \equiv \dot{\theta}_{11} - \dot{\theta}_{10}$, $\Delta \dot{\theta}_p \equiv \dot{\theta}_{p1}$, and $\Delta \dot{\theta}_e \equiv \dot{\theta}_{e1}$.

Step 3: This corresponds to unlocking. The escape wheel recoils through a small angle under the push of the pallet fork. The motion is governed by

$$J_1\ddot{\theta}_b + K\theta_b = -\sigma T_1 \tag{3.10}$$

where

$$J_1 = J_b + J_p \frac{r_4}{r_3}\frac{l_4}{l_3} - J_e \frac{r_4(r_6 + \mu r_8)}{r_3(r_5 - \mu r_7)}\frac{l_2 l_4}{l_1 l_3}$$

$$T_1 = \tau \frac{r_4(r_6 + \mu r_8)}{r_3(r_5 - \mu r_7)}$$

Here, μ is the friction coefficient. This step ends when the unlocking finishes at the moment t_2, generating the displacements and velocities $\theta_{b2}, \theta_{p2}, \theta_{e2}, \dot{\theta}_{b2}, \dot{\theta}_{p2}, \dot{\theta}_{e2}$.

Step 4: Because of the inertia, the escape wheel disengages from the pallet fork and continues to rotate under the only external torque τ. Then

$$J_e\ddot{\theta}_e = -\tau \tag{3.11}$$

It turns counter-clockwise until the kinematic energy is exhausted, and turns clockwise again. At the same time, for the balance wheel and pallet fork, one has

$$J_2\ddot{\theta}_b + K\theta_b = 0 \tag{3.12}$$

where

$$J_2 = J_b + J_p \frac{r_4}{r_3}\frac{l_4}{l_3}$$

This step ends when the escape wheel tooth reaches the pallet again at the time instant t_3, generating the displacements and velocities $\theta_{b3}, \theta_{p3}, \theta_{e3}, \dot{\theta}_{b30}, \dot{\theta}_{p30}, \dot{\theta}_{e30}$.

Step 5: The escape wheel impacts the pallet, and corresponds to the second shock. Almost simultaneously, the other side of the fork notch impacts the impulse pin, and corresponds to the third shock. In our model, these two shocks are considered to happen at the same time. The governing equation is

$$\frac{J_b}{r_2 r_4}\dot{\theta}_{b30} - \frac{J_p}{r_2 r_3}\dot{\theta}_{p30} - \sigma\frac{J_e}{r_1 r_4}\dot{\theta}_{e30} = \frac{J_b}{r_2 r_4}\dot{\theta}_{b31} - \frac{J_p}{r_2 r_3}\dot{\theta}_{p31} - \sigma\frac{J_e}{r_1 r_4}\dot{\theta}_{e31} \tag{3.13}$$

Because

$$\frac{\dot{\theta}_{b31}}{\dot{\theta}_{p31}} = -\frac{l_3}{l_4} \tag{3.14}$$

$$\frac{\dot{\theta}_{p31}}{\dot{\theta}_{e31}} = \sigma\frac{l_1}{l_2} \tag{3.15}$$

Equation 3.12 can be written as follows:

$$\frac{J_b}{r_2 r_4}\dot\theta_{b30} - \frac{J_p}{r_2 r_3}\dot\theta_{p30} - \sigma\frac{J_e}{r_1 r_4}\dot\theta_{e30} = \left(\frac{J_b}{r_2 r_4} + \frac{J_p l_4}{r_2 r_3 l_3} + \frac{J_e l_2 l_4}{r_1 r_4 l_1 l_3}\right)\dot\theta_{b31}$$

$$= -\left(\frac{J_b l_3}{r_2 r_4 l_4} + \frac{J_p}{r_2 r_3} + \frac{J_e l_2}{r_1 r_4 l_1}\right)\dot\theta_{p31} \qquad (3.16)$$

$$= \sigma\left(\frac{J_b l_1 l_3}{r_2 r_4 l_2 l_4} + \frac{J_p l_1}{r_2 r_3 l_2} + \frac{J_e}{r_1 r_4}\right)\dot\theta_{e31}$$

Also,

$$\Delta\dot\theta_b = \frac{J_e r_2 r_3 l_2 l_4}{J_b r_1 r_3 l_1 l_3 + J_p r_1 r_4 l_1 l_4 + J_e r_2 r_3 l_2 l_4}\dot\theta_{b30}$$

$$- \frac{\sigma J_e r_2 r_3 l_1 l_3}{J_b r_1 r_3 l_1 l_3 + J_p r_1 r_4 l_1 l_4 + J_e r_2 r_3 l_2 l_4}\dot\theta_{e30} \qquad (3.17)$$

$$\Delta\dot\theta_p = -\frac{J_e r_2 r_3 l_2 l_4}{J_b r_1 r_3 l_1 l_3 + J_p r_1 r_4 l_1 l_4 + J_e r_2 r_3 l_2 l_4}\frac{l_4}{l_3}\dot\theta_{b30}$$

$$+ \frac{\sigma J_e r_2 r_3 l_1 l_4}{J_b r_1 r_3 l_1 l_3 + J_p r_1 r_4 l_1 l_4 + J_e r_2 r_3 l_2 l_4}\dot\theta_{e30} \qquad (3.18)$$

$$\Delta\dot\theta_e = \sigma\frac{J_e r_2 r_3 l_2 l_4}{J_b r_1 r_3 l_1 l_3 + J_p r_1 r_4 l_1 l_4 + J_e r_2 r_3 l_2 l_4}\frac{l_2 l_4}{l_1 l_3}\dot\theta_{b30}$$

$$- \sigma\frac{\sigma J_e r_2 r_3 l_2 l_4}{J_b r_1 r_3 l_1 l_3 + J_p r_1 r_4 l_1 l_4 + J_e r_2 r_3 l_2 l_4}\dot\theta_{e30} \qquad (3.19)$$

Step 6: The escape wheel pushes the pallet fork and the balance wheel together and makes up the energy exhausted during the previous impulses. For the balance wheel, one has

$$J_3\ddot\theta_b + K\theta_b = \sigma T_3 \qquad (3.20)$$

where

$$J_3 = J_b + J_p\frac{r_4}{r_3}\frac{l_4}{l_3} + J_e\frac{(r_2 - \mu r_{10})r_4}{(r_1 + \mu r_9)r_3}\frac{l_2 l_4}{l_1 l_3}$$

$$T_3 = \tau\frac{(r_2 - \mu r_{10})r_4}{(r_1 + \mu r_9)r_3}$$

This step ends when the escape tooth leaves the pallet, fork, generating the displacements and velocities $\theta_{b4}, \theta_{p4}, \theta_{e4}, \dot\theta_{b4}, \dot\theta_{p4}, \dot\theta_{e4}$.

Step 7: This step is almost the same as Step 4. For the balance wheel, one has

$$J_2\ddot\theta_b + K\theta_b = 0 \qquad (3.21)$$

For the escape wheel

$$J_e \ddot{\theta}_e = -\tau \qquad (3.22)$$

This step ends when one escape wheel tooth reaches the other pallet, giving the displacements and velocities $\theta_{b5}, \theta_{p5}, \theta_{e5}, \dot{\theta}_{b50}, \dot{\theta}_{p50}, \dot{\theta}_{e50}$.

Step 8: The escape wheel impacts the pallet fork, giving the forth shock. This step is similar to Step 5:

$$\frac{J_b}{r_4 r_6}\dot{\theta}_{b50} - \frac{J_p}{r_3 r_6}\dot{\theta}_{p50} - \sigma\frac{J_e}{r_4 r_5}\dot{\theta}_{e50} = \frac{J_b}{r_4 r_6}\dot{\theta}_{b51} - \frac{J_p}{r_3 r_6}\dot{\theta}_{p51} - \sigma\frac{J_e}{r_4 r_5}\dot{\theta}_{e51} \qquad (3.23)$$

$$\Delta\dot{\theta}_b = \frac{J_e r_3 r_6 l_2 l_4}{J_b r_3 r_5 l_1 l_3 + J_p r_4 r_5 l_1 l_4 + J_e r_3 r_6 l_2 l_4}\dot{\theta}_{b50}$$
$$- \frac{\sigma J_e r_3 r_6 l_1 l_3}{J_b r_3 r_5 l_1 l_3 + J_p r_4 r_5 l_1 l_4 + J_e r_3 r_6 l_2 l_4}\dot{\theta}_{e50} \qquad (3.24)$$

$$\Delta\dot{\theta}_p = -\frac{J_e r_3 r_6 l_2 l_4}{J_b r_3 r_5 l_1 l_3 + J_p r_4 r_5 l_1 l_4 + J_e r_3 r_6 l_2 l_4}\frac{l_4}{l_3}\dot{\theta}_{b50}$$
$$+ \frac{\sigma J_e r_3 r_6 l_1 l_4}{J_b r_3 r_5 l_1 l_3 + J_p r_4 r_5 l_1 l_4 + J_e r_3 r_6 l_2 l_4}\dot{\theta}_{e50} \qquad (3.25)$$

$$\Delta\dot{\theta}_e = \sigma\frac{J_e r_3 r_6 l_2 l_4}{J_b r_3 r_5 l_1 l_3 + J_p r_4 r_5 l_1 l_4 + J_e r_3 r_6 l_2 l_4}\frac{l_2 l_4}{l_1 l_3}\dot{\theta}_{b50}$$
$$- \sigma\frac{\sigma J_e r_3 r_6 l_2 l_4}{J_b r_3 r_5 l_1 l_3 + J_p r_4 r_5 l_1 l_4 + J_e r_3 r_6 l_2 l_4}\dot{\theta}_{e50} \qquad (3.26)$$

Step 9: The torque on the escape wheel drives it to lock the pallet fork.

$$J_4 \ddot{\theta}_e = -\tau \qquad (3.27)$$

where

$$J_4 = J_e + J_p \frac{(r_5 + \mu r_7)\, l_1}{(r_6 - \mu r_8)\, l_2}$$

The balance wheel starts the descending supplementary arc, governed by

$$J_b \ddot{\theta}_b + K\theta_b = 0 \qquad (3.28)$$

This step ends when the pallet fork reaches the banking pin, giving the displacements and velocities $\theta_{b6}, \theta_{p6}, \theta_{e6}, \dot{\theta}_{b60}, \dot{\theta}_{p60}, \dot{\theta}_{e60}$. Note that

$$\theta_{b6} = -\theta_{b1} \qquad (3.29)$$

Step 10: The last shock occurs when the pallet fork collides with the banking pin. The governing equation is

$$\dot{\theta}_{p61} = 0, \dot{\theta}_{e61} = 0 \tag{3.30}$$

Therefore,

$$\Delta\dot{\theta}_b = 0 \tag{3.31}$$

$$\Delta\dot{\theta}_p = -\dot{\theta}_{e60} \tag{3.32}$$

$$\Delta\dot{\theta}_e = -\dot{\theta}_{e60} \tag{3.33}$$

Step 11: The balance wheel continues the descending motion along the supplementary arc while the pallet fork and the escape wheel are stationary. The governing equation is

$$J_b\ddot{\theta}_b + K\theta_b = 0 \tag{3.34}$$

This step ends when the angle reaches the maximum θ_{b7}, on which the angular velocity is zero. Subsequently, the balance wheel goes back in the opposite direction and the resulting motion is the same as that during the previous half-period.

3.3.2 Impulsive Differential Equation

Because of the shocks that are present in the system (i.e., the collisions involved during locking and unlocking), the ordinary differential equations in the developed model cannot be solved by ordinary numerical methods. Hence, the concept of impulsive differential equation is introduced. It is particularly designed for simulation of the processes in which the parameters undergo relatively long periods of smooth variation followed by a short-term rapid change in their values. There are three components in an impulsive differential equation: a continuous-time-differential equation, an impulse equation, and a jump criterion [9–12]. The first one determines the states between impulses. The second describes the jump by a jump function at the instant of the impulse. The last one is an event set by which the jump function works. These three components can be written by a general differential equation as follows:

$$\frac{dx}{dt} = g(t,x), \quad \text{if } \phi(t,x) \neq 0 \tag{3.35}$$

$$\Delta x = \rho(x), \quad \text{if } \phi(t,x) = 0 \tag{3.36}$$

where, $\phi(t,x) = 0$ is the jump set and $\rho(x)$ is the jump function. Define the state vector

$$x \equiv [\theta_b, \theta_p, \theta_e, \dot{\theta}_b, \dot{\theta}_p, \dot{\theta}_e]^T \tag{3.37}$$

The jump set is

$$S = S_1 \cup S_2 \cup S_3 \cup S_4 \tag{3.38}$$

where, $S_1 = \{x : x_1 = \theta_{b1}\}$, $S_2 = \{x : x_1 = \theta_{b3}\}$, $S_3 = \{x : x_1 = \theta_{b5}\}$, $S_4 = \{x : x_1 = \theta_{b6}\}$. The jump function is given by

$$
\rho(x) = \begin{cases}
\begin{bmatrix}
0 & 0 & 0 & 0 & 0 & 0 \\
0 & 0 & 0 & 0 & 0 & 0 \\
0 & 0 & 0 & 0 & 0 & 0 \\
0 & 0 & 0 & a_1 & 0 & 0 \\
0 & 0 & 0 & a_2 & 0 & 0 \\
0 & 0 & 0 & a_3 & 0 & 0
\end{bmatrix} x, & x \in S_1 \\[2em]
\begin{bmatrix}
0 & 0 & 0 & 0 & 0 & 0 \\
0 & 0 & 0 & 0 & 0 & 0 \\
0 & 0 & 0 & 0 & 0 & 0 \\
0 & 0 & 0 & a_4 & 0 & a_5 \\
0 & 0 & 0 & a_6 & 0 & a_7 \\
0 & 0 & 0 & a_8 & 0 & a_9
\end{bmatrix} x, & x \in S_2 \\[2em]
\begin{bmatrix}
0 & 0 & 0 & 0 & 0 & 0 \\
0 & 0 & 0 & 0 & 0 & 0 \\
0 & 0 & 0 & 0 & 0 & 0 \\
0 & 0 & 0 & a_{10} & 0 & a_{11} \\
0 & 0 & 0 & a_{12} & 0 & a_{13} \\
0 & 0 & 0 & a_{14} & 0 & a_{15}
\end{bmatrix} x, & x \in S_3 \\[2em]
\begin{bmatrix}
0 & 0 & 0 & 0 & 0 & 0 \\
0 & 0 & 0 & 0 & 0 & 0 \\
0 & 0 & 0 & 0 & 0 & 0 \\
0 & 0 & 0 & 0 & 0 & 0 \\
0 & 0 & 0 & 0 & -1 & 0 \\
0 & 0 & 0 & 0 & 0 & -1
\end{bmatrix} x, & x \in S_4
\end{cases}
\tag{3.39}
$$

where

$$
a_1 = -\frac{J_p r_4 r_5 l_1 l_4 + J_e r_4 r_6 l_2 l_4}{J_b r_3 r_5 l_1 l_3 + J_p r_4 r_5 l_1 l_4 + J_e r_4 r_6 l_2 l_4}
$$

$$
a_2 = -\frac{J_b r_3 r_5 l_1 l_4}{J_b r_3 r_5 l_1 l_3 + J_p r_4 r_5 l_1 l_4 + J_e r_4 r_6 l_2 l_4}
$$

$$
a_3 = \sigma \frac{J_b r_3 r_5 l_2 l_4}{J_b r_3 r_5 l_1 l_3 + J_p r_4 r_5 l_1 l_4 + J_e r_4 r_6 l_2 l_4}
$$

$$
a_4 = \frac{J_e r_2 r_3 l_2 l_4}{J_b r_1 r_3 l_1 l_3 + J_p r_1 r_4 l_1 l_4 + J_e r_2 r_3 l_2 l_4}
$$

$$
a_5 = -\frac{\sigma J_e r_2 r_3 l_1 l_3}{J_b r_1 r_3 l_1 l_3 + J_p r_1 r_4 l_1 l_4 + J_e r_2 r_3 l_2 l_4}
$$

$$a_6 = -\frac{J_e r_2 r_3 l_2 l_4}{J_b r_1 r_3 l_1 l_3 + J_p r_1 r_4 l_1 l_4 + J_e r_2 r_3 l_2 l_4}$$

$$a_7 = \frac{\sigma J_e r_2 r_3 l_1 l_4}{J_b r_1 r_3 l_1 l_3 + J_p r_1 r_4 l_1 l_4 + J_e r_2 r_3 l_2 l_4}$$

$$a_8 = \sigma \frac{J_e r_2 r_3 l_2 l_4}{J_b r_1 r_3 l_1 l_3 + J_p r_1 r_4 l_1 l_4 + J_e r_2 r_3 l_2 l_4} \frac{l_2 l_4}{l_1 l_3}$$

$$a_9 = -\sigma \frac{\sigma J_e r_2 r_3 l_2 l_4}{J_b r_1 r_3 l_1 l_3 + J_p r_1 r_4 l_1 l_4 + J_e r_2 r_3 l_2 l_4}$$

$$a_{10} = \frac{J_e r_3 r_6 l_2 l_4}{J_b r_3 r_5 l_1 l_3 + J_p r_4 r_5 l_1 l_4 + J_e r_3 r_6 l_2 l_4}$$

$$a_{11} = -\frac{\sigma J_e r_3 r_6 l_1 l_3}{J_b r_3 r_5 l_1 l_3 + J_p r_4 r_5 l_1 l_4 + J_e r_3 r_6 l_2 l_4}$$

$$a_{12} = -\frac{J_e r_3 r_6 l_2 l_4}{J_b r_3 r_5 l_1 l_3 + J_p r_4 r_5 l_1 l_4 + J_e r_3 r_6 l_2 l_4} \frac{l_4}{l_3}$$

$$a_{13} = \frac{\sigma J_e r_3 r_6 l_1 l_4}{J_b r_3 r_5 l_1 l_3 + J_p r_4 r_5 l_1 l_4 + J_e r_3 r_6 l_2 l_4}$$

$$a_{14} = \sigma \frac{J_e r_3 r_6 l_2 l_4}{J_b r_3 r_5 l_1 l_3 + J_p r_4 r_5 l_1 l_4 + J_e r_3 r_6 l_2 l_4} \frac{l_2 l_4}{l_1 l_3}$$

$$a_{15} = -\sigma \frac{\sigma J_e r_3 r_6 l_2 l_4}{J_b r_3 r_5 l_1 l_3 + J_p r_4 r_5 l_1 l_4 + J_e r_3 r_6 l_2 l_4}$$

The supplementary set of the jump set is

$$\bar{S} = S_5 \quad S_6 \quad S_7 \quad S_8 \quad S_9 \tag{3.40}$$

where

$$S_5 = \{x : |x_1| > \theta_{b1}\}, \quad S_6 = \{x : (x_1 - \theta_{b1})(x_1 - \theta_{b2}) < 0\}$$

$$S_7 = \{x : (x_1 - \theta_{b2})(x_1 - \theta_{b3}) < 0\} \cup \{x : (x_1 - \theta_{b4})(x_1 - \theta_{b5}) < 0\}$$

$$S_8 = \{x : (x_1 - \theta_{b3})(x_1 - \theta_{b4}) < 0\} \quad \text{and} \quad S_9 = \{x : (x_1 - \theta_{b5})(x_1 - \theta_{b6}) < 0\}$$

In summary, the continuous-time differential equation is

$$\dot{x} = f(x) = \begin{cases} \begin{bmatrix} 0 & 0 & 0 & 1 & 0 & 0 \\ 0 & 0 & 0 & 0 & 1 & 0 \\ 0 & 0 & 0 & 0 & 0 & 1 \\ -\dfrac{K}{J_b} & 0 & 0 & 0 & 0 & 0 \\ 0 & 0 & 0 & 0 & 0 & 0 \\ 0 & 0 & 0 & 0 & 0 & 0 \end{bmatrix} x, & x \in S_5 \\[3em] \begin{bmatrix} 0 & 0 & 0 & 1 & 0 & 0 \\ 0 & 0 & 0 & 0 & 1 & 0 \\ 0 & 0 & 0 & 0 & 0 & 1 \\ -\dfrac{K}{J_1} & 0 & 0 & 0 & 0 & 0 \\ 0 & 0 & 0 & 0 & 0 & 0 \\ 0 & 0 & 0 & 0 & 0 & 0 \end{bmatrix} x + \begin{bmatrix} 0 \\ 0 \\ 0 \\ -\sigma \dfrac{1}{J_1} \dfrac{r_4(r_6 + \mu r_8)}{r_3(r_5 - \mu r_7)} \\ \sigma \dfrac{1}{J_1} \dfrac{r_4(r_6 + \mu r_8)}{r_3(r_5 - \mu r_7)} \dfrac{l_4}{l_3} \\ \dfrac{1}{J_1} \dfrac{r_4(r_6 + \mu r_8)}{r_3(r_5 - \mu r_7)} \dfrac{l_2 l_4}{l_1 l_3} \end{bmatrix} \tau, & x \in S_6 \\[3em] \begin{bmatrix} 0 & 0 & 0 & 1 & 0 & 0 \\ 0 & 0 & 0 & 0 & 1 & 0 \\ 0 & 0 & 0 & 0 & 0 & 1 \\ -\dfrac{K}{J_2} & 0 & 0 & 0 & 0 & 0 \\ 0 & -\dfrac{K}{J_2} & 0 & 0 & 0 & 0 \\ 0 & 0 & 0 & 0 & 0 & 0 \end{bmatrix} x + \begin{bmatrix} 0 \\ 0 \\ 0 \\ 0 \\ 0 \\ -\dfrac{1}{J_e} \end{bmatrix} \tau, & x \in S_7 \\[3em] \begin{bmatrix} 0 & 0 & 0 & 1 & 0 & 0 \\ 0 & 0 & 0 & 0 & 1 & 0 \\ 0 & 0 & 0 & 0 & 0 & 1 \\ -\dfrac{K}{J_3} & 0 & 0 & 0 & 0 & 0 \\ 0 & -\dfrac{K}{J_3} & 0 & 0 & 0 & 0 \\ \sigma \dfrac{K}{J_3} \dfrac{l_2 l_4}{l_1 l_3} & 0 & 0 & 0 & 0 & 0 \end{bmatrix} x + \begin{bmatrix} 0 \\ 0 \\ 0 \\ \sigma \dfrac{1}{J_3} \dfrac{(r_2 - \mu r_{10})r_4}{(r_1 + \mu r_9)r_3} \\ -\sigma \dfrac{1}{J_3} \dfrac{(r_2 - \mu r_{10})r_4}{(r_1 + \mu r_9)r_3} \dfrac{l_4}{l_3} \\ -\dfrac{1}{J_3} \dfrac{(r_2 - \mu r_{10})r_4}{(r_1 + \mu r_9)r_3} \dfrac{l_2 l_4}{l_1 l_3} \end{bmatrix} \tau, & x \in S_8 \\[3em] \begin{bmatrix} 0 & 0 & 0 & 1 & 0 & 0 \\ 0 & 0 & 0 & 0 & 1 & 0 \\ 0 & 0 & 0 & 0 & 0 & 1 \\ -\dfrac{K}{J_b} & 0 & 0 & 0 & 0 & 0 \\ 0 & 0 & 0 & 0 & 0 & 0 \\ 0 & 0 & 0 & 0 & 0 & 0 \end{bmatrix} x + \begin{bmatrix} 0 \\ 0 \\ 0 \\ 0 \\ -\sigma \dfrac{1}{J_4} \dfrac{l_1}{l_2} \\ -\dfrac{1}{J_4} \end{bmatrix} \tau, & x \in S_9 \end{cases} \tag{3.41}$$

These equations can be solved numerically, giving the dynamics of the Swiss lever mechanism.

TABLE 3.1 The Simulation Parameters

r_1	1.381 mm	r_9	1.84 mm	J_e	1×10^{-8} kg.m^2
r_2	0.996 mm	r_{10}	0.98 mm	θ_{b0}	1.2
r_3	2.263 mm	l_1	2.394 mm	θ_{b1}	0.87
r_4	0.545 mm	l_2	1.576 mm	θ_{b6}	0.87
r_5	2.368 mm	l_3	2.27 mm	τ.	1.5×10^{-6} N.m
r_6	0.122 mm	l_4	0.64 mm	μ	0.15
r_7	0.354 mm	J_h	2×10^{-6} kg.m^2		
r_8	1.572 mm	J_p	8×10^{-8} kg.m^2		

3.4 Simulation Studies

Simulations are carried out based on the model presented in the previous section. The parameters of the escapement are summarized in Table 3.1.

Figures 3.5–3.7 show the time history of the angular displacement, velocity, and acceleration of the balance wheel, pallet fork, and the escape wheel. From the figures it is seen that the motion of the balance wheel is mainly determined by the supplementary arc. However, the push steps change the magnitude

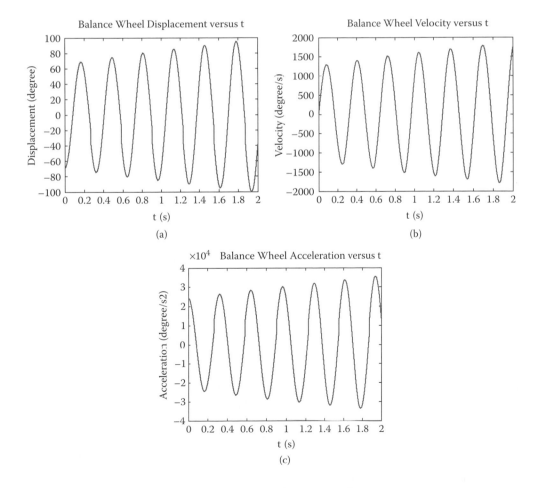

FIGURE 3.5 Motion of balance wheel: (a) θ_b versus t, (b) $\dot{\theta}_b$ versus t, (c) $\ddot{\theta}_b$ versus t.

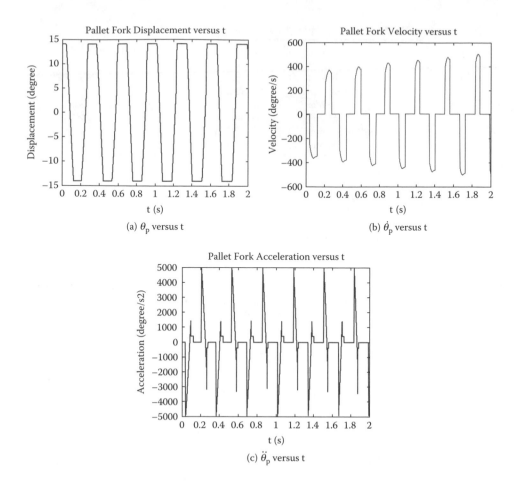

FIGURE 3.6　Motion of pallet fork: (a) θ_p versus t, (b) $\dot{\theta}_p$ versus t, (c) $\ddot{\theta}_p$ versus t.

of oscillation until it is stabilized. There are three impulses that occur within a half period. For the pallet fork, the motion follows a step function. The angle of displacement changes between two extremes as determined by the banking pins. When the displacement reaches the extremes, it remains there for a period of time. From Figure 3.6(c) it is clear that four impulses incur in a half-period. The last one takes place when the pallet fork impacts the guard pin. The angle of the escape wheel changes step by step. The width of the steps is determined by the dynamic system and hence changes in a small neighborhood of a fixed value. The small variation, however, has a significant effect as seen in Figure 3.7(c).

Figure 3.8 shows the velocity of the pallet fork and the escape wheel in a half-period. From the figure it is seen that the first shock and the fifth shock are critical for the pallet fork, whereas the second and the fourth shocks are critical for the escape wheel.

Figure 3.9 shows the relationship between the displacement and velocity. It is seen that the motion is almost periodic. It should be pointed out that, if the dynamics are not considered, the motion will be exactly periodic. However, owing to the dynamics, the initial condition is different each time. It is the escapement mechanism that constrains the dynamic effect, thereby making the escapement accurate in timekeeping. This is shown in Figure 3.9(b). It is seen that regardless of the variation, the pallet fork always comes to the exact position with exact velocity. This is very important in a watch mechanism.

Based on the simulation of 10 periods, the standard deviation of the half-period is found to be 0.0002781 s. The standard deviation of the amplitude is 1.3685 degrees.

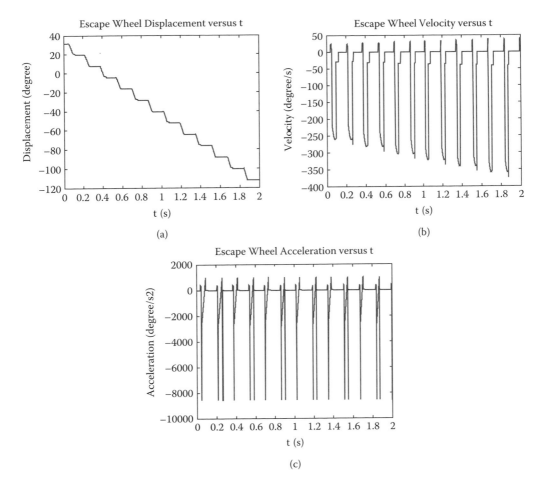

FIGURE 3.7 Motion of escape wheel: (a) θ_e versus t, (b) $\dot{\theta}_e$ versus t, (c) $\ddot{\theta}_e$ versus t.

It should be emphasized here that the use of the impulsive differential equation is necessary. The use of ordinary differential equations has led to mathematical and numerical difficulty. Therefore, it is suggested that any mechanisms involving impacts or abrupt changes should be modeled using impulsive differential equations.

3.5   Conclusion

From the study presented in the chapter, the following conclusions can be made:

1. The Swiss level escapement is a complex mechanical system that can be treated as a mechatronic device. It is necessary to use impulsive differential equations to model this system.
2. The Swiss level escapement mechanism has a complex pattern of motion. It can be regarded as a type of mechanical switch. With an applied torque, it can self-start and run almost periodically. Within a half period, the balance wheel experiences three shocks, whereas the pallet fork and the escape wheel experience four shocks.
3. Owing to the dynamics of the system, the motion of the balance wheel is near periodic, and the variation is noticeable. It is the pallet fork that constrains the effect of the dynamics, making the escapement accurate for the purpose of timekeeping.

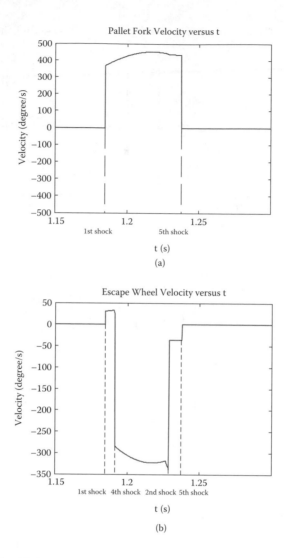

FIGURE 3.8 Critical shocks: (a) shocks on pallet fork, (b) shocks on escape wheel.

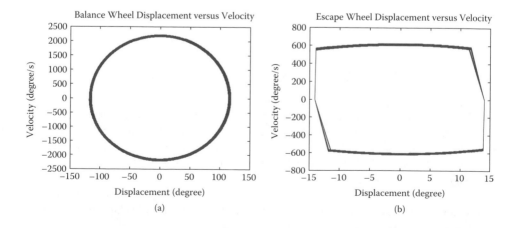

FIGURE 3.9 Phase diagram: (a) for balance wheel, (b) for pallet fork.

The impulsive differential equation is a powerful tool. It can be applied to model other escapements, mechanisms, or mechatronic devices that involve shocks.

Acknowledgment

This project is supported by the Hong Kong Watch Manufacturers Association (HKWMA), Federation of Hong Kong Watch Industry and Trade (FHKWIT), and Hong Kong Innovation and Technology Commission (ITC).

References

1. Lepschy, A.M., Mian, G.A., and Viaro, U., Feedback control in ancient water and mechanical clocks, *IEEE Transactions on Education*, Vol. 35, No. 1, 3–10, February 1992.
2. Roup, A.V. and Bernstein, D.S., On the dynamics of the escapement mechanism of a mechanical clock, *Proceedings of the 38th IEEE Conference on Decision and Control*, Phoenix, Arizona, Vol. 3, December 1999, pp. 2599–2604.
3. Roup, A.V., Bernstein, D.S., Nersesov, S.G., Haddad, W.M., and Chellaboina, V., Limit cycle analysis of the verge and foliot clock escapement using impulsive differential equations and poincare maps, *Proceedings of the 2001 American Control Conference*, Arlington, VA, Vol. 4, June 2001, pp. 3245–3250.
4. Schwartz, C. and Gran, R., Describing function analysis using MATLAB and Simulink, *IEEE Control Systems Magazine*, Vol. 21, No. 4, 1927, August 2001.
5. Headrick, M.V., Origin and evolution of the anchor clock escapement, *IEEE Control Systems Magazine*, Vol. 22, No. 2, April 2002.
6. Daniels, G., *Watchmaking*, Sotheby's Publications, London, 1985.
7. Headrick, M.V., *Clock and Watch Escapement Mechanics*, 1997 [available on line at: http://www.abbeyclock.com/TToc.htm].
8. Reymondin, C., Monnier, G., Jeanneret, D., and Pelaratti, U., *The Theory of Horology*, The Swiss Federation of Technical Colleges, Geneva, Switzerland, 1999.
9. Lakshmikantham, V., Bainov, D.D., and Simeonov, P.S., *Theory of Impulsive Differential Equations*, World Scientific, Singapore, 1988.
10. Bainov, D.D. and Simeonov, P.S., *Impulsive Differential Equations: Periodic Solutions and Applications*, Longman Scientific, Essex, U.K. 1993.
11. Samoilenko, A.M. and Perestyuk, N.A., *Impulsive Differential Equations*, World Scientific, Singapore, 1995.
12. Bainov, D.D. and Simeonov, P.S., *System with Impulse Effect*, Ellis Horwood, New York, 1989.

4

Instrumented Wheel for Wheelchair Propulsion Analysis

M. Mallakzadeh

F. Sassani

Summary

In this chapter the development of an instrumented wheel for use in a kinetic and kinematic study of manual wheelchair propulsion is presented. The instrumented wheel is a mechatronic device that is composed of a six-component load transducer and modules of signal conditioning, processing, and amplification. A slip ring is used for continuous operation and data transmission. The angular position of the wheel is measured by an encoder, and its resolution is mechanically amplified through a gear system. Two PCs are used in the system and are activated via a common mouse for consistent data acquisition in various tests. The loads applied by the wheelchair user on the handrim are calculated. The angular position of the wheelchair user's hand on the handrim during the pushing phase (φ) is calculated by means of kinetic parameters without using a camera or a motion analysis system. The propulsion moment with respect to the hand coordinate system (M_{hz}) is calculated using φ. Next, a general uncertainty analysis is performed to determine the uncertainty equations for the local and global forces and moments, the local hand forces and moments, and the hand-contact angular position. The numerical uncertainty values for these parameters are calculated using the related equations. The results provide an estimate of the errors and uncertainties for the output of the instrumented wheel. Finally, a complete experimental procedure is performed to determine the specifications of the instrumented wheel. The static and dynamic test set-ups are designed, and tests are performed under different conditions. The results of the static and dynamic tests are used for both qualitative and quantitative verification of the system specifications.

4.1 Introduction

Manual wheelchair propulsion (MWP) is an inefficient and physically straining process. Manual wheelchair users (MWUs) depend on wheelchairs for most of their daily activities. It is essential for them to be comfortable when they use their wheelchairs. Manual wheelchair users do, however, benefit greatly from the cardiovascular exercise involved in propulsion [1]. Actually, the nature of wheelchair propulsion is such that manual wheelchair users are essentially walking with their arms [2]. The propulsion of the standard and commonly used handrim wheelchair is a form of ambulation, whose mechanical efficiency is about 10% at best [3–7]. As a consequence, MWP is associated with a high mechanical load on the joints of the upper limbs, which may lead to overuse injuries in the shoulder, elbow, and wrist.

Researchers and designers have made many attempts to improve the efficiency of MWP, explain its dynamics, and determine the causes of the injuries associated with the use of manual wheelchairs. They have tried to better understand this activity and improve it by using new propulsion techniques, new wheelchair designs, or both. One category of related research focuses on the kinetics of wheelchair propulsion [7–9]. A number of research groups have fabricated and instrumented wheels to measure the forces and moments applied on the handrim by the wheelchair user [10–12].

Knowing the forces and moments that a wheelchair user exerts on the handrim is necessary for an inverse dynamics solution and the calculation of the forces and moments in the upper limb joints. Collecting reliable three-dimensional (3-D) kinetic data, especially the forces and moments applied to the handrim of a manual wheelchair, is one of the challenging aspects of gaining an in-depth understanding of the biomechanics of wheelchair propulsion.

Although it is advantageous to develop an in-house system, which allows greater flexibility in adding hardware and obtaining various signals, it is important to determine the specifications of a fabricated instrument. Such information determines the system's level of reliability and usefulness. The collected data cannot be useful if they are not reliable. Uncertainty analysis is a method that can help the researcher to calculate the level of uncertainty of the acquired data. This method can estimate the expected errors for different results obtained from the system. Cooper et al. [13] determined the uncertainty levels of the data acquired using their instrumented wheel (Smart^{wheel}). They determined the uncertainty for the forces and moments as 1.1–2.5 N and 0.03–0.19 N.m in the plane of the handrim, and 0.93 N and 2.24 N.m in the wheel axle direction, respectively. As uncertainty analysis is an analytical method, one will have greater confidence when an experimental technique uses the actual output of the system to determine the specification of the instrumented device. Wu et al. [11] performed static and dynamic analyses for their fabricated instrumented wheel and determined the linearity and drift of their system.

In the present work, an instrumented wheel system is fabricated and validated by using general uncertainty analysis. Also, the specifications of the designed and instrumented wheel system are determined by using both static and dynamic experiments. This system enables the forces and moments applied by the wheelchair user on the handrim to be determined. It is important to understand how these forces and moments are generated and what factors influence them. The applied loads are part of the data required to calculate propulsion efficiency and analyze the optimum seat position for wheelchair users, in order to improve performance, identify the probable causes of injuries, and develop prevention techniques.

Apart from describing these developments, this chapter also presents a procedure for calculating the essential dynamic variables used in the study of manual wheelchair propulsion. An important feature of the force/moment calculation procedure is that, together with encoder data analysis, it allows one to determine the angular position of the contact point between the hand and the handrim without cameras. This angular position is a critical factor in determining moments and the effective tangential force counteracting on the wheelchair user's hands and upper limbs, which can result in discomfort or injury. A complete and general uncertainty analysis is performed here for different outputs of the instrumented wheel, and the system's level of reliability is determined. The results indicate that the uncertainty levels for the forces and moments of interest are in the range of 1.4–1.7 N and 0.58–0.68 N.m in the plane of the handrim, and about 3.40 N and 0.25 N.m in the wheel axle direction, respectively. For the developed system, however, the uncertainty values for the important load components, namely the planar forces and axial moment, are low. The resulting uncertainties represent an estimation of the expected errors in future data gathering and analysis.

The static and dynamic test protocols are designed to cover all loading conditions. To determine the specifications of the system, the linearity, repeatability and the mean error of the measurement system in static and dynamic situations are calculated. These specifications allow one to determine the level of the system's reliability, and gain confidence in the results and future applications.

4.2   Instrumentation

To conduct an uncertainty analysis, as well as static and dynamic verification for manual wheelchair propulsion, stationary tests are performed to measure the forces and moments applied by the wheelchair user on the handrim, and the angular position of the wheel during propulsion. Currently, there is no sensor available that can measure the required forces and moments directly. Instrumented wheels are mechatronic systems (a combination of hardware and software) that process the data acquired by the sensors and calculate the desired values. In the work presented in this chapter, an instrumented wheel assembly is designed and fabricated and the required set-up for the tests is prepared to accurately determine the uncertainties and system specifications. The set-up for this study consists of a wheelchair, the instrumented wheel, a platform with two rollers to allow stationary tests (roller-rig), an AC motor, two personal computers (PCs), an analog/digital (A/D) data acquisition board, and four different static and dynamic loading set-ups.

4.2.1   Wheelchair

A Quickie (Sunrise Medical, Inc.) 40-cm-wide wheelchair with standard spokes is used in this study. The wheelchair has solid gray rubber tires 58.25 cm in diameter and 3 cm in width. The handrim is 54 cm in diameter, and the positions of the backrest and axle of the wheels (thus, the seat) are adjustable.

4.2.2   Instrumented Wheel

The instrumented wheel system itself consists of a standard-spoke wheel from Quickie wheelchair, a six-component load transducer (Model PY6-500, Bertec, Inc., Measurement Excellence™), an AM 6500 external amplifier, a signal conditioning circuit, a power supply, a handrim assembly, an encoder (S1 360 IB), a slip ring (AC 6373), two gears, and insulated shielded cables (Figure 4.1). As some of the data acquisition components had to be mounted within the wheelchair, counterweights were used to maintain the wheel's rotational balance.

The PY6 load transducer comes with a digitally stored calibration matrix and uses 16-bit digital signal acquisition and conditioning with a sampling rate of 1 kHz. The digital signal output can be plugged directly into the standard USB port of a personal computer via the AM 6500 amplifier without requiring an additional PC board for analog-to-digital (A/D) signal conversion.

The transducer has sensitivity levels of 2 mV/N and 2 mV/(N.m); an accuracy of 99.5%; ±1.0% cross-talk; and a full mechanical load rating of 1250 N for in-plane forces F_x and F_y, 2500 N for the out-of-plane force F_z, 60 N.m for moments M_x and M_y, and 30 N.m for moment M_z with respect to the first local

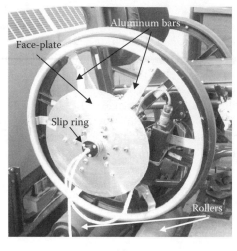

(a)

(b)

FIGURE 4.1 Instrumented wheel: (a) side view, (b) front view.

coordinate system (Figure 4.2). The load cell is mounted on the wheel with its z-axis aligned with the axle of the wheel. The origins of the global and first local coordinate systems are concentric.

The handrim assembly has three parts: a handrim, a 3-mm-thick round aluminum face-plate, and six reinforced aluminum spokes. The encoder has three input-output channels, and its resolution is 1°. Gears are used to transfer the wheel rotation, eliminate the wheel-slipping effect on the rollers, and, with a gear ratio of 8 to 1 increase the resolution of the measured angular position (θ) to 0.125°. The first (large) gear is mounted on the wheel shaft and the second (small) gear is mounted on the encoder shaft (Figure 4.3). A slip ring aligned with the wheel axle allows continuous transfer of the transducer signals to the computer during wheel rotation (Figure 4.1).

4.2.3 Roller Rig

The roller rig is a platform with two parallel rollers to allow stationary tests. The entire wheelchair is placed on the roller rig, which has adjustable legs to maintain a horizontal position (Figures 4.1b and 4.4).

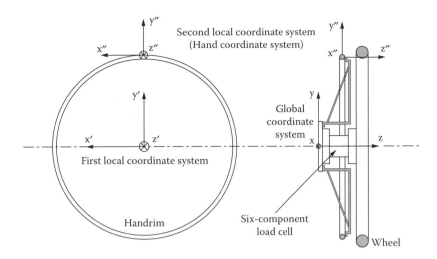

FIGURE 4.2 Initial position and orientation of global and two local coordinate systems on the instrumented wheel.

4.2.4 AC Motor

The AC motor is connected to the shaft of one of the rollers to rotate the instrumented wheel during some of the tests. The motor speed is adjustable, and three different wheelchair angular velocities (3.0, 3.8, and 4.8 rad/s) are used for the dynamic tests (Figure 4.4). The equivalent linear velocities are 3.15, 3.98, and 5.03 km/h, respectively.

4.2.5 Computers

The experimental set-up includes two personal computers. The transducer interface software Digital Acquire™ uses one computer to record the load exerted on the handrim by the user. The LabVIEW™ software on the second computer calculates the wheel's angular position as the load data is collected on the first computer. The output of the encoder is an analog voltage. So, the angular position of the wheel is calculated by using the LabVIEW software and its counting option, which increases the value of the output by 360 per each revolution of the shaft of the encoder. A single PC is not used for both measurements

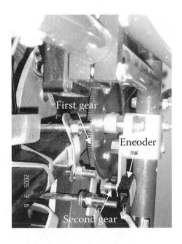

FIGURE 4.3 Encoder gear system.

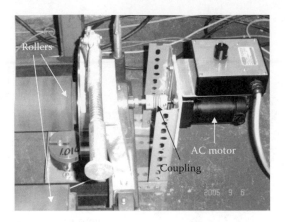

FIGURE 4.4 AC motor and its coupling to a shaft of the roller rig.

because running two different software programs at the same time on one PC causes the measured times to be incompatible and shifted.

4.2.6 Data Acquisition Board

A 12-bit analog-to-digital (A/D) signal conversion board (PCI-6025E) transfers the encoder's analog output data (Note: the used encoder has built-in hardware that converts the encoder pulses into an analog output) to the computer. For consistency, when tests are set up, a specially wired push-button is used to activate both PCs at the same instance.

Figures 4.5 and 4.6 give global views of the system as a schematic sketch and block diagrams, respectively.

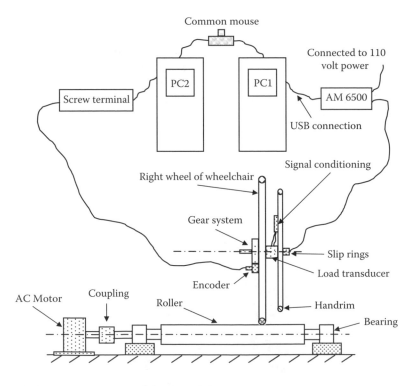

FIGURE 4.5 Global schematic rear view of the physical data acquisition system.

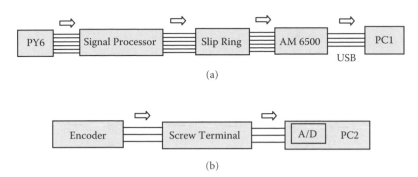

FIGURE 4.6 Measurement signal flow diagram for (a) load and (b) angular position.

4.2.7   Static and Dynamic Loading Set-Ups

To perform the tests, three different static loading set-ups and one dynamic loading set-up are used. These set-ups are elaborated in Section 4.5.1.

In the next section, three coordinate systems for the instrumented wheel are introduced and the required dynamic equations are generated to determine the applied forces and moments of the user's hand on the handrim.

4.3   Generation of Dynamic Equations

To characterize and measure the forces and moments applied by the user's hand on the handrim, three different coordinate systems are used (Figure 4.2). The global and first local coordinate systems have the same origin at the center of the wheel and the same direction in the beginning of the propulsion, which is the direction of the transducer coordinate system. The first local coordinate rotates with the wheel. The origin of the second local coordinate system (hand-coordinate system) is at the contact point between the hand and the handrim and moves with the handrim, but its axes stay parallel to the global coordinate axes.

4.3.1   System Calibration

As indicated before, the PY6 load transducer comes with the calibration matrix digitally stored in it and interfaces to the PC through software (Digital Acquire™). Therefore, there is no need to determine this matrix, due to the automatic conversion of voltages to forces and moments, and the cross-coupling involved. Using this transducer, the digital signal output is plugged directly into the standard USB port of a personal computer without requiring an A/D board. During the propulsion phase, in addition to the loads produced by the user's hand, the system experiences dynamic preloads due to the rotating weight of the measurement system and the balancing weights, which should be taken into consideration for the purpose of eliminating their effects.

4.3.2   Preload Equations

The instrumented wheel is mechanically turned by the AC motor on the rollers without applying any force on the handrim to measure the net preloads. As the preloads change sinusoidally with the rotation of the wheel, their values are calculated and their equations are determined with respect to the global coordinate system.

All of the preloads follow a generic periodic equation given by

$$P = a \, \mathrm{Sin}(\theta + \alpha) + b \qquad (4.1)$$

TABLE 4.1   Constants for Different Preload Equations

P (preload)	a [N, N.m]	b [N, N.m]	α [radian]
F_{px}	−26.5	2.5	0
F_{py}	25.5	24.5	−π/2
F_{pz}	1	~0	0
M_{px}	1.05	−1.05	π/2
M_{py}	−1.2	−0.2	0
M_{pz}	−0.1	~0	0

In this equation, P represents preload forces or moments, a and b are constants with dimensions [N] or [N.m], θ is the wheel's angular position [radian] (which is a function of time), and α is the phase difference [radian] for the output of the x, y, and z channels. Table 4.1 shows the preload equation constants for various load components. They are obtained by using the measured preloads, F_{px}, F_{py}, F_{pz}, M_{px}, M_{py}, and M_{pz}, with respect to the first local coordinate system.

4.3.3   Local and Global Forces and Moments

Forces and moments produced by the data acquisition software are not the values required for our analysis. The effects of preloads must be considered, and by using the following equations, one can calculate the net local forces and moments with respect to the first local coordinate system:

$$F_{Lx} = F_x - F_{px}$$

$$F_{Ly} = F_y - F_{py}$$

$$F_{Lz} = F_z - F_{pz}$$

$$M_{Lx} = M_x - M_{px}$$

$$M_{Ly} = M_y - M_{py}$$

$$M_{Lz} = M_z - M_{pz}$$

(4.2)

where F_{Lx}, F_{Ly}, F_{Lz}, M_{Lx}, M_{Ly}, and M_{Lz} are the forces and moments applied by the wheelchair user, and F_x, F_y, F_z, M_x, M_y, and M_z are the measured forces and moments. All values are with respect to the first local coordinate system at the center of the wheel.

The first local coordinate system is fixed to the wheel and rotates with it. The global coordinate system must therefore be used to calculate the forces and moments with respect to a fixed reference system. Thus, the next step is to transform the values from the first local to the global coordinate system. It should be emphasized that the origin of the global coordinate system coincides with that of the first local coordinate system, and that their z-axes are aligned. To calculate forces and moments in the global coordinate system, the following transformation relations, with reference to Figure 4.7, are used:

$$F_{gx} = \text{Cos}\theta \times F_{Lx} - \text{Sin}\theta \times F_{Ly}$$

$$F_{gy} = \text{Sin}\theta \times F_{Lx} + \text{Cos}\theta \times F_{Ly}$$

$$F_{gz} = F_{Lz}$$

$$M_{gx} = \text{Cos}\theta \times M_{Lx} - \text{Sin}\theta \times M_{Ly}$$

$$M_{gy} = \text{Sin}\theta \times M_{Lx} + \text{Cos}\theta \times M_{Ly}$$

$$M_{gz} = M_{Lz}$$

(4.3)

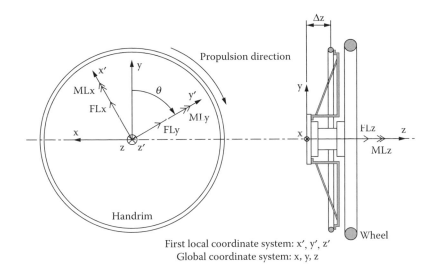

First local coordinate system: x′, y′, z′
Global coordinate system: x, y, z

FIGURE 4.7 Illustration of local loads after $\theta°$ of wheel rotation.

These relations can be expressed in the matrix form as

$$
\begin{bmatrix} F_{gx} \\ F_{gy} \\ F_{gz} \\ M_{gx} \\ M_{gy} \\ M_{gz} \end{bmatrix} = \begin{bmatrix} \text{Cos}\theta & -\text{Sin}\theta & 0 & 0 & 0 & 0 \\ \text{Sin}\theta & \text{Cos}\theta & 0 & 0 & 0 & 0 \\ 0 & 0 & 1 & 0 & 0 & 0 \\ 0 & 0 & 0 & \text{Cos}\theta & -\text{Sin}\theta & 0 \\ 0 & 0 & 0 & \text{Sin}\theta & \text{Cos}\theta & 0 \\ 0 & 0 & 0 & 0 & 0 & 1 \end{bmatrix} \begin{bmatrix} F_{Lx} \\ F_{Ly} \\ F_{Lz} \\ M_{Lx} \\ M_{Ly} \\ M_{Lz} \end{bmatrix}
\tag{4.4}
$$

and in the following compact form:

$$
Q_g = \lambda \cdot Q_L
\tag{4.5}
$$

where λ is the transformation matrix for transforming the local into global values, Q_g represents the vector of global forces and moments, and Q_L is the vector of local forces and moments (Figures 4.8 and 4.9).

Using equations (4.1), (4.2), and (4.3), the global forces and moments during the propulsion phase are calculated. The global forces are the same as the local (hand-coordinate system) forces. Figure 4.8 shows the forces produced by the wheelchair user during the pushing phase on the handrim with respect to the global coordinate system. It is postulated that the dip on the curve for F_{gy} (the vertical force in the global coordinate system) during the primary time of the propulsion phase is due to the contact impact between the hand and the handrim. This dip appears in the results because an able-bodied subject (inexperienced wheelchair user) was used in the experiments. The presence or absence of the dip has also been reported by other researchers who employed inexperienced or experienced wheelchair users in their investigations [6,12].

Figure 4.9 shows the moments produced by the wheelchair user's hand with respect to the global coordinate system. These moments are calculated using equations (4.1), (4.2), and (4.3). The curve of M_{gx} (the moment about the global coordinate system's x-axis) shows a spike in the early phase of the

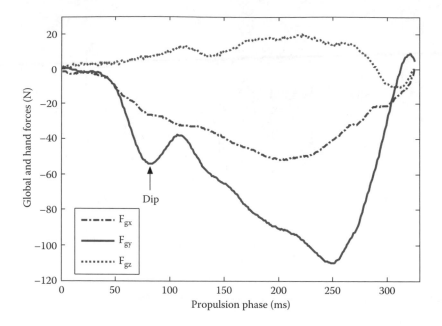

FIGURE 4.8 Propulsion force components with respect to global and hand local coordinate systems.

propulsion, which is due to the dip of F_{gy}. The only important moment for manual wheelchair propulsion is M_{gz}, which is the effective moment. The other two moments are undesirable and reduce propulsion efficiency.

As it is required to determine the forces and moments at the contact point between the hand of the wheelchair user and the handrim during the pushing phase, another transformation from the global coordinate system to the parallel-moving local hand (second local) coordinate system is needed. These

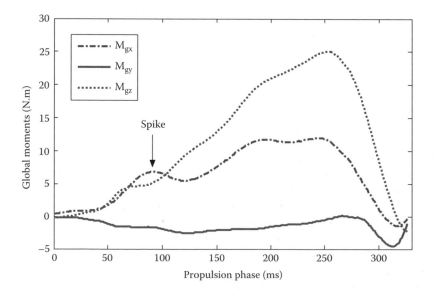

FIGURE 4.9 Propulsion moment components with respect to global coordinate system.

forces and moments, with reference to Figures 4.7 and 4.10, are as follows:

$$F_{hx} = F_{gx}$$

$$F_{hy} = F_{gy}$$

$$F_{hz} = F_{gz}$$

$$M_{hx} = M_{gx} - F_{gz} \times r_h \times \operatorname{Sin}\varphi + F_{gy} \times \Delta z$$

$$M_{hy} = M_{gy} + F_{gz} \times r_h \times \operatorname{Cos}\varphi - F_{gx} \times \Delta z$$

$$M_{hz} = M_{gz} + r_h \times (F_{gx} \times \operatorname{Sin}\varphi - F_{gy} \times \operatorname{Cos}\varphi)$$

(4.6)

where r_h is the handrim radius, and Δz is the offset distance between the plane of the handrim and the origin of the global coordinate system in the z direction. Also, the angle φ is the instantaneous position of the hand on the handrim in the global coordinate system (x-y plane) measured clockwise with respect to the $+x$ axis.

4.3.4 Important Kinetic Factors

The total force (F_{total}) applied on the handrim is obtained by using hand force components (F_{hx}, F_{hy}, and F_{hz}) and equation (4.7). The total effective force (*TEF*), which is the virtual force required to produce propulsion, is obtained by using M_{gz}, the moment around the z-axis and the handrim radius (equation 4.8). The fractional effective force (*FEF*) is an important factor because it shows the ratio of the required force for propulsion and the force produced by the wheelchair user during the propulsion phase. *FEF* (in percentage) is related to F_{total} and *TEF* as in equation (4.9) [12,14,15]. Figure 4.10 illustrates the above moment and forces.

The hand force components and F_{total} are calculated with respect to the second local coordinate system (these components are the same as in the global coordinate system). All other factors are calculated with respect to the global coordinate system.

$$F_{total} = \sqrt{\left(F_{hx}^{\ 2} + F_{hy}^{\ 2} + F_{hz}^{\ 2}\right)}$$

(4.7)

$$TEF = M_{gz}.r_h^{\ -1}$$

(4.8)

$$FEF = \left(\frac{TEF}{F_{total}}\right) \times 100$$

(4.9)

Figure 4.11 shows the total force, the total effective force, and the fractional effective force, which are calculated using equations (4.7), (4.8), and (4.9) and the data from the main test. The total force is related to its components (F_{hx}, F_{hy}, and F_{hz}).

The figure shows a spike on the curve for F_{total} during the early part of the propulsion phase. It is postulated that this spike has resulted from the contact impact between the hand of an able-bodied (inexperienced) wheelchair user and the handrim. The hand of the wheelchair user is in contact with the handrim for the entire propulsion phase, but the nature of the grip changes and affects the level of the loads transmitted. At the beginning and end of the propulsion phase the grip is partial and soft, but during the rest of the propulsion phase the user has a good grip. During part of the early period of the propulsion phase, the strength of the grip increases sharply for able-bodied (inexperienced) wheelchair users, which produces the spike.

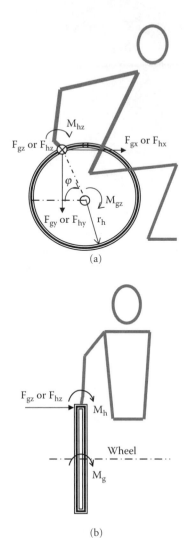

FIGURE 4.10 Illustration of forces and moments applied on the handrim: (a) side view, (b) front view.

The location or the time of the spike is not exactly the same during different tests. The shape of the spike depends on the propulsion style of the wheelchair user. This spike represents a loss of energy, which the user should learn to avoid.

Total effective force is a virtual force that produces the propulsive moment. Considering the generally low levels of efficiency for manual wheelchair propulsion, it is reasonable to expect a lower value for the total effective force compared with the total force produced during the propulsion phase. To improve manual wheelchair propulsion, one should attempt to reduce the total force as much as possible, closer to the total effective force, by choosing the proper wheelchair size and seating position for each user.

Fractional effective force is an important factor in determining the effectiveness of manual wheelchair propulsion and is used as an alternative to efficiency [12,14,15]. Figure 4.11 shows that *FEF* has low values at approximately the first 30% and the last 5% of the propulsion phase. We do not have high reliability in the first and last 5% of the propulsion phase because of the vibrations due to the initial contact between the hand and the handrim, and releasing the handrim. So, except for the first and last 5% of the propulsion phase, *FEF* has its lowest value at the time the spike is produced, which verifies what has been stated before.

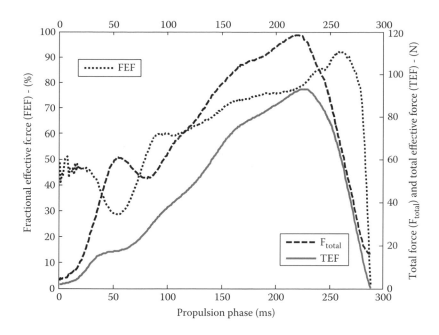

FIGURE 4.11 Total force (F_{total}) and total effective force (*TEF*) in global coordinate system, and fractional effective force (*FEF*) during propulsion.

4.3.5 Hand Position on the Handrim

The angle φ, which represents the hand position on the handrim, can be determined in a number of ways, but some assumptions must be made [12]. Considering the fourth and fifth relations in equations (4.6), (4.10) is obtained. First, M_{hx} and M_{hy} are assumed to be zero, yielding equation (4.11). Although this is a viable approach, it is not used because this equation is based on five parameters, of which only Δz is directly measurable. This poses a high risk of accumulation and propagation of error within the different equations that are needed to calculate φ.

$$\varphi = \mathrm{Tan}^{-1}\left(\frac{M_{gx} - M_{hx} + F_{gy} \times \Delta z}{-M_{gy} + M_{hy} + F_{gx} \times \Delta z}\right) \tag{4.10}$$

With $M_{hx} \approx M_{hy} \approx 0$, one has

$$\varphi = \mathrm{Tan}^{-1}\left(\frac{M_{gx} + F_{gy} \times \Delta z}{-M_{gy} + F_{gx} \times \Delta z}\right) \tag{4.11}$$

Instead, the fifth relation of equations (4.6) is used to obtain equation (4.12). By assuming that only M_{hy} is zero, equation (4.13) is derived for φ (Figure 4.12). This equation is also based on five parameters, but three are directly measurable (Δz, F_{gz}, r_h). Therefore, the risk of error accumulation and propagation is less.

$$\varphi = \mathrm{Cos}^{-1}\left(\frac{M_{hy} - M_{gy} + F_{gx} \times \Delta z}{F_{gz} \times r_h}\right) \tag{4.12}$$

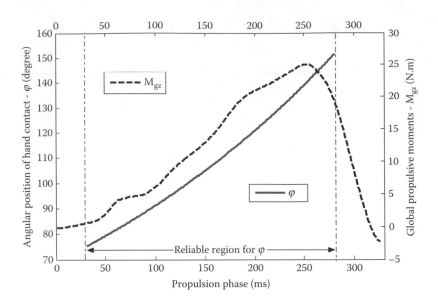

FIGURE 4.12 Global propulsive moments and angular position of the hand in global coordinate system.

With $M_{hy} \approx 0$, which is a reasonable assumption (see later for a discussion on this assumption), φ_c is calculated as

$$\varphi_c = \mathrm{Cos}^{-1}\left(\frac{-M_{gy} + F_{gx} \times \Delta z}{F_{gz} \times r_h}\right) \qquad (4.13)$$

φ_c is the "calculated φ," but for simplicity we will continue to use φ in the derivation. Now, using φ, M_{hz} is calculated as

$$M_{hz} = M_{gz} + r_h \times (F_{gx} \times \mathrm{Sin}\varphi - F_{gy} \times \mathrm{Cos}\varphi) \qquad (4.14)$$

Figure 4.12 shows the global propulsive moment (M_{gz}) and the hand's angular position on the handrim (φ) with respect to the global coordinate system. The results show that one does not have sufficient reliability at the beginning and end of the propulsion phase for the calculated φ, as similarly reported by Cooper et al. [13]. This is likely due to the lack of stability during the initial period of the propulsion phase (roughly the first 10%) when the hand impacts the handrim. During the later part of the propulsion phase (roughly the last 15%) the grip on the handrim becomes soft, and the propulsive moment begins to decrease. So, it is reasonable to attribute these instabilities to the making and breaking of hand contact with the handrim. During much of the propulsion phase, there is a reasonable relation between the behavior of φ and the global propulsive moment.

An iterative method is used to determine the value of M_{hz}. The first iteration is performed by setting M_{hy} to zero, and calculating and M_{hz} from equations (4.13) and (4.14). The new value for M_{hy} is then determined by using the calculated φ and the fifth relation of equation (4.6). In the second iteration, φ and M_{hz} are recalculated by using the new value for M_{hy} and equations (4.12) and (4.14), respectively. M_{hy} is also recalculated using the new value for φ and by performing another iteration with the equations. After three iterations, a comparison of the resulting values for M_{hz} shows very little difference (Figure 4.13). Whereas the iteration may be deemed superfluous, it was included to minimize the calculation error that was observed due to the non-linear form of the equations.

Figure 4.14 illustrates the behavior of the global propulsive moment and hand moment in the z direction during the propulsion phase. It can be seen that they act in opposite directions, meaning that M_{hz} reduces

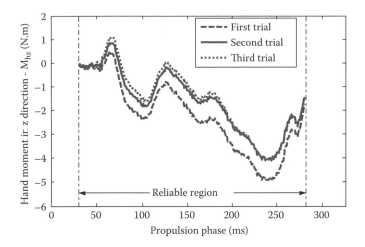

FIGURE 4.13 Comparison of hand moments in *z* direction for three different iterations.

the propulsive moment. This relationship is unavoidable and necessary for natural stability of the propulsion.

Microsoft® Excel, MATLAB® and LabVIEW™ software are used to calculate all forces, moments and φ.

In this section, the transformation matrix between the local and global values has been determined, and the applied forces and moments between the wheelchair user's hand and the handrim were calculated. The angular position of the hand on the handrim during the pushing phase (φ) was calculated by means of the kinetic parameters and the iteration method without using cameras or a motion analysis system. The periods with high estimated errors, due to making and breaking of hand contact (corresponding to the first 10%

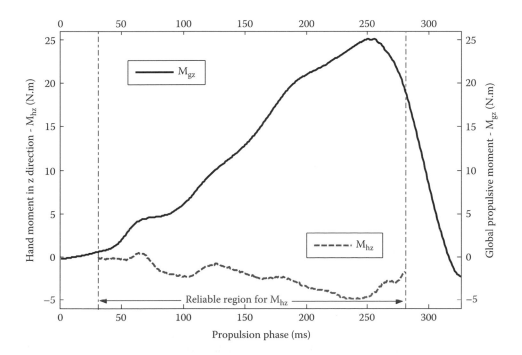

FIGURE 4.14 Global propulsive and hand moments in *z* direction after three trials.

and the last 15% of the propulsion phase), were truncated. If further verification is deemed necessary, a camera system may be used. The propulsion moments with respect to the hand coordinate system (M_{hz}) were calculated using the determined φ. The negative value of M_{hz} is significant and has been addressed by other researchers [4,16]. This negative value shows that M_{hz} is against the propulsive moment; from another point of view, however it is believed to stabilize the transmission of loads to the handrim.

In the following section, the experimental error of the system will be estimated using the general uncertainty method.

4.4   Uncertainty Analysis

The concept of uncertainty describes the degree of goodness of a measurement or experimental results [17]. Kline defines uncertainty as "what we think the error would be if we could and did measure it by calibration" [18]. Uncertainty is thus an estimate of experimental error.

Uncertainty analysis is a necessary and powerful tool, particularly when used in the planning and design of experiments. There are cases in which all of the measurements in an experiment can be made with 1% uncertainty, yet the uncertainty in the final experimental result could be greater than 50% [17]. Uncertainty analysis, used in an experiment's initial planning phase can identify such situations and save the researcher much time.

4.4.1   General Uncertainty Analysis

In the planning phase of an experimental program, one focuses on the general or overall uncertainties. Consider a general case in which an experimental result, f, is a function of n measured variables X_i [17]:

$$f = f(X_1, X_2, \ldots, X_n) \tag{4.15}$$

Equation (4.15) is the data reduction equation used for determining f from the measured values of the variables X_i. The overall uncertainty in the result is then given by

$$U_f^2 = \left(\frac{\partial f}{\partial X_1}\right)^2 U_{x_1}^2 + \left(\frac{\partial f}{\partial X_2}\right)^2 U_{x_2}^2 + \quad + \left(\frac{\partial f}{\partial X_n}\right)^2 U_{x_n}^2 \tag{4.16}$$

where U_{X_i} are the uncertainties in the measured variables X_i.

It is assumed that the relationship given by (4.15) is continuous and has continuous derivatives in the domain of interest, that the measured variables X_i are independent of one another, and that the uncertainties in the measured variables are also independent of one another.

If the partial derivatives are interpreted as absolute sensitivity coefficients such that

$$\delta_i = \frac{\partial f}{\partial X_i} \tag{4.17}$$

then Equation (4.16) can be written as

$$U_f^2 = \sum_{i=1}^{n} \delta_i^2 U_{X_i}^2 \tag{4.18}$$

In Equations (4.16) and (4.18), all of the absolute uncertainties (U_{Xi}) should be expressed with the same odds or level of confidence. In most cases, 95% confidence level (20:1 odds) is used, with the uncertainty in the result also being at the same level [17].

4.4.2 Experimental Protocol for Uncertainty Analysis

One able-bodied subject was used for the stationary tests. The main idea was to propel the wheelchair, measure the loads, and calculate uncertainties to verify the system rather than acquiring specific information related to the subjects themselves. The subject tried the wheelchair for 5 min per day for one week to become familiar with the experimental set-up. For the main test, the subject propelled the wheelchair for 2 min, increasing his speed as what is conveniently possible and maintaining it steady for 1 min [12]. The data were then collected for the last minute of the test. MATLAB® software was used for data filtering. A tenth-order linear-phase digital Equiripple-type filter (FIR) from MATLAB® was used for signal conditioning.

4.4.3 Uncertainty of Preloads

The general uncertainty equation for preloads is obtained by using Equations (4.16) and (4.1) as

$$U_p = \left[(\text{Sin}(\theta + \alpha))^2 U_a^2 + (a \text{Cos}(\theta + \alpha))^2 U_\theta^2 + U_b^2 \right]^{1/2} \tag{4.19}$$

where U_p represents the uncertainty for different preloads [N or N.m], U_a and U_b are the primary uncertainties for the constants a and b [N or N.m.] from Equation (4.1), and U_θ is the uncertainty for the wheel's angular position [radian]. Uncertainty equations for different preload components are then calculated by using Equation (4.19) according to

$$
\begin{aligned}
U_{F_{px}} &= \left[(\text{Sin}(\theta + \alpha_{F_x}))^2 U_{a_{F_x}}^2 + (a_{F_x} \text{Cos}(\theta + \alpha_{F_x}))^2 U_\theta^2 + U_{b_{F_x}}^2 \right]^{1/2} \\
U_{F_{py}} &= \left[(\text{Sin}(\theta + \alpha_{F_y}))^2 U_{a_{F_y}}^2 + (a_{F_y} \text{Cos}(\theta + \alpha_{F_y}))^2 U_\theta^2 + U_{b_{F_y}}^2 \right]^{1/2} \\
U_{F_{pz}} &= \left[(\text{Sin}(\theta + \alpha_{F_z}))^2 U_{a_{F_z}}^2 + (a_{F_z} \text{Cos}(\theta + \alpha_{F_z}))^2 U_\theta^2 + U_{b_{F_z}}^2 \right]^{1/2} \\
U_{M_{px}} &= \left[(\text{Sin}(\theta + \alpha_{M_x}))^2 U_{a_{M_x}}^2 + (a_{M_x} \text{Cos}(\theta + \alpha_{M_x}))^2 U_\theta^2 + U_{b_{M_x}}^2 \right]^{1/2} \\
U_{M_{py}} &= \left[(\text{Sin}(\theta + \alpha_{M_y}))^2 U_{a_{M_y}}^2 + (a_{M_y} \text{Cos}(\theta + \alpha_{M_y}))^2 U_\theta^2 + U_{b_{M_y}}^2 \right]^{1/2} \\
U_{M_{pz}} &= \left[(\text{Sin}(\theta + \alpha_{M_z}))^2 U_{a_{M_z}}^2 + (a_{M_z} \text{Cos}(\theta + \alpha_{M_z}))^2 U_\theta^2 + U_{b_{M_z}}^2 \right]^{1/2}
\end{aligned} \tag{4.20}
$$

Tables 4.2, 4.3, and 4.4 show the values for primary uncertainties. These values are determined in static tests and on the basis of parameter resolution as reported by the manufacturer of the transducer.

The values of $U_{F_{px}}, U_{F_{py}}, U_{F_{pz}}, U_{M_{px}}, U_{M_{py}}$, and $U_{M_{pz}}$ are calculated by using Equations (4.20) and Tables 4.2 and 4.4, with φ varied between 0 and 180°, the probable interval for the pushing phase.

4.4.4 Uncertainty of Local Loads

The uncertainties of the local forces and moments are obtained by using Equations 4.2 and 4.20, and Table 4.3 as follows:

$$
\begin{aligned}
U_{F_{Lx}} &= \left[U_{F_{px}}^2 + U_{F_{primx}}^2 \right]^{1/2} \\
U_{F_{Ly}} &= \left[U_{F_{py}}^2 + U_{F_{primy}}^2 \right]^{1/2} \\
U_{F_{Lz}} &= \left[U_{F_{pz}}^2 + U_{F_{primz}}^2 \right]^{1/2} \\
U_{M_{Lx}} &= \left[U_{M_{px}}^2 + U_{M_{primx}}^2 \right]^{1/2} \\
U_{M_{Ly}} &= \left[U_{M_{py}}^2 + U_{M_{primy}}^2 \right]^{1/2} \\
U_{M_{Lz}} &= \left[U_{M_{pz}}^2 + U_{M_{primz}}^2 \right]^{1/2}
\end{aligned} \tag{4.21}
$$

TABLE 4.2   Primary Uncertainties for Measured Variables

Variable uncertainties	U_θ [radian]	U_{r_h} [m]	$U_{\Delta z}$ [m]
Value used	0.001745	0.001	0.001

These uncertainties are shown in Figures 4.15 and 4.16.

The uncertainties for different outputs of the system are determined with respect to the local and global coordinate systems. The uncertainties for the preloads and local loads are calculated with the primary uncertainties for the measured variables and with Equations (4.19), (4.20) and (4.21), and the values presented in Tables 4.2, 4.3, and 4.4. The results show that there is no significant difference between the uncertainties of the preloads and local loads. Therefore, only the uncertainties of the local loads appear in the presented results.

Figure 4.15 shows the uncertainties for the local forces in the interval during which the propulsion phase can occur (from $+x$ to $-x$ direction of the global coordinate system). It indicates that F_{Lx} and F_{Lz} have their highest uncertainties (about 1.75 N for F_{Lx} and about 3.40 N for F_{Lz}) in the $+y$ direction of the global coordinate system, and that F_{Ly} has its highest uncertainty (about 1.75 N) in the $+x$ and $-x$ directions of the global coordinate system. F_{Lx} and F_{Ly} are the components of the applied force, which produces the propulsive moments. This information indicates that the highest uncertainties for F_{Lx} and F_{Ly} are low and acceptable.

Figure 4.16 shows the uncertainties for local moments in the interval during which the propulsion phase can take place (from $+x$ to $-x$ direction of the global coordinate system). It shows that M_{Ly} and M_{Lz} have their highest uncertainties (about 0.70 [N.m] for M_{Ly} and about 0.30 N.m for M_{Lz}) in the $+y$ direction of the global coordinate system and that M_{Lx} has its highest uncertainty (about 0.70 N.m) in the $+x$ and $-x$ directions of the global coordinate system. M_{Lz} is the moment, which produces the propulsion; its highest uncertainty is very low.

4.4.5   Uncertainty of Global Loads

The following relations are obtained by calculating F_{Lx}, F_{Ly}, F_{Lz}, M_{Lx}, M_{Ly}, M_{Lz}, $U_{F_{Lx}}$, $U_{F_{Ly}}$, $U_{F_{Lz}}$, $U_{M_{Lx}}$, $U_{M_{Ly}}$, and $U_{M_{Lz}}$ using Equations (4.2), (4.16), and (4.21), respectively, and employing equations (4.3) and (4.16):

$$U_{F_{gx}} = \left[(-\mathrm{Sin}\theta \times F_{Lx} - \mathrm{Cos}\theta \times F_{Ly})^2 U_\theta^2 + \mathrm{Cos}\theta^2 \times U_{F_{Lx}}^2 + \mathrm{Sin}\theta^2 \times U_{F_{Ly}}^2\right]^{1/2}$$

$$U_{F_{gy}} = \left[(\mathrm{Cos}\theta \times F_{Lx} - \mathrm{Sin}\theta \times F_{Ly})^2 U_\theta^2 + \mathrm{Sin}\theta^2 \times U_{F_{Lx}}^2 + \mathrm{Cos}\theta^2 \times U_{F_{Ly}}^2\right]^{1/2}$$

$$U_{F_{gz}} = U_{F_{Lz}}$$

$$U_{M_{gx}} = \left[(-\mathrm{Sin}\theta \times M_{Lx} - \mathrm{Cos}\theta \times M_{Ly})^2 U_\theta^2 + \mathrm{Cos}\theta^2 \times U_{M_{Lx}}^2 + \mathrm{Sin}\theta^2 \times U_{M_{Ly}}^2\right]^{1/2} \qquad (4.22)$$

$$U_{M_{gy}} = \left[(\mathrm{Cos}\theta \times M_{Lx} - \mathrm{Sin}\theta \times M_{Ly})^2 U_\theta^2 + \mathrm{Sin}\theta^2 \times U_{M_{Lx}}^2 + \mathrm{Cos}\theta^2 \times U_{M_{Ly}}^2\right]^{1/2}$$

$$U_{M_{gz}} = U_{M_{Lz}}$$

These uncertainties for the global forces and moments are shown in Figures 4.14 and 4.15. F_{Lx}, F_{Ly}, M_{Lx}, M_{Ly}, and θ are the parameters calculated from the data measured in the tests.

TABLE 4.3   Primary Uncertainties for Measured Loads

Load uncertainties	$U_{F_{primx}}$ [N]	$U_{F_{primy}}$ [N]	$U_{F_{primz}}$ [N]	$U_{M_{primx}}$ [N.m]	$U_{M_{primy}}$ [N.m]	$U_{M_{primz}}$ [N.m]
Value used	1	1	2	0.4	0.4	0.2

TABLE 4.4 Primary Uncertainties for Constants

Constant uncertainties	$U_{a_{F_x}}, U_{b_{F_x}}$ $U_{a_{F_y}}$ & $U_{b_{F_y}}$ [N]	$U_{a_{F_z}}$ & $U_{b_{F_z}}$ [N]	$U_{a_{M_x}}, U_{b_{M_x}}$ $U_{a_{M_y}}$ & $U_{b_{M_y}}$ [N.m]	$U_{a_{M_z}}$ & $U_{b_{M_z}}$ [N.m]
Value used	1	2	0.4	0.2

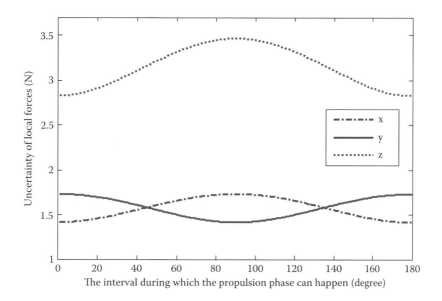

FIGURE 4.15 Uncertainties for local force components during propulsion phase.

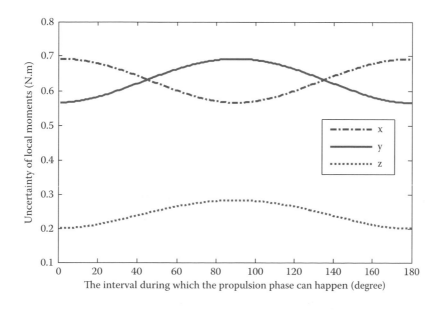

FIGURE 4.16 Uncertainties for local moment components during possible range of propulsion phase.

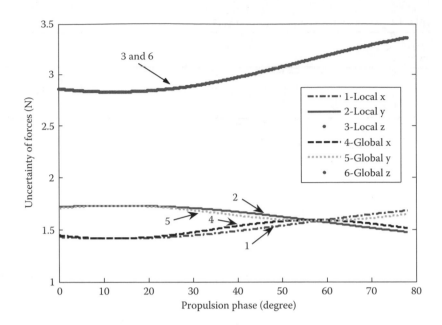

FIGURE 4.17 Uncertainties for local and global force components during propulsion phase.

Figure 4.17 reflects the uncertainties for the local and global forces. These uncertainties were calculated for the normal propulsion phase of 80°, covering a range from 75 to 155° of the possible propulsion phase. The local uncertainties were compared with the global uncertainties in the same graph and for the same period. This figure shows that the global uncertainty of F_z is the same as its local uncertainty. The global uncertainty of F_x shows a small increase compared with the local value, but its highest value of about 1.60 N is not near the end of the propulsion phase. It reaches its highest point at about 60° into the propulsion phase. The global uncertainty for F_y shows a small decrease compared with the local value. Its highest value of about 1.70 N is near 10° after the beginning of the contact between the hand and the handrim. It decreases to a minimum at about 60°.

Figure 4.18 shows the uncertainties for the local and global moments. These uncertainties were also calculated for the normal propulsion phase. The local uncertainties were compared with the global uncertainties in the same graph and for the same period. This figure shows that the global uncertainty of M_z is the same as its local uncertainty. The global uncertainty of M_y shows a modest increase compared with the local values. It starts to decrease after its peak of about 0.63 N.m at about 60° after the beginning of the contact between the hand and the handrim. The global uncertainties for M_x decrease to some extent compared with the local value, whose peak value of about 0.70 N.m occurs near 10° after the beginning of the contact between the hand and the handrim. It drops to a minimum at about 60°.

4.4.6 Uncertainty of Angular Position of Hand on Handrim (U_φ)

The uncertainty of the angular position of the hand on the handrim (φ) is obtained by using Equations (4.12) and (4.16), as

$$U_\varphi = \sqrt{(A+B)}$$

$$A = (\partial\varphi/\partial M_{hy})^2 U^2_{M_{hy}} + (\partial\varphi/\partial M_{gy})^2 U^2_{M_{gy}} + (\partial\varphi/\partial F_{gx})^2 U^2_{F_{gx}} \qquad (4.23)$$

$$B = (\partial\varphi/\partial F_{gz})^2 U^2_{F_{gz}} + (\partial\varphi/\partial \Delta z)^2 U^2_{\Delta z} + (\partial\varphi/\partial r_h)^2 U^2_{r_h})$$

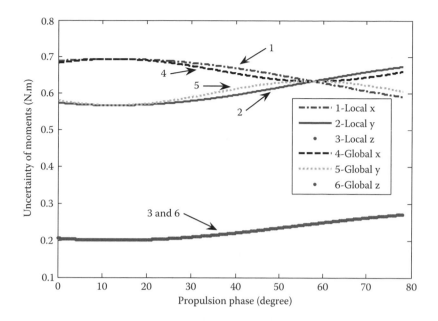

FIGURE 4.18 Uncertainties for local and global moment components during propulsion phase.

The derivatives of φ with respect to different variables are calculated as

$$D = \partial\varphi/\partial M_{hy} = 1/C$$

$$E = \partial\varphi/\partial M_{gy} = 1/C$$

$$F = \partial\varphi/\partial F_{gx} = -\Delta z/C$$

$$G = \partial\varphi/\partial F_{gz} = (M_{hy} - M_{gy} + F_{gx} \times \Delta z)/C \times F_{gz}$$

$$H = \partial\varphi/\partial \Delta z = -F_{gx}/C$$

$$I = \partial\varphi/\partial r_h = (M_{hy} - M_{gy} + F_{gx} \times \Delta z)/C \times r_h$$

(4.24)

The lumped parameter C in equation (4.24) is used to simplify some repetitive terms, which is given as

$$C = \sqrt{(F_{gz} \times r_h)^2 - (M_{hy} - M_{gy} + F_{gx} \times \Delta z)^2}$$

(4.25)

The term U_φ is determined on the basis of equations (4.23) and (4.24), as

$$U_\varphi = \sqrt{(D \times U_{M_{hy}})^2 + (E \times U_{M_{gy}})^2 + (F \times U_{F_{gx}})^2 + (G \times U_{F_{gz}})^2 + (H \times U_{\Delta z})^2 + (I \times U_{r_h})^2}$$

(4.26)

Using these relations and data from measured tests, one can obtain the time-dependent uncertainty for φ, as shown in Figure 4.19.

The uncertainty for the hand contact angular position for the reliable region of φ was calculated to be less than 10° for much of the region (Figure 4.19). The uncertainty for φ at the middle span of the reliable region is about 5°. Compared with the reported results by other researchers [19], and given that we did not use cameras in our measurements, this level of maximum uncertainty for φ is good.

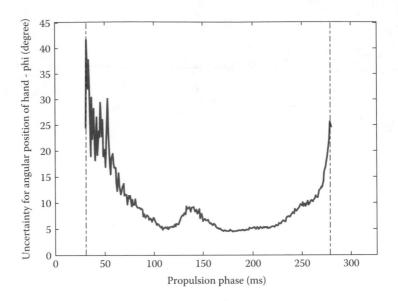

FIGURE 4.19 Uncertainty for angular position of hand during propulsion.

4.4.7 Uncertainty for Hand Contact Loads

Equations (4.6) show that the forces are the same in both the second local and global coordinate systems for the hand contact with the handrim. Using equation (4.16), the uncertainties for forces in the second local coordinate system are determined as

$$U_{F_{hx}} = U_{F_{gx}}$$

$$U_{F_{hy}} = U_{F_{gy}} \tag{4.27}$$

$$U_{F_{hz}} = U_{F_{gz}}$$

Knowing the uncertainties for r_h and Δz, and calculating the other required uncertainties, one can obtain the uncertainties for the moments with respect to the local hand coordinate system as follows:

$$U_{M_{hx}} = \left[U_{M_{gx}}^2 + (r_h \text{Sin}\varphi)^2 U_{F_{gz}}^2 + (F_{gz}\text{Sin}\varphi)^2 U_{r_h}^2 + (F_{gz}r_h \text{Cos}\varphi)^2 U_{\varphi}^2 + \Delta z^2 U_{F_{gy}}^2 + F_{gy}^2 U_{\Delta z}^2 \right]^{1/2}$$

$$U_{M_{hy}} = \left[U_{M_{gy}}^2 + (r_h \text{Cos}\varphi)^2 U_{F_{gz}}^2 + (F_{gz}\text{Cos}\varphi)^2 U_{r_h}^2 + (F_{gz}r_h \text{Sin}\varphi)^2 U_{\varphi}^2 + \Delta z^2 U_{F_{gx}}^2 + F_{gx}^2 U_{\Delta z}^2 \right]^{1/2} \tag{4.28}$$

$$U_{M_{hz}} = \left[U_{M_{gz}}^2 + (r_h \text{Sin}\varphi)^2 U_{F_{gx}}^2 + (F_{gx}\text{Sin}\varphi)^2 U_{r_h}^2 + (F_{gx}r_h \text{Cos}\varphi)^2 U_{\varphi}^2 + (r_h \text{Cos}\varphi)^2 U_{F_{gy}}^2 \right.$$

$$\left. + (F_{gy}\text{Cos}\varphi)^2 U_{r_h}^2 + (F_{gy}r_h \text{Sin}_{\varphi})^2 U_{\varphi}^2 \right]^{1/2} \right]$$

Among the moments in the hand coordinate system, M_{hz} is of considerable interest because it acts against the propulsion moment, which is produced by the applied forces F_{gx} and F_{gy} [4,16]. The uncertainty for M_{hz} is shown in Figure 4.20.

This figure shows the uncertainty for propulsive hand moment with respect to the local hand coordinate system. M_{hz} is the last parameter that is calculated by using the present hierarchy of equations for uncertainty propagation. Given the cumulative nature of the uncertainty of the equations, it is not

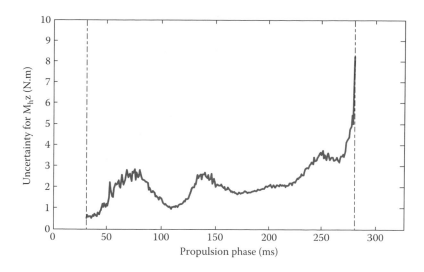

FIGURE 4.20 Uncertainty for hand moment in *z* direction of local hand coordinate system during propulsion.

surprising to see a high uncertainty for M_{hz}. This figure shows that the uncertainty varies between 0.50 and 3 N.m. For most of the reliable region of the propulsion phase, the uncertainty is less than 2 N.m.

In this section, a general uncertainty analysis is performed to determine the uncertainty equations for the local and global forces and moments, the local hand forces and moments, and the hand-contact angular position. The uncertainty values for these parameters are calculated from the respective equations. The results provide an estimation for the errors and uncertainty in the output of the instrumented wheel. The uncertainties are found to vary from 1.40 to 3.40 N for the local forces and from 0.20 to 0.70 N.m for the local moments. The maximum and minimum of the uncertainties for global values are about the same as the uncertainties for the local values, but the pattern of variation is different. The uncertainties are found to vary from 5 to 10° and from 0.5 to 3 N.m for φ and M_{hz}, respectively, for about 65% of the propulsion period. Uncertainties determined by Cooper et al. [13] for the forces and moments are in the range of 1.1–2.5 N and 0.03–0.19 N.m in the plane of the handrim, and 0.93 N and 2.24 N.m in the wheel axle direction, respectively. Our results show uncertainty levels for the forces and moments in the range of 1.40–1.70 N and 0.58–0.68 N.m in the plane of the handrim, and about 3.40 N and 0.25 N.m in the wheel axle direction, respectively. For our system, however, the uncertainty values for the important load components, namely the planar forces and the axial moment, are low.

In the next section, the instrumented wheel system is verified by using an experimental technique, and system specifications are determined by applying statistical methods.

4.5 System Verification

To obtain the degree of reliability of the results obtained from the designed and fabricated instrumented wheel, an experimental technique is used to determine the system specifications by performing static and dynamic verification tests. Four different set-ups are used for these static and dynamic tests, and both qualitative and quantitative analyses are performed. Pearson correlation and coefficient of variation techniques are used to determine linearity and repeatability, respectively, as key system specifications. Also, the error for quantitative analysis is estimated. Three different angular velocities are used in the dynamic tests. The static and dynamic tests are performed at different levels of loading on the handrim for all channels (different sensor signals) at four different loading positions.

FIGURE 4.21   First vertical loading set-up for static tests.

4.5.1   Set-Up Designs

To verify the system, the following four different loading set-ups are used: a first vertical loading set-up for static tests, a second vertical loading set-up for static tests, a horizontal loading set-up for static tests, and a dynamic loading set-up.

4.5.1.1   First Vertical Loading Set-Up

The first vertical loading set-up for static tests is used to apply the selected vertical loads at four loading points (one at a time) when they are placed in turn at the loading position of point 1 (Figures 4.21 and 4.22) on the handrim. Points 1 and 2 are at the intersections of a horizontal line passing through the handrim and its center. A loading disk is connected via a wire cable to the handrim with a clamp. Six different weights (22.27, 44.48, 66.76, 89.043, 111.50, and 133.30 N) are used in this set-up. The level of resolution for the weights depends on the resolution of the sensor—in our case, 0.01 N. The range of the weights covers the typical loads applied on the handrim during wheelchair propulsion. The first local coordinate system, which is attached to the load transducer, will turn in unison with points 2, 3, or 4, when they turn into the position of point 1. Rotation of the wheels during the static loading is prevented by locking the shaft of one of the rollers.

4.5.1.2   Second Vertical Loading Set-Up

The second vertical loading set-up for static tests applies the selected vertical loads at the four points when they are placed at the position of point 3 (Figure 4.22) on the handrim. Points 3 and 4 are at the intersections of the handrim and a vertical line passing through its center. A load-holding disk is hung

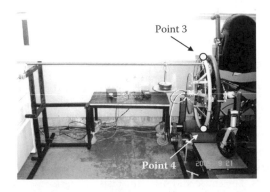

FIGURE 4.22   Second vertical loading set-up for static tests.

FIGURE 4.23 Horizontal loading set-up for static tests.

from a horizontal bar which is 2 m in length and 1.5 cm in diameter. One side of the bar is hung from the handrim using a rope so that only normal load is transmitted. The other side of the bar rests perpendicularly on a bar that is 2 cm in length and 0.5 cm in diameter, which itself rests on a smooth horizontal surface. This combination provides a rolling effect and eliminates horizontal frictional loads on the long bar due to deformation and shortening of its span after loading. The set of six weights used in this set-up is the same as the set used in the initial vertical loading set-up.

4.5.1.3 Horizontal Loading Set-Up

The horizontal loading set-up is used to apply horizontal static test loads at four different points located 90° apart from each other on the handrim's outer circumference. In fact, these are the same four points used in the previous tests. Four loading points are used to cover the entire circumference of the handrim. These points are at the intersections of the x and y axes of the first local coordinate system and the handrim, and are used to apply pure axial loads. A loading disk is connected to the loading point on the handrim through a pulley using a 2-mm wire cable (Figure 4.23). Six different weights (4.50, 9.02, 13.49, 16.41, 19.31, and 22.23 N) are used in this set-up for each point. During manual wheelchair propulsion, the subject applies a lower load in the direction of the axle of the wheel compared with the loads in the plane of the wheel. Therefore, a new set of the loads is used in the horizontal static tests.

4.5.1.4 Dynamic Loading Set-Up

The dynamic loading set-up applies the centrifugal test forces at four loading points, which are the same as the loading points in the static set-ups. Three different weights (4.50, 8.95, and 13.39 N) are used as loads. The loads are attached individually to the handrim's lateral surface with a very powerful magnet (Figure 4.24). The AC motor is used to mechanically turn the instrumented wheel at three different speeds.

Increasing the number of weights, loading points, and tests can yield more from different parts of the system, but it can also increase the calculation time. So, the number of weights, loading points, and tests are chosen such that proper statistical analysis can be performed within a reasonable time and with sufficient accuracy. The test loads are not meant to reproduce the loads applied by the wheelchair user on the handrim.

4.5.2 Test Protocol

After the design and fabrication of the instrumented wheel and the calculation of the transformation equations for the applied forces and moments [20], the system has to be verified. For this purpose, both

FIGURE 4.24 Dynamic loading set-up.

qualitative and quantitative analyses are performed for the output of the experiments. Two of the most important system specifications for qualitative analysis—linearity and repeatability—are determined by using statistical techniques such as Pearson correlation and descriptive analysis, respectively. The error for the quantitative analysis is estimated as well. Given the dynamic nature of the real situations, both dynamic and static conditions must be used to validate the system.

4.5.2.1 Static Verification

To verify the system under static conditions, the wheelchair is placed on and securely strapped to the roller rig. Three different test set-ups, described in Section 4.5.1, are used to apply loads in three different directions (x, y, and z) of the first local coordinate system (Figure 4.2). For vertical loading in the static tests, six different weights (22.27, 44.48, 66.76, 89.04, 111.50, and 133.30 N) are suspended independently from points 1, 2, 3, and 4 on the handrim circumference by using two vertical loading set-ups (Figures 4.21 and 4.22). The loading positions are 90° apart in the $-x'$, $-y'$, x', and y' directions of the first local coordinate system (Figure 4.2). The baseline of the load holding disk's own weight is measured by performing a no-load test, and the resultant loads are subtracted from the measured loads accordingly. Measurements are repeated three times at four different loading points with respect to the first local coordinate system.

To determine the specifications for qualitative analysis, the Pearson correlation coefficient (r) is used, which is defined as

$$r = b \times (s_x/s_y) \tag{4.29}$$

where s_x and s_y are the standard deviations of the independent and dependent variables and the value b is determined as

$$b = \sum_{i=1}^{N} \left((x_i - \bar{x})(y_i - \bar{y}) \middle/ (N-1)s_x^2 \right) \tag{4.30}$$

where x_i is the case value for the independent variable, $\bar{x}$ is the mean of the independent variable, y_i is the case value for the dependent variable, $\bar{y}$ is the mean of the dependent variable, N is the number of cases, and s_x^2 is the variance of the independent variable [21]. In this study, dependent variables are the measured forces and moments, and independent variables are the applied loads at different loading points.

The Pearson correlation coefficient (r) is used to determine the linearity of the system. The coefficient of variation is used for all different tests to determine the system repeatability and compare the variability

TABLE 4.5 Pearson Correlation Coefficient r (Static Verification)

Channel Position	F_x	F_y	F_z	M_x	M_y	M_z
1	1.000	1.000	0.999	1.000	1.000	1.000
2	1.000	0.994	0.998	0.993	1.000	0.985
3	1.000	1.000	0.997	1.000	1.000	1.000
4	1.000	1.000	0.999	1.000	1.000	1.000

of different parameters with different units. The coefficient of variation expresses the standard deviation as a percentage of the mean. This allows one to compare the variability of different parameters. The coefficient of variation is given by

$$Coefficient\ of\ variation = \left(\frac{standard\ deviation}{|mean|} \right) \times 100 \qquad (4.31)$$

where *mean* is the mean value of the variable of interest.

To determine the specifications for the quantitative analysis, the actual values are compared with the measured values. SPSS® 11.0 and Microsoft Excel® software are used to analyze the data and calculate the system specifications. All the r values are calculated by using the results of the first series of tests.

Table 4.5 shows the Pearson correlation coefficient (r) due to static verification. The "position" column gives the different load application points, and the "channel" row gives different measurements. The values of r show high linearity (above 0.9) at different loading points and for different measuring channels in the static situation.

Table 4.6 shows the mean value of the percentage of the coefficient of variation for different measured loads at the four loading points. The "load" column gives the different loading forces used during the tests. The loads differ for channel F_z because they do not reach high values during propulsion. These values indicate a low coefficient of variation (less than 2%), which are calculated using the measured values of the three different tests. The entries in Table 4.6 show high repeatability of the instrumented wheel. Tables 4.5 and 4.6 present the results for the qualitative analysis and collectively show reliable values for system specification. The average of the results from three series of the repeated tests has been used to calculate the mean errors.

Table 4.7 presents the results of the quantitative analysis and lists the mean errors of the measured forces and moments as a percentage of the loads. The values indicate low mean error (mostly less than 5%) for different loads on all channels. Some errors are expected because of the effect of other sources, such as human or experimental errors. The low levels of errors indicate that the parameters measured by the instrumented wheel are reliable.

The results of qualitative and quantitative analyses for the instrumented wheel in the static situation show a reliable range of values for system specifications.

TABLE 4.6 Mean Coefficient of Variation of Measured Loads (%; Static Verification)

Channel Load (N)	F_x	F_y	M_x	M_y	M_z	Channel Load (N)	F_z
22.27	0.110	0.166	1.547	1.784	0.133	4.50	1.736
44.48	0.045	0.147	1.545	0.398	0.166	9.02	1.293
66.76	0.059	0.174	1.401	0.289	0.082	13.49	0.863
89.04	0.070	0.053	1.314	0.355	0.067	16.41	0.747
111.50	0.102	0.117	1.192	0.381	0.563	19.31	0.895
133.30	0.096	0.106	0.234	0.485	0.117	22.23	1.325

TABLE 4.7   Mean Errors as Percentage of Loads (%; Static Verification)

Channel Load (N)	F_x	F_y	M_x	M_y	M_z	Channel Load (N)	F_z
22.273	0.857	0.070	1.608	2.969	0.344	4.504	3.572
44.482	0.583	0.291	8.422	4.556	0.113	9.015	1.374
66.755	0.640	0.144	2.586	1.088	0.179	13.489	2.037
89.043	0.576	0.170	1.922	1.420	0.081	16.406	2.680
111.504	0.726	0.074	3.187	0.881	0.259	19.308	3.718
133.299	0.666	0.097	3.383	0.614	0.128	22.225	7.401

4.5.2.2 Dynamic Verification

Dynamic verification is more challenging than static verification. The local coordinate system of the transducer spins with the wheel and the loadings are weights, so the loads (in the global coordinate system) cannot be measured directly. An encoder is used to determine the position of the load attached to the wheel with respect to the global coordinate system. The wheelchair is placed on the roller rig, and the AC motor rotates the driving roller. Three different angular velocities (3, 3.8, and 4.8 rad/s) are used for the dynamic tests to cover the wheeling speeds of the user. Three different weights (4.50, 8.95, and 13.39 N) and one powerful magnet are used for loading at points 1 to 4 (Figures 4.21, 4.22, and 4.23) on the handrim lateral surface. The loading positions are the same as for the static verification tests. The measured forces and moments of three successive cycles are used to verify the system repeatability. The baseline of the attachment's own weight is set to zero by using the method mentioned in the static verification tests. The actual values are compared with the measured values to obtain the specifications for quantitative analysis. The actual values are determined using the inverse dynamics method. The angular motion of the loaded wheel is considered in the vertical plane, where the centripetal force F is determined as

$$F = mr_h\omega^2 + mg \tag{4.32}$$

Here m is the mass of the weight that is attached to the wheel, r_h is the moment arm (handrim radius), ω is the wheel angular velocity, and g is the acceleration due to gravity. There is no force component in the z direction because the object has a planar motion (x-y) and the wheel camber angle is zero with respect to the global coordinate system. The x and y planar components are as follows:

$$F_x = mr_h\omega^2 \mathrm{Sin}\theta$$
$$F_y = mr_h\omega^2 \mathrm{Cos}\theta - mg \tag{4.33}$$

where θ is the angular position of the load.

Equations (4.29), (4.31) and (4.33) are used to determine the specifications for qualitative analysis in the dynamic tests with three different angular velocities.

As the nature of the manual wheelchair propulsion is dynamic, qualitative and quantitative analyses are performed for the instrumented wheel under dynamic conditions. These analyses are carried out for three different angular velocities (3, 3.8, and 4.8 rad/s).

Table 4.8 shows the Pearson correlation coefficient (r) for the tests conducted. These values mostly show high accuracy (r above 0.9) at different angular velocities and loadings. Channel F_z is not considered for dynamic verification because there is no appreciable load on this channel, due to the nature of dynamic loading.

The values in Table 4.9 show a low mean coefficient of variation for different measured loads (less than 4%) at four loading points, and high repeatability of the instrumented wheel. Tables 4.8 and 4.9 show the results of the qualitative analysis. They indicate reliable values for system specification in the dynamic verification tests.

TABLE 4.8 Pearson Correlation Coefficient r
(Dynamic Verification)

Position	Channel F_x	F_y	M_x	M_y	M_z
	$\omega = 3$ rad/s				
1	1.000	1.000	0.999	1.000	1.000
2	1.000	1.000	1.000	1.000	0.992
3	1.000	1.000	1.000	0.997	0.998
4	1.000	1.000	0.993	1.000	0.987
	$\omega = 3.8$ rad/s				
1	1.000	1.000	0.999	1.000	1.000
2	0.999	1.000	0.999	0.996	0.982
3	1.000	1.000	0.994	1.000	0.995
4	1.000	1.000	0.996	1.000	0.989
	$\omega = 4.8$ rad/s				
1	1.000	1.000	0.993	0.985	1.000
2	0.998	1.000	0.993	0.989	0.991
3	1.000	1.000	0.998	1.000	0.989
4	0.997	1.000	0.999	0.999	0.991

The mean errors produced by the instrumented wheel as a percentage of loads are presented in Table 4.10 (quantitative analysis). The low mean error values (mostly less than 6%) indicate that the parameters measured by the instrumented wheel are reliable for the dynamic situation too.

Given the actual performance for the instrumented wheel and its measurements, Figures 4.25 and 4.26 show the measured and predicted values for F_x, F_y, M_x, M_y, and M_z with respect to the global coordinate system. As mentioned previously, F_z is not considered in the dynamic measurements because there is no significant load on this channel due to the nature of loading for dynamic tests. These figures show that the patterns of the measured and predicted curves of the data for forces and moments are highly compatible with typical results measured by other researchers [12].

TABLE 4.9 Mean Coefficient of Variation of
Measured Loads (%; Dynamic Verification)

Load (N)	Channel F_x	F_y	M_x	M_y	M_z
	$\omega = 3$ rad/s				
4.50	2.170	1.467	1.547	0.967	0.842
8.95	1.871	1.752	1.236	1.226	0.144
13.39	1.230	0.954	0.638	0.761	0.115
	$\omega = 3.8$ rad/s				
4.50	2.511	3.196	2.892	1.475	0.445
8.95	2.144	1.498	2.781	1.915	0.300
13.39	1.054	0.968	0.896	0.937	0.078
	$\omega = 4.8$ rad/s				
4.50	3.986	3.404	2.191	3.051	0.844
8.95	2.529	1.604	2.247	3.695	0.456
13.39	1.691	0.856	3.380	1.721	1.347

TABLE 4.10 Mean Errors as Percentage of Loads (Dynamic Verification)

Load (N)	F_x	F_y	M_x	M_y	M_z
			$\omega = 3$ rad/s		
4.50	5.675	7.724	3.276	4.514	4.395
8.95	6.532	6.820	1.924	5.227	6.005
13.39	6.723	7.038	2.944	6.878	5.638
			$\omega = 3.8$ rad/s		
4.50	4.115	4.833	3.762	4.116	3.134
8.95	5.581	6.636	4.798	2.867	5.530
13.39	7.430	6.909	5.579	3.259	5.640
			$\omega = 4.8$ rad/s		
4.50	5.837	6.200	3.079	6.189	4.643
8.95	5.675	7.211	5.794	4.329	4.300
13.39	5.590	8.079	2.281	3.467	4.639

A complete experimental technique has been designed and carried out under different static and dynamic conditions to determine the specification of an instrumented wheel for manual wheelchair propulsion analysis. The verification techniques, which are highlighted and demonstrated step-by-step, can be implemented in similar wheelchair instrumentation set-ups. The results of the static and dynamic tests were used for both qualitative and quantitative analyses to determine the system specifications. The static tests showed high linearity (r above 0.9), very low standard deviation (mostly close to zero), and a low mean coefficient of variation for measured loads (less than 2%). These results indicate high

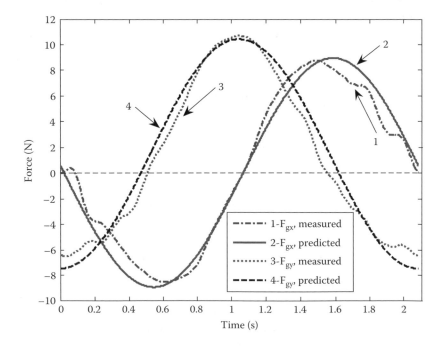

FIGURE 4.25 Measured and predicted global sample force components.

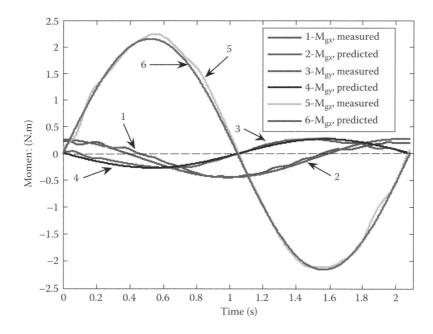

FIGURE 4.26 Measured and predicted global sample moment components.

repeatability and low mean error (mostly less than 5%) due to the different loading conditions for all load channels. Two cells of Table 4.6 show mean errors above 5%. One of them is at the maximum horizontal load. Usually, the horizontally applied loads are not so high during manual wheelchair propulsion, and the main idea is that the users try to apply planar loads. So, the system is not proportionally responsive to higher horizontal loading due to its structure, but it has mean errors less than 5% for all other lower horizontal loadings. The other mean error value above 5% corresponds to M_x. All of the mean error values in this column are close to 2% or more. Generally, one has higher values of mean error (%) for M_x compared with the mean error values (%) for the other channels. Dynamic tests were performed at three angular velocities and at four loading positions for all measuring channels. The results also show high linearity (r above 0.9). The low mean coefficient of variation for measured loads (less than 4%) confirm high repeatability (reliability) of the instrumented wheel. The results show that most of the mean errors are around 5%. The resultant specifications show high linearity, repeatability, and a low percentage of errors.

Conclusion

In this chapter the development of an instrumented wheel, which is a mechatronic device, for use in a kinetic and kinematic study of manual wheelchair propulsion was presented. The results collectively show that it was possible to reliably obtain the essential information required for manual wheelchair propulsion analysis, including the applied forces and moments, using the designed and fabricated instrumented wheel. The tests with an able-bodied subject reproduced patterns and overall behavior comparable to the available data, indicating that the system can be used for the designed and planned experiments. Using the introduced design and experimental technique under different static and dynamic conditions, the specification of the instrumented wheel were determined. The results of the static and dynamic tests were then used both for qualitative and quantitative analysis to determine the system specifications. It is noted that a mechatronic system developed in-house has enabled flexibility in enhancing the experimental scope. The instrumented and verified wheel can be used to conduct a host of other experiments for studying the kinetic aspects of wheelchair propulsion. For instance, varying the positions of the wheel axle (seat) and the backrest of the wheelchair affects all the forces and moments as well as the mechanics of propulsion. Therefore, determining and prescribing optimum positions can reduce pain and help prevent injury for manual wheelchair users, and improve the gross mechanical efficiency of propulsion.

References

1. Glaser, R.M., Sawka, M.N., Brune, M.F., and Wilde, S.W., Physiological responses to maximal effort wheelchair and arm crank ergometry, *Journal of Applied Physiology*, Vol. 48, No. 6, 1060–1064, 1980.
2. Boninger, M.L., Cooper, R.A., Shimada, S.D., and Rudy, T.E., Shoulder and elbow motion during two speeds of wheelchair propulsion: a description using a local coordinate system, *Spinal Cord*, Vol. 36, 418–426, 1998.
3. van der Woude, L.H.V., de Groot, G., and Hollander, A.P., Wheelchair ergonomics and physiological testing of prototypes, *Journal of Ergonomics*, Vol. 29, No. 12, 1561–1573, 1986.
4. de Groot, S., Veeger, H.E.J., Hollander, A.P., and van der Woude, L.H.V., Consequence of feedback-based learning of an effective handrim wheelchair force production on mechanical efficiency, *Clinical Biomechanics*, Vol. 17, 219–226, 2002.
5. van der Woude, L.H.V., Formanoy, M., and de Groot, S., Hand rim configuration: effect on physical strain and technique in unimpaired subjects?, *Journal of Medical Engineering and Physics*, Vol. 25, 765–774, 2003.
6. Veeger, H.E.J., van der Woude, L.H.V., and Rozendal, R.H., Effect of handrim velocity on mechanical efficiency in wheelchair propulsion, *Medicine and Science in Sports and Exercise*, Vol. 24, No. 1, 100–107, 1992.
7. Guo, L., Su, F., Wu, H., and An, K., Mechanical energy and power flow of the upper extremity in manual wheelchair propulsion, *Clinical Biomechanics*, Vol. 18, 106–114, 2003.
8. Wei, S., Huang, S., Chuan-Jiang, J., and Chiu, J., Wrist kinematic characterization of wheelchair propulsion in various seating position: implication to wrist pain, *Clinical Biomechanics*, Vol. 18, S46–S52, 2003.
9. Hintzy, F. and Tordi, N., Mechanical efficiency during hand-rim wheelchair propulsion: effects of base-line subtraction and power output, *Clinical Biomechanics*, Vol. 19, 343–349, 2004.
10. Asato, K.T., Cooper, R.A., Robertson, and Ster, J.F., SMARTWheel: development and testing of a system for measuring manual wheelchair propulsion dynamics, *IEEE Transaction on Biomedical Engineering*, Vol. 40, No. 12, 1320–1324, 1993.
11. Wu, H.W., Berglund, L.J., Su, F.C., Yu, B., Westreich, A., Kim, K.J., and An, K.N., An instrumented wheel for kinetic analysis of wheelchair propulsion, *Journal of Biomechanical Engineering*, Vol. 120, 534–535, 1998.
12. Cooper, R.A., Boninger, M.L., VanSickle, D.P., and DiGiovine, C.P., Instrumentation for measuring wheelchair propulsion biomechanics, in van der Woude, L., Hopman, M.T.E., and van Kemenade, C.H. (Eds.), *Biomedical Aspects of Manual Wheelchair Propulsion*, IOS Press, Amsterdam, 1999, pp. 104–114.
13. Cooper, R.A., Boninger, M.L., VanSickle, D.P., Robertson, R.N., and Shimada, S.D., Uncertainty analysis for wheelchair propulsion dynamics, *IEEE Transactions on Rehabilitation. Engineering*, Vol. 5, No. 2, 130–139, June 1997.
14. Veeger, H.E.J., Lute, E.M.C., Roeleveld, K., and van der Woude, L.H.V., Differences in performance between trained and untrained subjects during a 30-s sprint test in a wheelchair ergometer, *European Journal of Applied Physiology*, Vol. 64, 158–164, 1992.
15. van Kemenade, C.H., te Kulve, K.L., Dallmeijer, A.J., van der Woude, H.E.V., and Veeger, H.E.J., Changes in wheelchair propulsion technique during rehabilitation, in van der Woude, L., Hopman, M.T.E., and van Kemenade, C.H. (Eds.), *Biomedical Aspects of Manual Wheelchair Propulsion*, IOS Press, Amsterdam, 1999, pp. 104–114.

16. VanSickle, D.P., Cooper, R.A., Boninger, M.L., Robertson, R.N., and Shimada, S.D., A unified method for calculating the center of pressure during wheelchair propulsion, *Annals of Biomedical Engineering*, Vol. 26, 328–336, 1998.

17. Coleman, H.W. and Steele, W.G., *Experimentation and Uncertainty Analysis for Engineers*, John Wiley & Sons, New York, 1999, pp. 47–79.

18. Kline, S.J. and McClintock, F.A., Describing uncertainties in single-sample experiments, *Mechanical Engineering*, Vol. 75, 3–8, 1953.

19. Rozendal, L.A. and Veeger, H.E.L., Force direction in manual wheel chair propulsion: balance between effect and cost, *Clinical Biomechanics*, Vol. 15, No. 1, S39–S41, 2000.

20. Mallakzadeh, M.R., Sassani, F., and Oxland, T., Development of an instrumented wheel for three-dimensional analysis of manual wheelchair propulsion dynamics, *Proceedings of International Symposium on Collaborative Research in Applied Science (ISOCRIAS)*, Vancouver, Canada, 2005, pp. 22–29.

21. Norusis, M.J., *SPSS® 11.0 Guide to Data Analysis*, Prentice-Hall, Upper Saddle River, NJ, 2002.

5

MEMS-Based Ultrasonic Devices

T. Siu

M. Chiao

Summary

This chapter investigates the use of micro-ultrasonic-transducers (MUTs) for therapeutic applications such as enhancing the transfection efficacy of antisense oligonucleotides (ASOs) for cancer treatment. After a brief description of the two subcategories of MUT and their respective fabrication techniques, some experimental results are discussed. Finally, other potential applications of MUT are outlined and the advantages of MUTs over conventional ultrasonic transducers used in mechatronic applications are discussed.

5.1   Introduction

Ultrasonic transducers are simple yet typical examples of mechantronic devices. Besides, ultrasonic transducers are commonly presented as integral components in mechatronic systems. Acoustic waves, vibration, and movement of molecules in an organized manner are all mechanical phenomena. Yet, to produce these vibrations at high frequency (>20 kHz), one often looks to an electrical source as the driving signal. Furthermore, as MUTs are fabricated using micro-electromechanical systems (MEMS) technology, which evolved from photolithographic techniques of the microelectronic industry, they have inherited a great potential to be integrated with electronic circuits.

5.2   Micro-Ultrasonic-Transducers (MUTs)

Most ultrasonic transducers today use piezoelectric ceramic as their active elements. In recent years, researchers have used micromachining techniques to fabricate ultrasonic transducers that have active acoustic elements with the size of a few microns. Proposed applications for these MUTs include medical ultrasonography [1], acoustic anemometry [2], aerial ranging [3], and so on. In medical ultrasonography in particular, MUTs have many advantages over conventional ultrasonic transducers—for example, greater precision in fabrication and better acoustic matching [4]. In spite of the numerous MUT research around the world, by far no one has looked into the possibility of using MUTs for sonodynamic therapy.

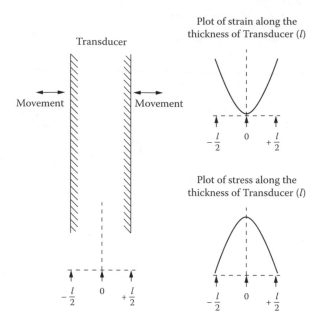

FIGURE 5.1 A conventional ultrasonic transducer made of piezoelectric ceramic is operated at the fundamental thickness-mode resonant frequency.

There are currently two approaches in fabricating MUTs. The first approach requires the deposition of a piezoelectric thin film on a micromachined diaphragm. The resultant transducers are called piezoelectric micro-ultrasonic transducers (pMUTs). Piezoelectric materials change their physical dimension when an electric field is applied across them. The dimension change occurs in the direction along the electric field and also in the directions perpendicular to the electric field.

In a conventional ultrasonic transducer, only the piezoelectric effect along the electric field is typically used to generate ultrasound. For power ultrasound applications, such as sonochemistry and sonodynamic therapy, the objective is to obtain maximal ultrasonic intensity. In such cases, the transducer is usually operated in a continuous wave fashion at the fundamental thickness-mode resonant frequency. At that frequency, a standing compressive wave is set up within the thickness of the transducer with the largest strain and therefore the least stress on its two surfaces [5]. Figure 5.1 shows a transducer at thickness mode resonant frequency.

Acoustic wave has a constant compressive velocity in a given material, the fundamental thickness-mode resonant frequency is related to its thickness and given by [5]:

$$f_1 = \frac{c_t}{2l} \tag{5.1}$$

where f_1 is the fundamental thickness-mode resonant frequency, c_t is the compressive acoustic wave velocity in the ceramic, and l is the thickness of the ceramic.

In a pMUT, however, the deposited piezoelectric thin film typically has a thickness of a few microns. The thickness-mode resonant frequency of such a thin film will be on the order of a few GHz. To allow the pMUT to operate at the MHz frequency that is more common in medical applications, the piezoelectric effect perpendicular to the applied electric field is employed, instead [4]. The micromachined diaphragm on which the thin piezoelectric film rests is made to buckle. When an electric field is applied across the piezoelectric film, the film will experience dimension changes in the directions perpendicular to the electric field. The change in area in the piezoelectric film will force the buckled diaphragm to

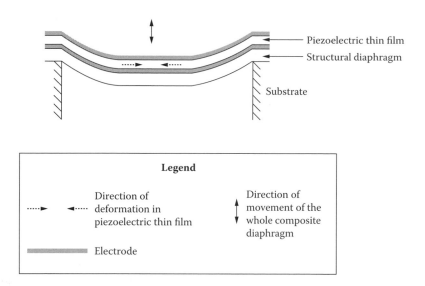

FIGURE 5.2 Basic design and operation of a pMUT.

flatten or further deform, generating the necessary motions to produce the ultrasound. Therefore, the resonant frequency of the pMUT will depend on the dimension and the material properties of the diaphragm. Figure 5.2 illustrates the operation of a pMUT. For detail fabrication processes of a pMUT, the readers may refer to the published papers of Muralt et al. [6] and Wang et al. [7].

The second way to fabricate MUTs is to utilize the electrostatic force between parallel plates with opposite charges. The resultant transducers are called capacitive micro-ultrasonic transducers (cMUTs). In general, a bottom electrode is first laid down, and a thin membrane is suspended on the bottom electrode with a cavity in between. On top of the thin membrane is the top electrode. When an alternating voltage is applied across the two electrodes, the electrostatic force between them will attract and then repel them from each other, thereby creating the necessary motions to generate ultrasound. The resonant frequency of a cMUT will also depend on the dimensions and the material properties of the membrane. The operation of a cMUT is illustrated in Figure 5.3.

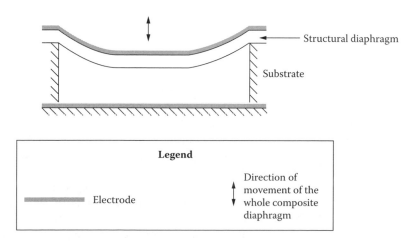

FIGURE 5.3 Basic design and operation of a cMUT.

A crucial step in the fabrication of cMUT is the incorporation of a precisely controlled cavity between the bottom electrode and the membrane. This cavity needs to be deep enough to allow free movement of the membrane but maintain an adequate electrostatic effect. Methods that are used to incorporate this cavity include: (1) use of sacrificial layer that is etched by chemical [8, 9] and (2) a wafer bonding process [2].

It is understood that cavitation, and therefore sonodynamic phenomenon, occurs only beyond a certain acoustic power density threshold [10]. Therefore, the most important design criterion for a MUT intended for sonodynamic therapy is the output acoustic power density.

The power density of an acoustic wave is defined as "the instantaneous power flowing through a unit area perpendicular to the direction of propagation of the wave as one elemental volume of the fluid acts on a neighboring element" [11]. Given that, one can write:

$$\text{Power density} = \frac{\text{power}}{\text{area}} = \frac{\text{work}}{\text{area} \times \text{time}} = \frac{\text{force} \times \text{distance}}{\text{area} \times \text{time}} = \text{pressure} \times \text{velocity} \qquad (5.2)$$

The acoustic impedance (Z) of a material is defined as the relationship between the pressure and velocity that a particle in the medium feels [11]:

$$Z = \frac{\text{pressure}}{\text{velocity}} \qquad (5.3)$$

Combining Equation 5.2 and Equation 5.3, one can write

$$I = Zu^2 \qquad (5.4)$$

where I is the power density of the acoustic wave and u is the velocity of a particle in the medium. Now let us consider what happens at the interface between the transducer surface and the liquid at a power density below the cavitation threshold. The velocity of a liquid particle that is adjacent to the transducer surface must be the same as the surface; otherwise one would have a void formation, which is not possible under the threshold power density. Therefore, the power emitted by an ultrasonic transducer is determined by the velocity of movement of its emitting face.

5.3 Acoustic Cavitation

Acoustics cavitation is a physical phenomenon in which cavities are created in the liquid by a large negative (i.e., rarefactional) pressure, produced either intentionally, such as by ultrasound, or unintentionally, such as by the movement of a ship's propellers. These cavities can disappear during the subsequent compression or remain to grow and oscillate for several cycles [10]. When these cavities finally collapse, they give out enormous amounts of energy in the form of intense pressure and temperature for a very brief moment [12]. This energy can produce free radicals by chemical bond cleavage [10] as well as physical disruption to biological membrane [13].

The occurrence of cavitation depends on a variety of parameters including temperature, hydrostatic pressure, and impurity in the solution as well as frequency and intensity of the applied ultrasound [10]. It should be noted that ultrasonic intensity (units: W/cm^2) is also called power density, and these two terms will be used interchangeably, depending on the context. With all parameters set as constant, there is a threshold power density that the applied ultrasound must achieve before cavitation occurs. This is because cavities will only be formed when the applied power density (force acting over a distance in a given time) is strong enough to overcome the attractive forces between the liquid molecules.

5.4 MUTs for Enhancing Antisense Oligonucleotides Efficacy

In the biomedical field, researchers are actively investigating the ability of ultrasound to enhance efficacy of a variety of pharmaceutical molecules including cancer drugs [14–16]. A particularly promising cancer drug is antisense oligonucleotides (ASOs). ASOs are short sequences of nucleotides that could inhibit a specific messenger ribonucleic acid (mRNA) in cells. It has great potential because it enables rational drug design based on genomic information [17]. However, like many other hydrophilic molecules, one of the major challenges of using ASOs is the inefficient cellular uptake, which limits their therapeutic efficacy [14].

The most efficient tools to introduce nucleic acids such as deoxyribonucleic acid (DNA) and mRNA into cells are viruses. However, concerns for safety limit the use of viral-mediated transfection. A widely used nonviral method to enhance transfection is the addition of cationic liposomes [18,19]. More recently, ultrasound was found to further enhance the transfection of DNA that are complexed with cationic liposomes [20–24]. Koch et al. reported that ultrasound enhancement of liposome-mediated cell transfection is caused by cavitational effects [23]. However, the precise mechanism involved remains unclear. Transfection could possibly be enhanced by sonoporation—the formation of transient pores on cell membranes due to cavitations. Cavitation might also disrupt the endosomes containing the uptaked nucleic acids and allow the nucleic acids to evade lysosomal degradation.

Despite the exciting possibility of using ultrasound to enhance the efficacy of ASOs, researchers have only been using conventional ultrasonic transducers. Treatment with conventional ultrasonic transducers would require multiple clinical visits for the ASOs injections as well as ultrasound exposures. Unwanted tissue heating and skin lesions due to the extracorporeal application of ultrasound also can present problems. Instead, an implantable MUT is proposed with an integrated controlled-release mechanism. Figure 5.4 shows a schematic diagram of the proposed device. Such a device could

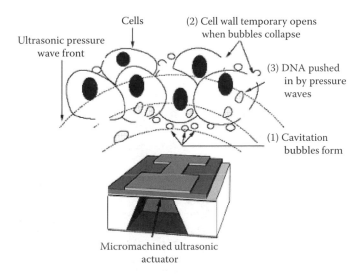

FIGURE 5.4 Proposed device for enhancing efficacy of cancer drugs including ASOs (not to scale). The operating sequence is marked in numerical order.

be implanted into/near a tumor with minimally invasive surgery. It could then automatically release the ASOs and transmit ultrasound over time. Therefore, it could reduce treatment cost as well as patient discomfort.

A series of preliminary experiments have been carried out in our laboratory to investigate the possible use of MUTs to increase the efficacy of ASOs. The ASOs used in the experiments target the bcl-xL mRNA, which has been investigated previously for cancer therapy [25,26]. The cell lines investigated by us were (1) human umbilical vein endothelial cells (HUVEC) and (2) human prostate cancer cells (PC3). In our experiments, three different controls were used. The first was the untreated wells, which served as a negative control and gave the number of viable cells without any treatment. The second control gave the number of viable cells after sonication. In the third control, Lipofectin, a state-of-the-art cationic lipid transfection agent, was added in conjunction with the ASOs but the wells were not sonicated.

For the HUVEC, a fair amount of cells were lysed due to sonication alone. After transfection of ASOs using lipofectin, only about 72% of cells remained viable. Sonication has enhanced the efficacy of ASOs+lipofectin and decreased surviving cells to 62%. The result of the experiment is summarized in Figure 5.5. Student's t-test has shown that the result from ASOs+lipofectin+sonication is significantly different ($p < 0.01$) from sonication and ASOs+lipofectin.

For the PC3, sonication and ASOs+lipofectin individually did not contribute to cell death. However, ASOs+lipofectin+sonication enhanced the ASOs efficacy and only 93% of the cells survived. Student's t-test again showed that the result from ASOs+lipofectin+sonication were significantly different ($p < 0.01$) from sonication and ASOs+lipofectin. Figure 5.6 illustrates the PC3 result.

Our experiments have shown that sonication with the prototype ultrasonic transducer enhances bcl-xL ASOs efficacy in HUVEC and PC3 cells. However, there existed a large number of parameters in our experiments, and they have not been systematically optimized. These parameters include reagent concentration, sonication frequency/power/duration, incubation time, and so on. In the future, we will systematically investigate the effect of each parameter on the results. In this manner we will be able to optimize the synergetic effect of ultrasound and possibly further enhance the ASOs efficacy. We are also continuing our work on the fabrication of miniaturized prototypes and the MUT. Once those devices are fabricated, we plan to use them for repeating the experiments.

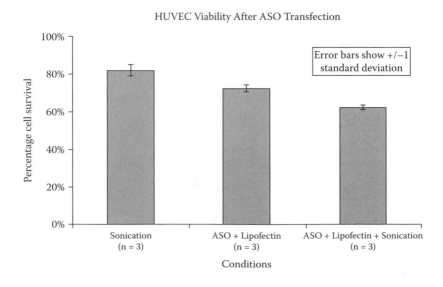

FIGURE 5.5 HUVEC viability after ASOs transfection in different conditions. ASOs+lipofectin+sonication resulted in a significant ($p < 0.01$) decrease of surviving cells compared to sonication and ASOs+lipofectin.

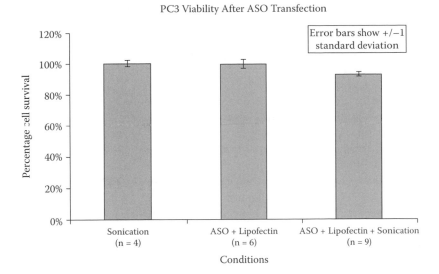

FIGURE 5.6 PC3 viability after ASOs transfection in different conditions. ASOs+lipofectin+sonication resulted in a significant $(p < 0.01)$ decrease of surviving cells compared to sonication and ASOs+lipofectin.

References

1. Oralkan, O., Ergun, A.S., Cheng, C.-H., Johnson, J.A., Karaman, M., Lee, T.H., and Khuri-Yakub, B.T., Volumetric ultrasound imaging using 2-D CMUT arrays, *IEEE Transactions on Ultrasonics, Ferroelectrics, and Frequency Control*, Vol. 50, 1581–1594, 2003.

2. Huang, Y., Capacitive micromachined ultrasonic transducers (CMUTs) built with wafer-bonding technology, in *Electrical Engineering*, Stanford University, 2005, p. 128.

3. Zhu, H., Miao, J., Wang, Z., Zhao, C., and Zhu, W., Fabrication of ultrasonic arrays with 7 um PZT thick films as ultrasonic emitter for object detection in air, *Sensors and Actuators A*, Vol. 123–124, 614–619, 2005.

4. Muralt, P. and Baborowski, J., Micromachined ultrasonic transducers and acoustic sensors based on piezoelectric thin films, *Journal of Electroceramics*, Vol. 12, 101–108, 2004.

5. Christensen, D.A., Transducers, beam patterns, and resolution, in *Ultrasonic Bioinstrumentation*, John Wiley, New York, 1998, pp. 69–118.

6. Muralt, P., Schmitt, D., Ledermann, N., Baborowski, J., Weber, P.K., Steichen, W., Petitgrand, S., Bosseboeuf, A., Setter, N., and Gaucher, P., Study of PZT coated membrane structures for micromachined ultrasonic transducers, presented at *IEEE Ultrasonics Symposium*, 2001.

7. Wang, Z., Zhu, W., Tan, O.K., Chao, C., Zhu, H., and Miao, J., Ultrasound radiating performances of piezoelectric micromachined ultrasonic transmitter, *Applied Physics Letters*, Vol. 86, 033508-1 to 3, 2005.

8. Knight, J., McLean, J., and Degertekin, F.L., Low temperature fabrication of immersion capacitive micromachined ultrasonic transducers on silicon and dielectric substrates, *IEEE Transactions on Ultrasonics, Ferroelectrics, and Frequency Control*, Vol. 51, 1324–1333, 2004.

9. Ladabaum, I., Jin, X., Soh, H.T., Pierre, F., Atalar, A., and Khuri-Yakub, B.T., Microfabricated ultrasonic transducers: towards robust models and immersion devices, presented at *IEEE Ultrasonics Symposium*, 1996.

10. Mason, T.J. and Lorimer, J.P., *Applied Sonochemistry: The Uses of Power Ultrasound in Chemistry and Processing*, John Wiley-VCH, Weinheim, 2002.

11. Christensen, D.A., Impedance, power and reflection, in *Ultrasonic Bioinstrumentation*, John Wiley, New York, 1998, pp. 21–38.

12. Leighton, T.G., Effects and mechanisms, in *The Acoustic Bubble*, Academic Press, London, 1994, pp. 439–566.
13. Deng, C.X., Sileling, F., Pan, H., and Cui, J., Ultrasound-induced cell membrane porosity, *Ultrasound in Medicine and Biology*, Vol. 30, 519–526, 2004.
14. Lysik, M.A. and Wu-Pong, S., Innovations in oligonucleotide drug delivery, *Journal of Pharmaceutical Sciences*, Vol. 92, 1559–1573, 2003.
15. Rosenthal, I., Sostaric, J.Z., and Riesz, P., Sonodynamic therapy — a review of the synergistic effects of drugs and ultrasound, *Ultrasonics Sonochemistry*, Vol. 11, 349–363, 2004.
16. Frairia, R., Catalano, M.G., Fortunati, N., Fazzari, A., Raineri, M., and Berta, L., High energy shock waves (HESW) enhance paclitaxel cytotoxicity in MCF-7 cells, *Breast Cancer Research and Treatment*, Vol. 81, 11–19, 2003.
17. Dove, A., Antisense and sensibility, *Nature Biotechnology*, Vol. 20, 121–124, 2002.
18. Felgner, P.L., Gader, T.R., Holm, M., Roman, R., Chan, H.W., Wenz, M., Northrop, J.P., Ringold, G.M., and Danielsen, M., Lipofectin: a highly efficient, lipid-mediated DNA-transfection procedure, *Proceeding of National Academy of Science USA*, Vol. 84, 7413–7417, 1987.
19. Malone, R.W., Felgner, P.L., and Verma, I.M., Cationic liposome-mediated RNA transfection, *Proceeding of National Academy of Science USA*, Vol. 86, 6077–6081, 1989.
20. Anwer, K., Kao, G., Proctor, B., Anscombe, I., Florack, V., Earls, R., Wilson, E., McCreery, T., Unger, E., Rolland, A., and Sullivan, S., Ultrasound enhancement of cationic lipid-mediated gene transfer to primary tumors following systemic administration, *Gene Therapy*, Vol. 7, 1833–1839, 2000.
21. Feril, L.B., Ogawa, R., Kobayashi, H., Kikuchi, H., and Kondo, T., Ultrasound enhances liposome-mediated gene transfection, *Ultrasonics Sonochemistry*, Vol. 12, 489–493, 2005.
22. Huang, S., Tiukinhoy, S., McPherson, D., and MacDonald, R., Combined use of ultrasound and acoustic cationic liposomes results in improved gene delivery into smooth muscle cells, *Molecular Therapy*, Vol. 5, S9, 2002.
23. Koch, S., Pohl, P., Cobet, U., and Rainov, N.G., Ultrasound enhancement of liposome-mediated cell transfection is caused by cavitation effects, *Ultrasound in Medicine and Biology*, Vol. 26, 897–903, 2000.
24. Unger, E., McCreery, T., and Sweiter, R.H., Ultrasound enhances gene expression of liposomal transfection, *Investigative Radiology*, Vol. 32, 723–727, 1997.
25. O'Neill, J., Manion, M., Schwartz, P., and Hockenbery, D.M., Promises and challenges of targeting Bcl-2 anti-apoptotic proteins for cancer therapy, *Biochimica et Biophysica Acta*, Vol. 1705, 43–51, 2004.
26. Piro, L., Apoptosis, Bcl-2 antisense, and cancer therapy, *Oncology (Williston Park, New York)*, Vol. 18(13 Suppl. 10), 5–10, 2004.

6

Polyaniline Nanostructures

H. Xia

H.S.O. Chan

Summary

Miniaturization of mechatronic devices has important uses and advantages. Nanostructures are relevant in this context. In this chapter, different nanostructures of polyaniline (PANI) and their preparation are briefly introduced. The novel *Y*-junction PANI nanorods and nanotubes are described in more detail. The morphology of *Y*-junction PANI nanostructures (rods and tubes with diameter in the range of 40–100 nm) is confirmed by transmission electron microscopy (TEM), and the effects of reaction conditions on the morphology of PANI nanostructures are investigated. Further characterization experiments including FTIR, UV-vis spectrometry, and x-ray diffraction are carried out to study the chemical and electronic structures of the PANI nanostructures. The magnetic property of the *Y*-junction PANI nanorods with Fe_3O_4 nanoparticles is investigated with a superconducting quantum interference device (SQUID).

6.1 Introduction

Mechatronics is a multidisciplinary field. In the effort to miniaturize mechatronic devices, micro- and nanotechnologies have received increased attention. Such technologies are particularly applicable in sensing, actuation, and control of miniature mechatronic devices. Since their discovery in mid-1970s, conducting polymers have been widely studied [1,2]. Among the conducting polymers, PANI is the most investigated because of its simple preparation, good environmental and thermal stability, structural versatility, and potential applications as electrical and optical materials [3,4]. The polymeric structure of

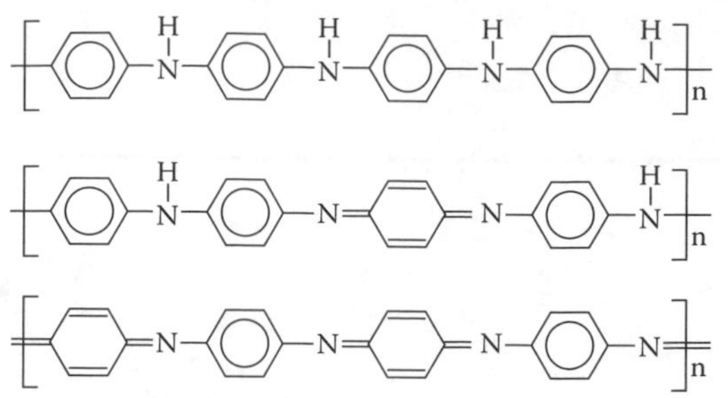

FIGURE 6.1 Three forms of PANI: (a) leucoemeraldine, (b) emeraldine, (c) pernigraniline.

PANI is now generally accepted (see Figure 6.1). It has three different oxidation states: leucoemeraldine, emeraldine, and pernigraniline. Emeraldine is the state with the highest conductivity.

Together with the electrical conductivity of the emeraldine salt form, the reversible redox and pH switching properties of the nanostructured PANI have been currently developed in a wide range of potential applications, which will be discussed in the next section.

Nanostructured PANIs (nanorods/-tubes/-fibers) offer the possibility of enhanced performance wherever a high interfacial area between PANI and its environment is important. For example, in sensor application for chemical gas, nanostructured (fibers) PANI has greater sensitivity and faster response time compared with their bulk counterparts due to higher effective surface area and shorter penetration depth for target molecules [5,6]. However, such sensors have not been widely exploited, probably due to the lack of facile and reliable methods for making high-quality conducting PANI nanostructures.

6.2 Synthetic Methods of PANI Nanostructures

Generally, the chemical synthesis of PANI uses monomer aniline and a strong oxidant in the presence of common mineral acids such as HCl, H_2SO_4, and H_3PO_4 as a dopant [7]. The polymerization of PANI in aqueous medium has been studied as a function of a wide variety of synthesis parameters such as pH, relative concentration of reactants, polymerization temperature, and time [8]. However, the resulting PANIs generally show a mixed morphology of granular flakes and fibers entangled together.

Template synthesis is an effective and common method to prepare the nanostructures of conducting polymers [9–24]. For example, PANI nanofibers with extraordinarily small diameter of 3 nm have been made by template-guided polymerization within channels of zeolites [25] or within nanoporous membranes [26,27]. Wallace et al. [28] also reported the formation of nanofibers by the use of insoluble solid template of opals.

Another way, known as the self-assembly method, synthesizes PANI nanostructures by adding structure-directing agents such as surfactants [29] or polyelectrolytes [30] into the chemical polymerization bath. When organic dopants (structure-directing agents) with surfactant functionalities are used, micelles and emulsions can be formed, leading to tubes, fibers, or rodlike structures [31–35]. For example, Huang and Wan [36] polymerized aniline using ammonium persulfate as an oxidant in the presence of naphthalene sulfonic acid (NSA) as a dopant by self-assembly. In particular, it is noted that the structure-directing agent NSA has acted as a soft template in the formation of the PANI structure. Obviously, the self-assembly method is simple and inexpensive as compared to template synthesis, as the former method avoids the use of a microporous membrane as a template and the need for expensive equipment.

6.3 PANI Nanofibers

Chen and Lee [3] reported the fibrous network morphology of PANI electropolymerized in a tetrafluoroboric acid (HBF_4) aqueous solution. A similar fibrous morphology of electrochemically polymerized PANI was also observed by Desilvestro and Scheifele [37], and Bedekar and coworkers [38]. Although these methods of electrochemical polymerization and other physical methods such as electrospinning [39] and mechanic stretching [40] also produce conducting polymer nanofibers, these materials have only been made in a very limited scale. Recently, using common mineral acids as dopants in an interfacial polymerization route, Huang et al. [41] have successfully synthesized PANI nanofibers with diameters close to 50 nm, which are among the smallest reported for PANI nanofibers and nanowires without using a template. An extremely simple "nanofiber seeding" method for the preparation of PANI nanofibers was reported by Zhang [42]. Seeding the reaction with very small amounts of nanofibers, regardless of their chemical nature, resulted in a precipitate with bulk fibrillar morphology.

6.4 *Y*-Junction PANI Nanorods and Nanotubes

Although micro- and nanotubes of PANI have been prepared by chemical and electrochemical oxidative polymerization of aniline within templates [21,25,43] or with structure-directing agents [32,36,44–47], few papers reported the preparation of *Y*-junction PANI nanotubes and nanorods [48].

Since the discovery of carbon nanotubes by Iijima [49], micro- and nanotubular materials have attracted a great deal of interest because of their potential applications in nanoelectronics and biomedical devices [50,51]. For such applications, it is important to be able to connect the nanotubes of different shapes and sizes [52–56]. The three-point nanotube junctions have been one of the most common building blocks in nanoelectronics, and especially the *Y*- and *T*-junctions have been considered as prototypes. With this aim in mind, a number of carbon nanotube junctions have been prepared by chemical vapor deposition (CVD) of products generated from a pyrolysis of metallocenes [57–59], template synthesis [60], and nanowelding of overlapping isolated nanotubes using high-intensity electron beams [61].

In the meantime, the "flash welding" technique [62,63] makes PANI an ideal material [64] for use as the "solder" to weld together the nanoscale building blocks of complex functional devices. The successful synthesis of the *Y*-shaped junction PANI nanostructures would certainly have an important impact on the development of nanoelectronic devices or miniature mechatronic devices.

6.5 Preparation of *Y*-Junction PANI Nanorods and Nanotubes

Recently, it was reported that molecular junctions for molecular electronics devices were fabricated by magnetically directed assembly of silica particles with deposition of nickel and gold films [65]. Magnetically directed assembly provides a wafer-level route for the fabrication of molecular junctions and opens up the potential for hybrid complementary metal-oxide semiconductor/molecular electronic applications.

We describe in this chapter a new way of making PANI nanostructures. Water-soluble Fe_3O_4 nanoparticles with a coating of polyethylene glycol(5), nonylphenyl ether (NP5), and cyclodextrin (CD) were synthesized and used as templates for the preparation of PANI nanostructures. NP5 was used to stabilize the magnetic nanoparticles by the formation of a suitable surface coating [66], and CD would improve their miscibility with water [67]. The preparation steps are given next.

The templates (A, Figure 6.2) are formed first in the solution. For the formation of *Y*-junction PANI nanorods, the initial pH of the reaction system is kept at 8. Aniline monomers first surround the templates to form micelles [68] while retaining the *Y*-shape (B, Figure 6.2). Polymerization starts after ammonium persulphate (APS) addition, and the growing PANI chains promote the formation of longer and more stable *Y*-junction structures [69] (C, Figure 6.2). Aggregates of *Y*-junction PANI nanorods (D, Figure 6.2) are formed eventually. On the other hand, *Y*-junction PANI nanotubes (E, Figure 6.2) are obtained when the initial pH of the reaction system is adjusted to about 5 by the addition of HCl solution.

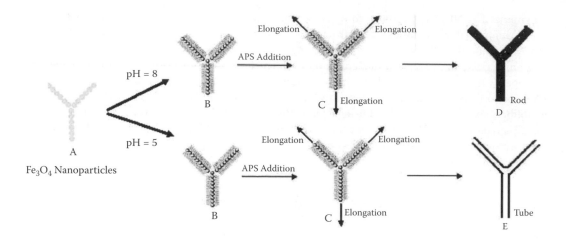

FIGURE 6.2 Schematic formation of PANI nanorods and nanotubes with Y-junctions. (Xia, H., Cheng, D., Xiao, C., Chan, H.S.O. *J. Mater. Chem.*, 2005, 15, 4161. Reproduced by permission of The Royal Society of Chemistry.)

6.6 Characterization of Templates

Figure 6.3a shows the aggregates of Fe_3O_4 nanoparticle chains and nanoparticles. Y-junctions made up of Fe_3O_4 nanoparticles (average diameter of 10 nm) are observed in the enlarged TEM image (indicated by arrows in Figure 6.3b). It is difficult to obtain a good TEM picture of more isolated Y-junctions because of the thermal motion experienced at ambient temperatures. The presence of many individual nanoparticles indicates that a snapshot of dynamic clusters was observed, which is consistent with that reported by Butter et al. [70].

The presence of Fe_3O_4 is identified from the electron diffraction pattern and the x-ray diffraction pattern (Figure 6.4). The patterns obtained confirm that the nanoparticles prepared in this study are Fe_3O_4 nanoparticles [71]. The Bragg reflection peaks are all relatively broad because of the extremely small dimensions of the Fe_3O_4 nanoparticles.

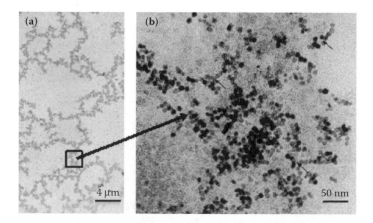

FIGURE 6.3 TEM images of assemblies of Fe_3O_4 nanoparticles and aniline: (a) before polymerization, (b) an enlargement of a part of (a). (Xia, H.; Cheng, D.; Xiao, C.; Chan, H.S.O. *J. Mater. Chem.*, 2005, 15, 4161. Reproduced by permission of The Royal Society of Chemistry.)

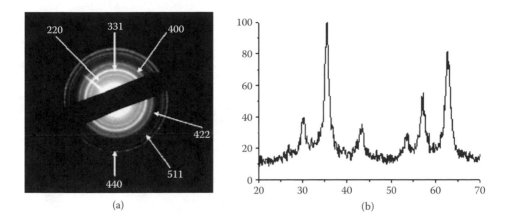

(a)   (b)

FIGURE 6.4 (a) Selected area electron diffraction pattern of Fe_3O_4 nanoparticles deposited from dispersion, (b) x-ray diffraction pattern of Fe_3O_4 nanoparticles on glass substrate. (Xia, H.; Cheng, D.; Xiao, C.; Chan, H.S.O. *J. Mater. Chem.*, 2005, 15, 4161. Reproduced by permission of The Royal Society of Chemistry.)

6.7 Morphology

In this section some important issues of morphology as related to PANI products are presented.

6.7.1 Effect of Aniline Concentration on Morphology of PANI Products

The TEM picture of PANI nanorods obtained at an optimal [An] of 0.10 mol/L is shown in Figure 6.5. The average diameter of the PANI nanorods is about 80 nm (Figure 6.5a). One clear case of a single *Y*-junction PANI nanorod is shown in Figure 6.5b. Figure 6.5c shows how multiple *Y*-junctions can

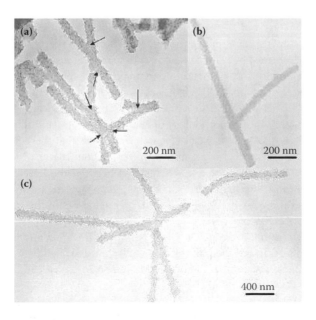

FIGURE 6.5 TEM photos of PANI nanorods with *Y*-junctions ([An] = 0.10 mol/L, [An]: [APS] = 1, reaction time = 18 h, temperature = 2.5°C, initial pH = 8). (Xia, H.; Cheng, D.; Xiao, C.; Chan, H.S.O. *J. Mater. Chem.*, 2005, 15, 4161. Reproduced by permission of The Royal Society of Chemistry.)

 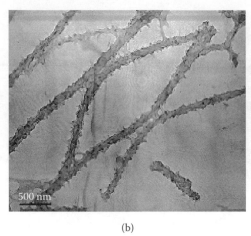

(a)   (b)

FIGURE 6.6 TEM images of PANI nanorods at other concentrations of aniline: (a) 0.20 mol/L, (b) 0.05 mol/L ([An]: [APS] = 1, reaction time = 18 h, temperature = 2.5°C, initial pH = 8). (Xia, H., Cheng, D., Xiao, C., Chan, H.S.O. *J. Mater. Chem.*, 2005, 15, 4161. Reproduced by permission of The Royal Society of Chemistry.)

coexist, which is consistent with the results of the self-assembly of Fe_3O_4 nanoparticles reported by Butter [70] and works based on computer simulations [72,73]. In addition, it is interesting to note that some X-junctions are also present due to the connection of the Y-junction to another nanorod. This result confirms that higher levels of magnetic particle organization are possible, and some of the PANI chains may form long, ringlike structures or connect to form an extended network [70,73,74]. About 30% of the products are well-fined Y-junction structures.

The morphology of PANI nanorods prepared at other concentrations of aniline is shown in Figure 6.6. When [An] was at a low of 0.05 mol/L, the dendritic structures of the PANI nanorods obtained (Figure 6.6a) are very similar to those predicted by analytical simulations [72,73]. When [An] is increased from 0.10 to 0.20 mol/L, the diameter of PANI nanorods increases only a little compared to the increase in length (shown in Figure 6.6b), but there are fewer Y-junctions. The increase in [An] may inhibit the formation of the Fe_3O_4 templates, and only normal PANI nanorods are formed because the polymerization of cylindrical micelles formed by aniline monomers in the reaction system is likely to dominate [68].

Accordingly, PANI nanotubes with Y-junctions were prepared at an initial pH of about 5 as compared to 8 for the nanorods. No Y-junction PANI nanotubes or nanorods can be obtained if the initial pH of reaction system is below 3. This is because the template is dissolved in this low pH before polymerization.

Similar to the PANI nanorods, the morphology of PANI Y-junction nanotubes is also affected by the aniline concentration (Figure 6.7). TEM image (Figure 6.7a) shows that the Y-junction PANI nanotube has an outer diameter of around 70 nm and an inner diameter of about 10 nm, which is equal to the average diameter of the Fe_3O_4 nanoparticles. The angle θ enclosed between the lower arms of the Y-junction nanotube is about 60°. Figure 6.7a shows that some Fe_3O_4 nanoparticles are still present, and perfectly hollow PANI nanotubes can be formed (Figure 6.7b) by reducing pH of the sample solution to about 1 and kept for 24 h before characterization.

When [An] is increased from 0.13 to 0.26 mol/L, for example, the outer and inner diameters of PANI nanotubes changed significantly, but the wall thickness only changed slightly (Figure 6.7c). As found in the PANI nanorods, Y-junction nanotubes disappeared (Figure 6.7d) when [An] is too high, for the same reasons outlined for the PANI nanorods.

6.7.2   Influence of Organic Solvent on Morphology of PANI Products

As previously mentioned, self-assembled flexible chains of magnetic particles are determined by a thermodynamic balance of forces [75,76]. If the self-assembled flexible chains of Fe_3O_4 nanoparticles acted

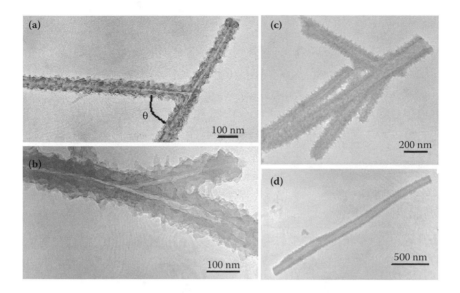

FIGURE 6.7 TEM images of PANI products at different concentrations of aniline: (a) 0.13 mol/L, (b) 0.13 mol/L after treatment in HCl (pH = 1) solution, (c) 0.20 mol/L, (d) 0.26 mol/L ([An] : [APS] = 1, reaction time = 18 h, temperature = 2.5°C, initial pH = 5).

as the templates in the formation of *Y*-junction PANI nanorods, the presence of an organic solvent is expected to have some influence on the shape of the PANI products. When an organic solvent, ethanol, is added into the reaction system (ethanol/Fe$_3$O$_4$ solution = 1:10 by volume), the *Y*-junction PANI nanorods are able to form rings (Figure 6.8a), and the chosen overlapping region of PANI nanorods is enlarged and clearly shown in Figure 6.8b.

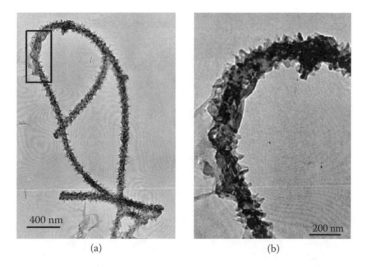

(a)   (b)

FIGURE 6.8 (a) TEM images of PANI rings with *Y*-junctions, (b) an enlargement of a part of (a) ([An] = 0.10 mol/L, [An] : [APS] = 1, reaction time = 18 h, temperature = 2.5°C, initial pH = 8, ethanol/Fe$_3$O$_4$ solution = 1:10 by volume). (Xia, H., Cheng, D., Xiao, C., Chan, H.S.O. *J. Mater. Chem.*, 2005, 15, 4161. Reproduced by permission of The Royal Society of Chemistry.)

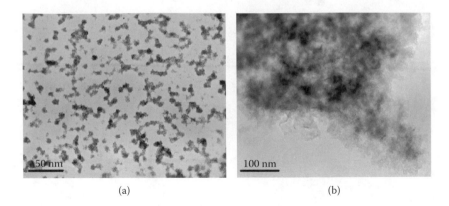

(a)   (b)

FIGURE 6.9 TEM images of Fe_3O_4 nanoparticles and aniline monomers in the reaction system (toluene/Fe_3O_4 solution = 1:10 by volume): (a) before polymerization, (b) PANI-Fe_3O_4 composite after polymerization.

Alternatively, aggregates of PANI-Fe_3O_4 composite are obtained when an organic solvent, toluene, immiscible with water, is added into the reaction system (toluene/Fe_3O_4 solution = 1:10 by volume). That is because the toluene solvent destroys the surfactant layer of the water-soluble Fe_3O_4 nanoparticles (Figure 6.9a), and then makes Fe_3O_4 nanoparticles hydrophobic. Consequently, the normal PANI-Fe_3O_4 composite is obtained (Figure 6.9b).

Controlled experiments were also carried out. Normal PANI nanostructures (rods or tubes) are obtained if aniline is polymerized under the same conditions without the Fe_3O_4 nanoparticles [68]. The incorporation of NP5 and CD produced nanoparticles only (Figure 6.10). These results confirm the important contribution of the Fe_3O_4 nanoparticles as the template. In addition, our results are significantly different from polyaniline-Fe_3O_4 composites [77,78], in which Fe_3O_4 nanoparticles are hydrophobic and there are no surfactant layers. When the concentration of Fe_3O_4 nanoparticles is very low (Figure 6.11) in our system, PANI nanofibers are obtained. This result is consistent with those reported by Huang et al. [79]. Under this condition, the Fe_3O_4 nanoparticles cannot act as a template, and aniline is polymerized into a fibril structure.

FIGURE 6.10 TEM image of PANI particles obtained in presence of NP5 and CD.

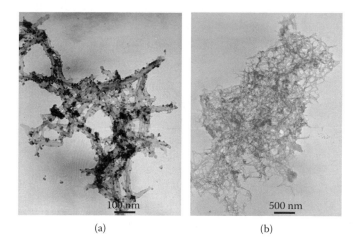

(a)   (b)

FIGURE 6.11 TEM image of PANI products obtained under the same conditions except the amount of Fe_3O_4 solution (< 5 mL): (a) initial pH = 8, (b) initial pH = 5.

6.8 Structural and Magnetic Characterization

The molecular structure of PANI nanostructure was studied by Fourier transform infrared (FTIR) spectroscopy. The FTIR spectra (Figure 6.12) show the characteristic peaks of PANI at around 1582 cm[1], 1489 cm[1] (C=C stretching deformation of quinoid and benzenoid ring, respectively), 1296 cm[1] (C-N stretching of secondary aromatic amine), 1115 cm[1], and 822 cm[1] (out-of-plane deformation of C-H in the 1,4-dissubstituted benzene ring) [80]. Compared with normal bulk PANI (figure 6.12a), other peaks are found in the FTIR spectra due to irregular structure of the PANI nanorods and nanotubes (figures 6.12b and 6.12c). The peak at about 695 cm[1] (out-of-plane deformation of C-H) indicates the formation of the 1,3-dissubstituted benzene ring [81–83]. The peaks around 1445 cm[1], 1040 cm[1], and 1415 cm[1] are

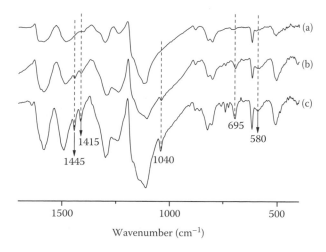

Wavenumber (cm^{-1})

FIGURE 6.12 (a) FTIR spectra of normal bulk PANI, (b) PANI nanotubes with Y-junctions (sample conditions are the same as in figure 6.7a), (c) PANI nanorods with Y-junctions (sample conditions are the same as in Figure 6.5). (Xia, H., Cheng, D., Xiao, C., Chan, H.S.O. *J. Mater. Chem.*, 2005, 15, 4161. Reproduced by permission of The Royal Society of Chemistry.)

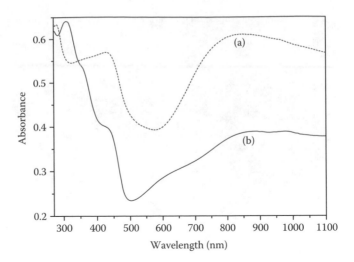

FIGURE 6.13 UV-vis spectra of (a) doped PANI nanotubes in water (sample conditions are the same as in Figure 6.7a), (b) normal bulk PANI in water.

attributed to the N=N stretching [83,84] due to the presence of some "head to head" [84–86] instead of the normal "head to tail" couplings [81,86,87]. The peak at about 580 cm¹ is attributed to the presence of Fe_3O_4 nanoparticles (Figure 6.12b and Figure 6.12c) [88].

In UV-vis spectrometry, the peaks around 280 and 420 nm (Figure 6.13a) are assigned to π-π* transition and the polaron band transition of PANI backbone, respectively [89,90]. The band at about 840 nm (Figure 6.13a) is characteristic of the emeraldine salt form of PANI [91]. Compared with the PANI prepared by conventional methods (Figure 6.13b), there is a blue shift of the π-π* transition band from 310 nm to 280 nm, which can be attributed to PANI in the nanoscale and a decrease of conjugation length in the *Y*-junction polymer chains [92].

The x-ray diffraction patterns of PANI-HCl, and *Y*-junction PANI nanorods and nanotubes, are shown in Figure 6.14. The broad peaks centered at $2\theta = 15–30°$ suggest that all the resulting PANI nanostructures are amorphous. Moreover, the two broad peaks, at about 21° and 25°, can be ascribed to the periodicity

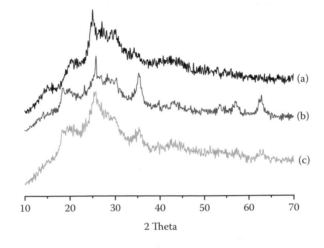

FIGURE 6.14 X-ray diffraction patterns of: (a) common PANI-HCl, (b) *Y*-junction PANI rods (sample conditions are the same as in Figure 6.5), (c) *Y*-junction PANI tubes (sample conditions are the same as in Figure 6.7a).

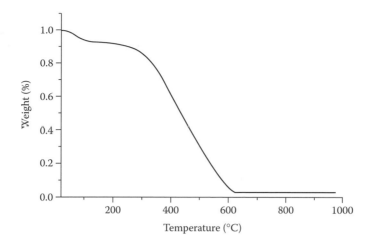

FIGURE 6.15 The content of Fe_3O_4 in the final PANI nanorods characterized by TGA under air environment (sample conditions are the same as in Figure 6.5).

parallel and perpendicular to the polymer chain [93,94], respectively. In particular, the diffraction peaks of the Fe_3O_4 nanoparticles are observed in Y-junction PANI nanorods and nanotubes (Figures 6.14b and 6.14c), compared with the normal bulk PANI-HCl (Figure 6.14a). These data are in good agreement with those of Fe_3O_4 nanoparticles (Figure 6.4b). In addition, the diffraction peaks are strong in PANI nanorods but weak in PANI nanotubes. These results show that there are considerable amounts of Fe_3O_4 nano-particles in the PANI nanorods and less in the PANI nanotubes. These results are also consistent with those obtained from FTIR.

The magnetic property of the PANI nanorods with Fe_3O_4 nanoparticles was investigated with SQUID. For Y-junction PANI nanotubes, the content of Fe_3O_4 nanoparticles in them is very close to zero, and no meaningful data of magnetic property could be extracted. For Y-junction PANI nanorods, the content of Fe_3O_4 nanoparticles is 2.74 wt%, which has been determined by thermogravimetric analysis (TGA) characterization (Figure 6.15). Figure 6.16 shows the room-temperature magnetization (300 K) of as-prepared

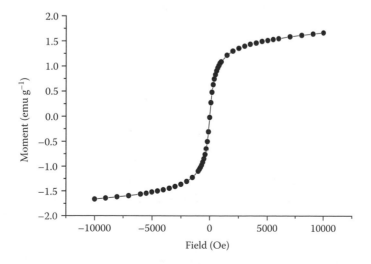

FIGURE 6.16 Room-temperature magnetization curve of obtained PANI nanorods. (Xia, H., Cheng, D., Xiao, C., Chan, H.S.O. *J. Mater. Chem.*, 2005, 15, 4161. Reproduced by permission of The Royal Society of Chemistry.)

PANI-Fe_3O_4 nanocomposite. The curve is perfectly superimposable as the field is cycled between 10 kOe and 10 kOe. The curves are consistent with superparamagnetic behavior and the nanoscale dimensions of the particles [95,96]. The saturation magnetization is 1.67 emu/g, which is attributed to the low content of Fe_3O_4 nanoparticles in the nanocomposites.

6.9 Outlook

It is possible to make polymeric nanostructures with the desired structure for applications in the nano-electronic devices by controlling the synthetic conditions. The method mentioned in this chapter can be extended to prepare a variety of magnetic metal-conducting polymer nanostructures under appropriate conditions. In addition, totally hollow polymer nanotubes can be obtained by removal of the core material under certain conditions.

References

1. Chandrasekhar, P. *Conducting polymers, fundamentals and applications: a practical approach.* Boston, MA: Kluwer Academic, 1999.
2. Wallace, G.G.; Spinks, G.M.; Teasdale, P.R. *Conductive electroactive polymers: intelligent materials systems,* 2nd ed. Boca Raton, FL: CRC Press, 2003.
3. Chen, S.A.; Lee, T.S. *J. Polym. Sci. Part C: Polym. Lett.* 1987, *25*, 455.
4. Desilvestro, J.; Scheifele, W. *J. Mater. Chem.* 1993, *3*, 263.
5. Huang, J.; Virji, S.; Weuller, B.H.; Kaner, R.B. *Chem. Eur. J.* 2004, *10*, 1314.
6. Virji, S; Huang, J.; Kaner, R.B.; Weuller, B.H. *Nano Letters* 2004, *4*, 491.
7. Huang, W.S.; Humphrey, B.D.; MacDiarmid, A.G. *J. Chem. Soc., Faraday Trans.* 1986, *82*, 2385.
8. Negi, S.; Adhyapak, P.V. *J. Macromol. Sci. Polymer Rev.* 2002, *C42*, 35.
9. Parthasarathy, R.V.; Martin, C.R. *Nature* 1994, *369*, 298.
10. Penner, R.M.; Martin, C.R. *J. Electrochem. Soc.* 1986, *133*, 2206.
11. Cai, Z.; Martin, C.R. *J. Am. Chem. Soc.* 1989, *111*, 4138.
12. Cai, Z.; Lei, J.; Liang, W.; Menon, V.; Martin, C.R. *Chem. Mater.* 1991, *3*, 960.
13. Martin, C.R.; Parthasarathy, R.; Menon, V. *Synth. Met.* 1993, *55*, 1165.
14. Van Dyke, L.S.; Martin, C.R. *Langmuir* 1990, *6*, 1123.
15. Liang, W.; Martin, C.R. *J. Am. Chem. Soc.* 1990, *112*, 9666.
16. Penner, R.M.; Martin, C.R. *Anal. Chem.* 1987, *59*, 2625.
17. Brumlik, C.J.; Martin, C.R.; Tokuda, K. *Anal. Chem.* 1992, *64*, 1201.
18. Foss, C.A., Jr.; Hornyak, G.L.; Stockert, J.A.; Martin, C.R. *J. Phys. Chem.* 1992, *96*, 7497.
19. Foss, C.A., Jr.; Hornyak, G.L.; Stockert, J.A; Martin, C.R. *Adv. Mater.* 1993, *5*, 135.
20. Foss, C.A., Jr.; Hornyak, G.L.; Stockert, J.A.; Martin, C.R. *J. Phys. Chem.* 1994, *98*, 2963.
21. Martin, C.R. *Adv. Mater.* 1991, *3*, 457.
22. Brumlik, C.J.; Martin, C.R. *J. Am. Chem. Soc.* 1991, *113*, 3174.
23. Brumlik, C.J.; Menon, V.P.; Martin, C.R. *J. Mater. Res.* 1994, *9*, 1174.
24. Klein, J.D.; Herrick, R.D.I.; Palmer, D.; Sailor, M.J.; Brumlik, C.J.; Martin, C.R. *Chem. Mater.* 1993, *5*, 902.
25. Wu, C.G.; Bein, T. *Science* 1994, *264*, 1757.
26. Wang, C.W.; Wang, Z.; Li, M.K.; Li, H.L. *Chem. Phys. Lett.* 2001, *341*, 431.
27. Wang, Z.; Chen, M.A.; Li, H.L. *Mater. Sci. Eng. A* 2002, *328*, 33.
28. Misoska, V.; Price, W.; Ralph, S.; Wallace, G. *Synth. Met.* 2001, *121*, 1501.
29. Yu, L.; Lee, J.I.; Shin, K.W.; Park, C.E.; Holze, R. *J. Appl. Polym. Sci.* 2003, *88*, 1550.
30. Liu, J.M.; Yang, S.C. *Chem. Commun.* 1991, 1529.
31. Kinlen, P.J.; Liu, J.; Ding, Y.; Graham, C.R.; Remsen, E.E. *Macromolecules* 1998, *31*, 1735.
32. Wei, Z.X.; Zhang, Z.M.; Wan, M.X. *Langmuir* 2002, *18*, 917.
33. Wei, Z.X.; Wan, M.X. *J. Appl. Polym. Sci.* 2003, *87*, 1297.

34. Langer, J.J.; Framski, G.; Joachimiak, R. *Synth. Met.* 2001, *121*, 1281.
35. Qiu, H.J.; Wan, M.X.; Matthews, B.; Dai, L.M. *Macromolecules* 2001, *34*, 675.
36. Huang, K.; Wan, M.X. *Chem. Mater.* 2002, *14*, 3486.
37. Desilvestro, J; Scheifele, W. *J. Mater. Chem.* 1993, *3*, 263.
38. (a) Bedekar, A.G.; Patil, S.F.; Pail, R.C.; Agashe, C. *Polymer* 1994, *35*, 2902; (b) Okampto, H.; Okampto, M.; Kotaka, T. *Polymer* 1998, *39*, 4359.
39. MacDiarmid, A.G.; Jones, W.E.; Norris, I.D.; Gao, J.; Johnson, A.T.; Pinto, N.J.; Hone, J.; Han, B.; Ko, F.K.; Okuzaki, H.; Llaguno, M. *Synth. Met.* 2001, *119*, 27.
40. He, H.X.; Li, C.Z.; Tao, N.J. *Appl. Phys. Lett.* 2001, *78*, 811.
41. Huang, J.; Kaner, R.B. *J. Am. Chem. Soc.* 2004, *126*, 851.
42. Zhang, X.; Goux, W.J.;. Manohar, S.K. *J. Am. Chem. Soc.* 2004, *126*, 4502.
43. Parthasarathy, R.V.; Martin, C.R., *Chem. Mater.* 1994, *6*, 1627.
44. Wei, Z.; Wan, M.X.; Lin, T.; Dai, L.M. *Adv. Mater.* 2003, *15*, 136.
45. Xia, H.; Chan, H.S.O.; Xiao, C.Y.; Cheng, D.M. *Nanotechnology* 2004, *15*, 1807.
46. Xia, H.; Janaky, N.; Cheng, D.; Xiao, C.; Liu, X.-Y.; Chan, H.S.O. *J. Phys. Chem.: B*, 2005, *109*, 12677.
47. Xia, H.; Cheng, D.; Lam, P.; Chan, H.S.O. *Nanotechnology*, 2006, *17*, 3957.
48. Xia, H.; Cheng, D.; Xiao, C.; Chan, H.S.O. *J. Mater. Chem.*, 2005, *15*, 4161.
49. Iijima, S. *Nature* 1991, *354*, 56.
50. Ajayan, P.M. *Handbook of Nanostructured Materials and Nanotechnology* Vol. 5 (Ed. H.S. Nalwa), Academic Press, San Diego, CA, 1999.
51. Lei, J.; Menon, V.P.; Martin, C.R. *Polym. Adv. Technol.* 1992, *4*, 124.
52. Dresselhaus, M.S.; Dresselhaus, G.; Ecklund, P.C. *Science of Fullerenes and Carbon Nanotubes*, Academic Press, San Diego, CA, 1996.
53. Chico, L.; Crespi, V.H; Benedict, L.X.; Louie, S.G.; Cohen, M.L. *Phys. Rev. Lett.* 1996, *76*, 971.
54. Menon, M.; Srivastava, D. *Phys. Rev. Lett.* 1997, *79*, 4453.
55. Kouwenhoven, L. *Science* 1997, *275*, 1896.
56. McEuen, P.L. *Nature* 1998, *393*, 15.
57. Satishkumar, B.C.; John Thomas, P.; Govindaraj, A.; Rao, C.N.R. *Appl. Phys. Lett.* 2000, *77*, 2530.
58. Li, W.Z.; Wen, J.G.; Ren, Z.F. *Appl. Phys. Lett.* 2001, *79*, 1879.
59. Deepak, F.L.; Govindaraj, A.; Rao, C.N.R. *Chem. Phys. Lett.* 2001, *345*, 5.
60. Papadopoulos, C.; Rakitin, A.; Li, J.; Vedeneev, A.S.; Xu, J.M. *Phys. Rev. Lett.* 2000, *85*, 3476.
61. Terrones, M.; Banhart, F.; Grobert, N.; Charlier, J.C.; Terrones, H.; Ajayan, P.M. *Phys. Rev. Lett.* 2002, *89*, 075505.
62. Huang, J; Kaner, R.B. *Nature Mater.* 2004, *3*, 783.
63. Li, D.; Xia, Y.N. *Nature Mater.* 2004, *3*, 753.
64. MacDiarmid, A.G. *Synth. Met.* 1997, *84*, 27.
65. Long, D.P.; Patterson, C.H.; Moore, M.H.; Seferos, D.S.; Bazan, G.C.; Kushmerick, J.G. *Appl. Phys. Lett.* 2005, *86*, 153105.
66. Ditsch, A.; Laibinis, P.E.; Wang, D.I.C.; Hatton, T.A. *Langmuir* 2005, *21*, 6006.
67. Wang, Y.; Wong, J.F.; Teng, X.; Lin, X.Z.; Yang, H. *Nano Lett.* 2003, *3*, 1555.
68. Zhang, Z.; Wei, Z.; Wan, M.X. *Macromolecules* 2002, *35*, 5937.
69. Gao, J.; Bender, C.M.; Murphy, C.J. *Langmuir*, 2003, *19*, 9065.
70. Butter, K.; Boman, P.H.; Fredrik, P.M.; Vroege, G.J.; Philipse, A.P. *Nature Mater.* 2003, *2*, 88.
71. Jin, J.; Iyoda, T.; Cao, C.; Song, Y.L.; Jiang, L.; Li, T.J.; Zhu, D.B. *Angew. Chem. Int. Ed.* 2001, *40*, 2315.
72. Camp, P.J.; Shelley, J.C.; Patey, G.N. *Phys. Rev. Lett.* 2000, *84*, 115.
73. Tlusty, T.; Safran, S.A. *Science* 2000, *290*, 1328.
74. Klokkenburg, M.; Vonk, C.; Claesson, E.M.; Meeldijk, J.D.; Erne´, Ben H.; Philipse, A.P *J. Am. Chem. Soc.* 2004, *126*, 16706.
75. Safran, S.A. *Nature Mater.* 2003, *2*, 71.
76. Tavares, J.M.; Weis, J.J.; Telo da Gama, M.M. *Phys. Rev. E* 1999, *59*, 4388.
77. Wan, M.X.; Zhou, W.; Li, J. *Synth. Met.* 1996, *78*, 27.

78. Zhang, Z.; Wan, M.X. *Synth. Met.* 2003, *132*, 205.

79. Huang, J.; Kaner, R.B. *Angew. Chem. Int. Ed.* 2004, *43*, 5817.

80. Chen, S.A.; Lee, H.T. *Macromolecules* 1995, *28*, 2858.

81. Manohar, S.K.; Macdiarmid, A.G.; Cromack, K.R.; Ginder, J.M.; Epstein, A.J. *Synth. Met.* 1989, *29*, 349.

82. (a) Sayyah, S.M.; Abd El-Khalek, A.A.; Bahgat, A.A.; Abd El-Salam, H.M. *Polym. Int.* 2001, *50*, 197; (b) Gök, A.; Sari, B.; Talu, M. *Synth. Met.* 2004, *142*, 41.

83. Pretsch, E.; Bühlmann, P.; Affolter, C. *Structure Determination of Organic Compounds: Tables of Spectral Data,* Springer, New York, 2000.

84. Harada, I.; Furukawa, Y.; Ueda, F. *Synth. Met.* 1989, *29*, 303.

85. Lacroix, J. Ch.; Garcia, P.; Audiere, J.P.; Clément, R.; Kahn, O. *New J. Chem.* 1990, *14*, 87.

86. Mora, M.A.; Alicia, L.; Vázquze, H. *Int. J. Quantum Chem.* 2000, *78*, 99.

87. Hagiwara, T; Demura, T.; Iwata, K. *Synth. Met.* 1987, *18*, 317.

88. (a) Cornell, R.M.; Schwertmann, U. *The iron oxide, VCH,* New York, 1996; (b) Kryszewski, M; Jeszka, J.K. *Synth. Met.* 1998, *94*, 99.

89. Yue, J.; Epstein, A.J. *J. Am. Chem. Soc.* 1990, *112*, 2800.

90. Yue, J.; Wang, Z.H.; Cromack, K.R.; Epstein, A.G.; MacDiarmid, A.G. *J. Am. Chem. Soc.* 1991, *113*, 2665.

91. Cao, Y.; Smith, P.; Heeger, A.J. *Synth. Met.* 1989, *32*, 263.

92. Kulkarni, M.V.; Viswanath, A.K.; Marimuthu, R.; Mulik, U.P. *J. Mater. Sci.: Materials in Electronics* 2004, *15*, 781.

93. Pouget, J.P.; J´ozefowicz, M.E.; Epstein, A.J.; Tang, X.; MacDiarmid, A.G. *Macromolecules* 1991, *24*, 779.

94. Moon, Y.B.; Cao, Y.; Smith, P.; Heeger, A. *J. Polym. Commun.* 1989, *30*, 196.

95. Bidan, G; Jarjayes, O; Fruchart, J.M; Hannecart, E. *Adv. Mater.* 1994, *6*, 152.

96. Tang, B.Z.; Geng, Y.; Lam, J.W.; Li, B.; Jing, X.; Wang, X.; Wang, F.; Pakhomov, A.B.; Zhang, X.X. *Chem. Mater.* 1999, *11*, 1581.

Communication
Technologies

7

TCP Connectivity Analysis in Mobile Ad Hoc Networks

Sasan Adibi

Summary

Communication link forms the backbone of a mechatronic system. When operating a group of mechatronic devices for accomplishing a common task, an interdevice communication takes place through appropriate network connection. Wireless communication is particularly useful in this context. This chapter deals with the transmission control protocol (TCP) connectivity and performance issues in mobile ad hoc networks (MANETs). The majority of applications running on wireless links require seamless TCP connectivity to ensure continuous service from the source to the destination. Continuity of services is of vital importance during mobile-IP handoff/handover procedure in which the mobile subscriber changes the service domain, especially if Quality of Service (QoS) is required for high-performance mechatronic applications. To study the TCP connectivity in MANETs, a closer look is taken here at TCP and its different flavors and the issue of retransmission. An introduction to mobile ad hoc networks is given, and a simulation environment is set up using ad hoc elements over Dynamic Source Routing (DSR) with IPv6 style option header as part of the network protocol. The simulation results reveal how TCP's retransmission pattern reacts during a lossy handoff/handover period.

7.1 Introduction to TCP

Communication protocols become relevant when two or more mechatronic devices are interconnected through a wired or wireless network. In this context, TCP plays an important role. It is located at layer 4 of the 7-layer open system interconnection (OSI) model of communication protocols. TCP is a connection-oriented protocol responsible for the end-to-end delivery of data from the source all the way to the destination. It guarantees data delivery by three-way handshaking and retransmission mechanisms so that whenever a packet is transmitted successfully, an acknowledgment (ACK) is issued by the receiver to inform the sender. If the packet is lost or encounters longer-than-expected delays, the sender will be informed with a negative acknowledgment (NACK), and it will try to retransmit the packet based on the previous successfully transmitted packet. A NACK could mean that there is congestion along the path to the destination. This way the sender could reduce the window size to avoid worsening of the congestion condition.

For the network layer, either IPv4 or IPv6 could be used. In the simulation configuration presented in this chapter, a DSR over IPv6 is chosen. The reason is that there is a government-based urge to shift from IPv4 to IPv6 by the year 2008, and the entire industry has been mandated to switch the Internet Protocol (IP)-driven technologies to IPv6 [1]. IPv6 by far has proven to address many micromobility issues and facilitated the mobile-IP operation by its simplified structure and operational units and would offer a competitive QoS to the end users.

The organization of the sections of the chapter is as follows: Section 2 provides a detailed overview of TCP, different TCP flavors, congestion control, and retransmission mechanisms. Section 3 introduces basics of ad hoc networks, QoS, mobile-IP, and handoff/handover processes. The simulation test bed and results are considered in Section 4. Section 5 gives conclusions. The chapter ends with a list of relevant references.

7.2 Congestion and Retransmission of TCP

TCP was first introduced in Request for Comment (RFC) 793 in September 1981. The frame format is shown in Figure 7.1 and is explained in the next section [2,3,4].

7.2.1 TCP Header Format

Fields specified in the TCP header (Figure 7.1), which will be discussed in detail later, are introduced in this section. These fields are the following:

Source and destination ports (32 bits each): These are numbers that identify specific TCP-based applications running on both sending and receiving endpoints, which enable TCP endpoints to perform multiplexing and demultiplexing. Multiple instances of TCP segments related to different applications could be transmitted between two endpoints on a single medium.

Sequence number (32 bits): A number that is used to track segments successfully sent across and is used in error detection by tracking down failed sent segments.

Acknowledgment number (32 bits): This indicates that the receiver has successfully received the previous segment and is expecting the next segment from the receiver.

Data offset (4 bits): This field indicates where the starting point of the data is.

Reserved (6 bits): This is reserved for future use and is always set to 0.

Code bits (6 bits): These are 6-bit flags that indicate the nature of the header. They are the following:

> **URG**—Urgent pointer.
> **ACK**—Acknowledgment.
> **PSH**—Push function; it causes the TCP sender to push all unsent data to the receiver.
> **RST**—Resets the connection.
> **SYN**—Synchronizes sequence numbers.
> **FIN**—End of data.

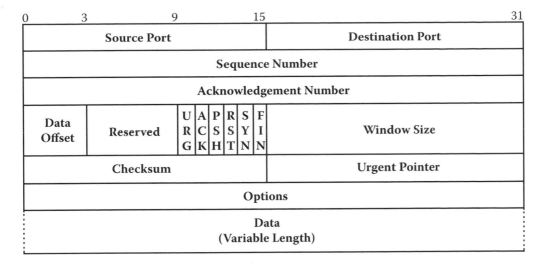

FIGURE 7.1 TCP header (RFC 793).

Window size (16 bits): This indicates the number of segments the receiver can process at once before the sender expects an acknowledgment.

Urgent pointer (16 bits): Points to the end of urgent data. When the URG bit is set, this data is given priority over other data streams.

Option (Variable): Mostly used to specify the TCP maximum segment size (MSS), which is sometimes called maximum window size or send maximum segment size (SMSS).

Padding (Variable): The TCP header padding is used to ensure that the TCP header ends and data begins on a 32-bit boundary. The padding is composed of only zeros.

Data (Variable): Data being sent by TCP are often set to the maximum transfer unit (MTU).

7.2.2 Port Numbers

According to Figure 7.1, the TCP header starts with the Source Port and Destination Port, which are 16-bit fields each [5,6]. These port numbers are specific numbers generally ranging from 0 to 65536; however, the port numbers 0 to 1024 (designated as well-known ports) are reserved for specific services and applications. A few well-known port numbers are given in Table 7.1. The Internet Assigned Number

TABLE 7.1 Some Well-Known TCP Port Numbers

Port Number	Description
20	File Transfer Protocol (FTP)—Data
21	File Transfer Protocol (FTP)—Control
22	Secure Shell (SSH) Remote Login Protocol
23	Telnet
25	Simple Mail Transfer Protocol (SMTP)
53	Domain Name System (DNS)
80	Hypertext Transfer Protocol (HTTP)
109	Post Office Protocol v2 (POP2)
110	Post Office Protocol v3 (POP3)
115	Simple File Transfer Protocol (SFTP)
161	Simple Network Management Protocol (SNMP)
179	Border Gateway Protocol (BGP)
546	Dynamic Host Configuration Protocol (DHCP)—client
547	Dynamic Host Configuration Protocol (DHCP)—server

Authority (IANA) maintains the assigned port numbers (referred to RFC 1700, issued in October 1994). It is worth mentioning here that the concept of port numbers is common to both TCP and UDP (User Datagram Protocol, RFC 768). For instance, port number 69 is used for TFTP (Trivial File Transfer Protocol), which runs on UDP. Note that UDP is particularly useful in non-real-time mechatronic applications.

7.2.3   Sequence Number

The sequence number is located in the third section of the TCP header with a 32-bit field. The reason for using this field is that, in most cases, there is a maximum number of bytes that a transmitter could handle in a single segment the TCP transmitting end points could send and receive. This is denoted by MTU, which is often dictated by the data link layer (usually 1500 octets). The data being transferred is often larger than the MTU size. Therefore, the TCP sender end point has to divide the data into several MTU-size segments and send the segments one by one.

Due to congestion, packet loss, packet drop at the router's queues, and packets traveling in different paths, it is quite possible that packets received at the destination are either not in the order that they were sent or some of them are lost along the path. The sequence number is used to track the number of received segments and to report the missing ones to the sender.

7.2.4   Three-Way Handshaking

For end point TCP transmitters to be able to send and receive data, a three-way handshaking mechanism is used (Figure 7.2) [3]:

- This mechanism starts with a SYN signal from the sender by setting the SYN bit-flag in the code-field containing a 32-bit sequence number X called the initial send sequence (ISS).
- The receiver receives the SYN with the sequence number X and sends an SYN segment with its own totally independent ISS number Y in the sequence number field. It also increments X to $X + 1$ and sends it back to the sender. This acknowledgment number informs the initial sender that its data was received and it expects the next segment with the sequence number of $X + 1$. This stage is often called the SYN-ACK.
- The initial sender receives this SYN-ACK segment and sends an ACK segment containing the next sequence number, $Y + 1$, which is called forward acknowledgment and is received by the initial receiver. The ACK segment is identified by the fact that the ACK field is set in the code-field. Segments that are not acknowledged within a certain time frame are retransmitted.

Incrementing the sequence numbers one by one is not an efficient way of transmitting data, as this slows down the communication channel. This is addressed in the next section.

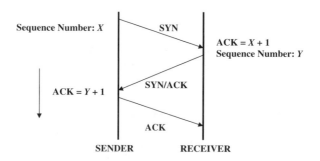

FIGURE 7.2 TCP three-way handshake.

7.2.5 Sliding Window

As discussed in the previous section, sending the segments one by one is not an efficient method for TCP, which poses as a bottleneck in the transmitting data path. Instead of sending segments one by one, the transmitting parties could agree on a number M, which is the number of segments that the sender sends one after another without waiting for any acknowledgment from the receiver. After sending M segments, it will wait for the receiver to acknowledge the receipt of all M segments, at which time the sender is triggered to send the next segments. The efficiency of this scheme requires good channel condition, low packet loss, and negligible congestion conditions. However, if these conditions are not met, the chances of receiving a NACK from the receiver will be high, and retransmitting all M segments would make this scheme less efficient than sending the segments one by one. The mechanism works as follows:

- Assume that the current sequence number of the TCP sender is X.
- The TCP receiver informs the current negotiated window size M in every packet. By default, M is 536 bytes unless specified by the operating system.
- The TCP sender sends M-byte segments one after another and will wait for the ACK.
- The receiver sends an ACK with the value $X + M$.
- After a successful transmission, the window size increases by an additional M, which is called a "slow start."
- The sender sends segments with $2M$ bytes, and then $3M$ bytes, and so on, up to the MSS, as indicated in the TCP options.

The following criteria affect the window size:

- If the segment is either not received by the receiver, which triggers a NACK, or the ACK signal sent by the receiver is not received by the sender in a specific amount of time, the sender believes that the previously sent segment has either failed or fallen into a congestion area. This results in the window size being cut by half (for example, from $4M$ to $2M$).
- If the segment retransmitted, due to the failure of the previous transmission, fails again, the window size is again cut by half. This continues until the window size becomes zero. This is a typical scenario when the receiver's buffer becomes full. In this case, the window is said to be "frozen," and the sender cannot send any more segments until it receives a datagram from the receiver with a window size greater than zero.
- If the next retransmission is successful, the slow ramp up starts again.

7.3 Different Flavors of TCP

Different applications running on TCP may require different characteristics in terms of how the transport layer should react to congestion scenarios. In the standard TCP, when facing congestion and when decreasing the sliding window, the slow start is often unacceptable for many applications requiring fast reactions and real-time scenarios. Due to these requirements, different flavors of TCP have been developed, such as the following [7,8]:

- **TCP Vanilla:** It incorporates the slow start TCP and congestion avoidance.
- **TCP Tahoe:** It is similar to TCP Vanilla as it contains a TCP congestion control algorithm that performs basic slow start and congestion avoidance; however, it has an integrated fast retransmission algorithm, in which the TCP sender is allowed to retransmit a segment when more than three duplicate acknowledgments have been received.
- **TCP Reno:** TCP Reno covers the basic operation of TCP Tahoe with an additional fast recovery algorithm in which it does not allow a slow start after a fast retransmission and avoids congestion. TCP Reno also adds some intelligence compared to Tahoe, which helps in detecting lost segments and fast recovery.

- **New Reno:** New Reno incorporates a slight modified version of the TCP Reno. It modifies the reaction after receiving new ACKs. For a fast recovery, the sender must receive an ACK for the highest sequence number sent.
- **TCP Vegas:** TCP Vegas is built on the basis of TCP Reno. It takes the congestion control mechanism into consideration and works as a rather proactive than a reactive one. TCP Vegas also includes a modified retransmission strategy (compared to TCP Reno), which is based on fine-grained measurements of the round-trip Time (RTT) as well as new mechanisms for congestion detection.
- **TCP Sack:** This is another TCP flavor with the option of selective acknowledgments, an extension of the TCP Reno. It tackles problems that are faced by TCP Reno and TCP New Reno. These problems include multiple lost package detection and retransmission of more than one lost packet per RTT. In general, TCP uses the round-trip delay estimates for its adaptive windowing scheme to transmit data reliably over an unreliable data path with variable network parameters (i.e., bandwidth, delays, etc.).

7.4 Mobile Ad Hoc Networks (MANETs)

Ad hoc networks usually have flexible and variable topologies in which nodes are not fixed to physical locations and are free to move about, and only nodes within the wireless range are included in the communication schemes. As soon as they leave the wireless domains, they will be automatically excluded from the routing scheme. MANETs are networks of mobile nodes in ad hoc routing protocol scenarios.

7.4.1 Different Flavors of Ad Hoc Networks

There are different categories of MANET protocols suitable for a variety of applications and scenarios. Figure 7.3 shows the main categories of MANETs [9]: Flat (proactive and reactive), hierarchical, and geographical position-assisted routing schemes. These schemes are further explained next:

Flat routing: In this type of routing scheme, all nodes have the same hierarchy, and there is no specific order for processing. Whenever a node receives a packet, it forwards it based on the routing scheme, not based on its location or position. There are two subcategories of flat routing schemes, proactive and reactive, as outlined here:

- **Proactive routing**—This routing scheme maintains frequent routing information updates. Therefore, all nodes have a rough knowledge of where they are. Thus, when a route is requested at a node, it already knows where the destination is, and finding destinations requires no additional time. However, maintaining these updates requires excessive processing power, large local memory, and extra bandwidth.

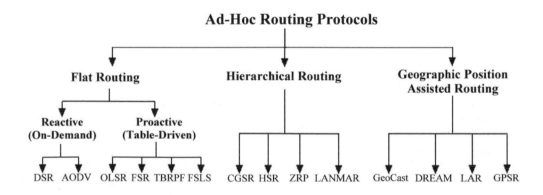

FIGURE 7.3 Main mobile ad hoc network protocols.

- **Reactive routing**—In this category, whenever a node receives a packet, it has to find the destination first. At that time the node usually has to send a route request (RREQ) message. As soon as the destination receives the RREQ message, it responds back with a route reply (RREP) message. As soon as the RREP packet is received by the requesting node, the packet is good to be sent.
- Reactive routing schemes do not require frequent updates and excessive processing time compared to proactive routing schemes. However, finding routes from the source to the destination requires additional time (tRREQ + tRREP).

Hierarchical routing: When nodes increase in number and the MANET entity population grows, frequent updates, RREQs, RREPs, and other types of interactive signals will flood the wireless domain. This could be remedied by keeping local traffic locally. The whole idea of having hierarchical schemes is to divide the complete routing domain into subgroups and to have local administrative entities oversee local traffic. Therefore, all local movement information and updates are kept locally. This further decreases overhead and unnecessary traffic.

Geographical position-assisted routing: These schemes use the geographical coordinates (i.e., Global Positioning System or GPS) to locate destinations. Therefore, through the use of geographical means, all nodes should be aware of other nodes within the routing range.

7.4.2 TCP in Ad Hoc Networks

TCP was introduced in the early 1980s when wireless communication was not used for data applications. Therefore, TCP and wireless applications do not usually go hand in hand without any problems. The most serious problem arises due to the fact that wireless links experience multipath and fading effects with higher BER (Bit Error Rate) compared to wired networks, which will cause TCP time-out and unnecessary retransmissions. One particular problem occurs during handoff and handover period when a mobile host being served by one entity has to change its domain. Therefore, the current connection has to tear down (handoff) and a new connection has to establish (handover) with a new Access Point. This is a frequent procedure in MANETs. In our test bed, we will study the effect of handoff/handover on the retransmission pattern.

7.4.3 Dynamic Source Routing (DSR)

DSR was created by David B. Johnson, David A. Maltz, and Josh Broch in 1994 and is a flat-reactive routing protocol. In DSR, each data packet carries the complete and ordered list of nodes from source to destination. There is a minimal update traffic that is required between nodes (usually less than 1% of the total traffic) and no periodic routing messages take place (i.e., no router advertisements and no link-level neighbor status messages). DSR is a self-organizing and self-configuring multihop protocol with low bandwidth overhead and very good battery power conservation. It quickly adapts to network changes and offers high mobility speeds.

There are two important mechanisms associated with DSR: route discovery and route maintenance. In route discovery (Figure 7.4), when a node (A) wants to communicate with another node (E) it first checks the route cache for routes previously learned. If there is no entity, then an RREQ message is

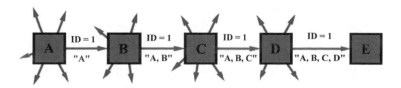

FIGURE 7.4 Route discovery process in DSR.

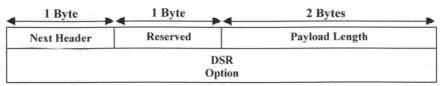

1 Byte	1 Byte	2 Bytes
Next Header	Reserved	Payload Length
DSR Option		

DSR Fixed Header

Option Type	Option Length	Identification
Target Address		
IN Index (*i*)		
Out Index (*i*)		
Address 1 ··· Address *n*		

Route Request Option

Option Type	Option Length	L	Reserved
Target Address			
Out Index (*i*)			
Address 1 ··· Address *n*			

Route Reply Option

FIGURE 7.5 DSR fixed and option headers.

broadcast to the entire routing range. Each RREQ message identifies the initiator and the target of the route discovery, and also contains a unique request ID. The target (E) sends back an RREP message to the initiator of the route discovery (A), giving a copy of the accumulated routerecord from the RREQ. Route maintenance finds broken and down links and uses the route discovery to find new paths if a required path is broken or down. Both unidirectional and bidirectional links are supported.

Figure 7.5 shows the DSR frame header specifications (DSR fixed header, route request, and route reply headers).

7.5   Simulation Environment, Test-Bed, and Results

In this section, a test bed is set up using a DSR-based MANET scenario. Figure 7.6 shows a typical ad hoc mobile-IP architecture with essential entities such as Home Agent (HA), Correspondence Node (CN), hosts, and ad hoc elements, and optional entities such as Access Points and ad hoc manager.

HA is a server located at the base location, the same location as the mobile host. When the mobile host is at its base location, only HA serves it, and no additional roaming, handoff, and handover procedures are involved. Whenever an entity requires to communicate with the mobile host, it sends its queries to the HA, and whenever the mobile host needs to communicate, it communicates through the HA as well. As soon as the mobile host leaves its home location, the HA is not able to serve the mobile host directly. In this situation, ad hoc managers and Access Points come to play their roles. When the mobile host is away from the base location, HA will send the related queries first to the ad hoc manager, which will be in touch with the current Access Point. An ad hoc manager manages a wireless domain of ad hoc entities, and an Access Point directly registers a mobile host as it is roaming inside a local region.

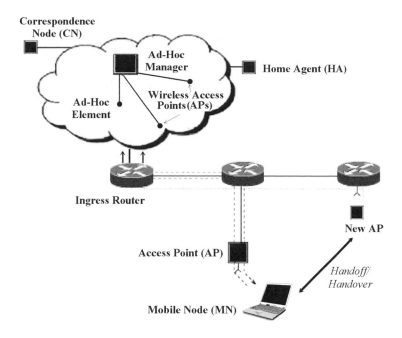

FIGURE 7.6 Typical ad hoc mobile-IP architecture.

7.5.1   Simulation Parameters

NS-II simulation software is used for the simulation purposes. Within NS-II, one is able to create close-to-real-life channel conditions. A realistic modeling is considered for the following physical layer-related factors:

- Free space and ground reflection propagation
- Transmission power
- Antenna gain
- Receiver sensitivity
- Propagation delay
- Carrier sense
- Discarding packets below carrier-sense threshold
- CSMA/CD (CA) collision leftovers (packets not discarded due to collisions)

An efficient and complete distributed coordination function (DCF) MAC layer model is used, which applies to IEEE 802.11-based wireless LAN protocol standards.

The following parameters specify the traffic conditions:

- Traffic rate of the channel is set to 2 Mbps.
- Ad hoc elements use DSR protocol for routing on top of IPv6.
- For traffic generation, file transfer protocol (FTP) is used over TCP for all the flows in the network.
- 802.11b is used for MAC layer protocol.

The simulation shows the retransmission pattern in the course of mobile host movements from one Access Point to the next (during a handoff/handover procedure). Figure 7.7 shows this pattern, in which a handoff/handover takes place between the 3rd and the 4th seconds. This results in a slight degradation

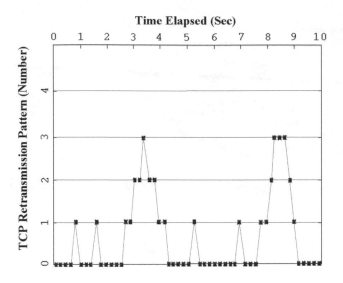

FIGURE 7.7 TCP retransmission pattern during mobile-IP activity.

of the connection due to the physical occurrence of handoff. The number of retransmissions rises to three in the worst case. The same happens between the 8th and 9th seconds.

7.6 Conclusion

This chapter has discussed the importance of TCP as a reliable transport layer in communication between mechatronic devices. TCP was shown to play an important role in error detection, forward error correction (retransmission), and congestion control. Different TCP flavors were introduced with descriptions on how each of them deals with the congestion issue. Different fields of TCP's frame header were studied, and special attention was given to the retransmission procedure. Furthermore, the definition of ad hoc networks was introduced. It was discussed how TCP mechanism could be affected due to longer delays and physical limitations incurred within ad hoc domains. The end of the chapter included a simulation environment and test bed to study the effect of handoff/handover procedure on the number of retransmissions of TCP. For most mechatronic applications, TCP seamless connectivity has to maintain connectivity for high-performance connection-oriented handshaking with peers. During handoff/handover, which is the worst case for TCP retransmissions, the maximum number of retransmissions rises to three, which, depending on the specific application, could be nontolerable.

References

1. Wilson, D.J. and Dragnea, R., *IPv6 in Fixed and Mobile Networks*, Technology White Paper, Alcatel Corp. Press, Alcatel Headquarters, Paris, France, November 2004.
2. Kurose, J., *Connection-Oriented Transport: TCP*, Lecture Notes, University of Massachusetts, Amherst, MA, May 2004.
3. Haden, R., *TCP*, The Data Network Resource, (http://www.rhyshaden.com/tcp.htm),1996–2006.
4. *Transmission Control Protocol*, RFC 793, IETF, Information Sciences Institute, University of Southern California, Marina del Rey, California, 1981.
5. Port Numbers and Services Database (www.sockets.com/services.htm).
6. List of TCP and UDP port numbers, Wikipedia Web site (http://en.wikipedia.org/wiki/List_of_TCP_and_UDP_port_numbers).

7. Xu, S. and Saadawi, T., Performance evaluation of TCP algorithms in multi-hop wireless packet networks, *Journal of Wireless Communications and Mobile Computing*, Vol. 2, No. 1, 85–100, 2002.
8. Stoica, I., *A Comparative Analysis of TCP Tahoe, Reno, New-Reno, SACK and Vegas*, Communication Networks, Student Project, University of California, Berkeley, CA, 2005.
9. Adibi, S., Different flavors of mobile ad-hoc networks (MANETs), *Photonic Networking*, ECE 710 Final Course Project, Department of Electrical and Computer Engineering, University of Waterloo, Waterloo, Ontario, Canada, July 2006.

8

i.LIGHT: Communication Using Visible Light

G.K.H. Pang

Summary

Communication among mechatronic devices can be achieved in many ways. The use of visible light is one option, which has several advantages. i.LIGHT is a technology using visible light from light-emitting diodes (LEDs) as a medium to transmit information in open space. It was originally developed with an aim to use LEDs for illumination and broadcasting of information concurrently. Through prototypes development it has been shown that visible light can indeed be used as a medium for short-range, wireless communication. Essentially, all LED light sources can become information sources or beacons, in addition to the usual function utilized in visual illumination or display. Both digital and audio information can be transmitted using visible light. In this chapter, an i.LIGHT prototype design, an EXIT sign to broadcast audio information for providing guidance to visually impaired people and others is discussed. The i.LIGHT technology has also been extended to the use of digital cameras as receivers.

8.1 Introduction

Visible light, which is used for illuminating spaces, may be used for information communication in mechatronic applications. This novel communication technology, termed i.LIGHT, is presented in this chapter. First, the background of i.LIGHT is given. Second, an overview in the system of communication based on i.LIGHT is presented. Then, a prototype system for the interaction with an exit sign is described.

Evaluations on distance reception and angle deviation are studied, as well. Further examples of applications are given and an extension to the use of small digital cameras as receivers is also described. Conclusions are drawn at the end of the chapter.

8.2 Background of i.LIGHT

LEDs are increasingly being used in many mechatronic applications, including headlights and taillights of vehicles, traffic lights, indoor lighting, and sign boards [1–3]. Their usual functions are to provide illumination or visual display. However, one very important characteristic of LEDs is that they are semiconductor devices that have the capability for high frequency switching. With this property of LEDs and the use of electronics to control the switching of LEDs, the emitted light can be modulated with information [4–6]. To a normal eye an LED would appear to be turned on all the time. However, in actual fact, an LED is being switched on and off at a very high frequency. This switching of lights can be detected by a photodetector. Again, with the use of appropriate electronics, the information embedded in the visible light can be retrieved through demodulation of the light signal. In this use of i.LIGHT, LEDs are not just devices for illumination or indication but communications devices as well. This enables the concurrent use of visible light for simultaneous signaling and communication. The technology of i.LIGHT leads to the development of alternate wireless optical communication systems [7,8] for mechatronic applications.

8.3 System Overview

An i.LIGHT system consists of one or more transmitters and receivers. Figure 8.1 gives an overview of the system. An i.LIGHT transmitter unit is attached to an LED display or illumination unit. A receiver is made up of a photo sensor and other electronic circuitry for the demodulation of the light signal to audio or digital information. The information is then displayed on a screen (if digital) or heard through a speaker (if audio).

One novel application of the i.LIGHT technology in Intelligent Transport Systems (ITS)—a complex mechatronic system—is the development of intelligent traffic lights. With the modulation of light emitted by a traffic light, a considerable amount of local traffic information can be broadcasted to the driver. Figure 8.2 illustrates this concept.

Essentially, an intelligent traffic light becomes an information beacon. The information that is broadcasted may consist of information on nearby facilities or warning messages. A location and heading direction can also be provided. In this manner each traffic light is converted into a location beacon.

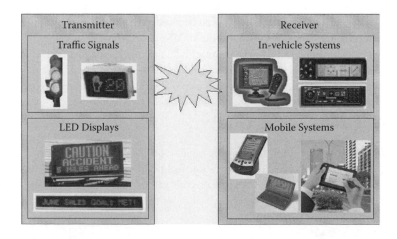

FIGURE 8.1 Overview of an i.LIGHT communications system.

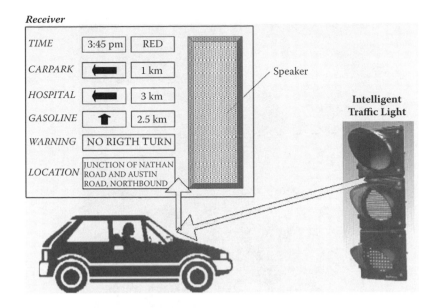

FIGURE 8.2 Intelligent traffic light.

8.4 A Prototype System

This section provides an overview of the hardware in a prototype i.LIGHT system that has been developed for an exit sign. The evaluation test was performed in an indoor environment that resembles a real situation of application. The objective of the evaluation test is to find out the maximum reception distance and angle deviation for an audio receiver. The prototype system is shown in Figure 8.3.

The exit sign display has been used for distance reception and angle deviation evaluation. The display was placed on the table vertically at a fixed position. The meter rule was fixed on the floor using a plastic sticker perpendicular to the center of the LED display. Power supply was provided for the sign. The transmitter was then connected to the sign for the transmission of audio information. The light meter was tightened on the 3-leg stand with an adjusted height matching the center of display, as shown in Figure 8.4. It was used for measuring the light intensity at different positions. Background brightness was also measured by the light meter for each measurement when the transmitter was turned off.

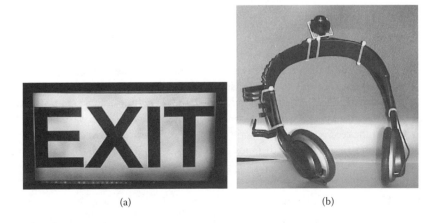

(a)   (b)

FIGURE 8.3 (a) A prototype i.LIGHT sign and (b) an i.LIGHT audio receiver.

FIGURE 8.4 Light meter fixed on a tripod.

8.5 Evaluations of an Exit Sign

The evaluation of the prototype i.LIGHT system for the exit sign is presented in this section. Pertinent results are presented and discussed.

8.5.1 Distance Reception Evaluation

The results are given in Table 8.1.

The tripod with the fixed light meter and receiver was placed directly in front of the center of the display. The measurements were taken at different distances to find out the maximum distance at which the audio receiver could successfully receive the signal satisfactorily from the sign.

The lower the light intensity from the LED display, the weaker the signal received by the receiver. If the detected light intensity is lower than a specified level, the receiver cannot function properly. Therefore, the receiver can only function properly if its distance from the sign is within a particular distance. From Table 8.1 it is seen that the receiver cannot function when the perpendicular distance is greater than 4.6 m. The light intensity due to LED light at a distance of 4.6 m is 5 lux. As a result, it can be concluded that the receiver can function properly if the detected light intensity is not less than 5 lux.

8.5.2 Angle Deviation Evaluation

The purpose of this test is to find out the maximum deviation in angle at which the receiver can still function properly even if it is not directly pointed towards the display. Note that the light sensor can

TABLE 8.1 Distance Reception of the Receiver

Distance (in meters)	Background Light Intensity	Light Intensity Due to LED and Background	Light Intensity Due to LED Only
0.5	30 lux	175 lux	145 lux
1.0	32 lux	85 lux	63 lux
1.5	29 lux	65 lux	46 lux
2.0	30 lux	49 lux	19 lux
2.5	32 lux	46 lux	14 lux
3.0	28 lux	39 lux	11 lux
3.5	29 lux	36 lux	7 lux
4.0	27 lux	33 lux	6 lux
4.5	27 lux	32 lux	5 lux
4.6	26 lux	31 lux	5 lux

TABLE 8.2 Angle Deviation of the Receiver

Perpendicular Distance to the LED Display (in meters)	Angle from Center Axis (in degrees)
0.5	17, +18
1.0	16, +18
1.5	14, +17
2.0	15, +14
2.5	16, +13
3.0	15, +15

detect the LED light within a deviation angle. Therefore, the smaller the deviation angle, the more exact the direction which the LED display can indicate. Measurements were taken up to a distance of 3 m from the display so that the reception is independent of the light intensity. The results are shown in Table 8.2.

From the measurement data in Table 8.2 it can be observed that the deviation angle for the receiver is about 15 degrees from the center axis and symmetric on both sides. As a result, when the LED sign comes from a direction within an angle of 15 degrees, the transmitted audio information can be detected by the light sensor and the receiver would be properly carrying out its function.

8.6 More Applications

The i.LIGHT technology can find many novel and interesting applications. For an average person the modulated light from visual signs can provide additional information. For visually impaired people, the visual signs can provide directional guidance. Although they cannot see the signs, they can hear the audio message demodulated from the light received from the signs. An i.LIGHT system is better than an infrared system [9–11] in this context for infrared can bounce back and forth and is not suitable for directional guidance. The line-of-sight property of an i.LIGHT system is well-suited for this application. Figures 8.5 and 8.6 provide some illustrations.

As the i.LIGHT technology is more suited for relatively short-range (around 5 to 10 meters) communication and indoor environments, applications can be developed for such environments as museums and exhibition halls. With the i.LIGHT technology, the existing lighting systems and displays can be converted for the purpose of information broadcasting, in addition to their normal functions as indoor lighting or indication devices. For example, in a museum setting, the information on individual exhibits can be sent via a plurality of LED lighting. A receiver could be a mobile phone

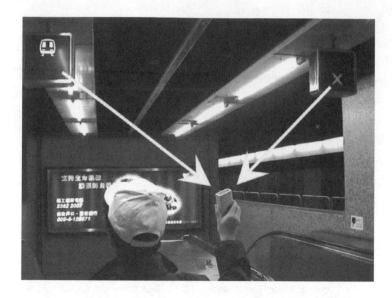

FIGURE 8.5　Application of i.LIGHT to provide platform directional guidance.

embedded with i.LIGHT technology. When a visitor points the receiver to the LED light (i.e., transmitter), with the ear jack attached to the mobile phone, he/she will be able to listen to the audio description of the specific exhibit in which he/she is interested. In this manner, there will be a dual use of the indoor lighting. The indoor environment would remain quiet while the visitors stroll in the museum. There would be no need for audio equipment rental for the purpose of exhibit description. Similar use can be found in an exhibition hall. Even if sales personnel or helpers are not available all the time, the booth lightings (when equipped with i.LIGHT) will broadcast product information continuously.

FIGURE 8.6　Application of i.LIGHT to provide station exit information.

A major advantage of this over a conventional broadcasting system is that an individual with a receiver has the freedom of choice in receiving specific messages and can selectively avoid unwanted advertisements or announcements.

8.7 Digital Cameras as i.LIGHT Receivers

The ideas for an innovative LED-based information beacon system have been developed. In the current implementation, the LED device serves as a direct information transmitter. The encoded information is broadcasted by the high frequency switching of all the LEDs in the device, which are turned on or off simultaneously. The receiver is constructed using a phototransistor and other demodulation electronics. In extending this technology to other practical situations of communication, the receiver can be constructed using a micro-digital camera typically found in a mobile phone. The LEDs in the transmitter could broadcast the modulated information by high frequency switching on–off of the LEDs arranged in a special pattern. With the digital camera of a typical mobile phone, through the processing of the acquired images by appropriate algorithms [12,13], one can recover the information embedded in the visible light.

The developed system is a combination of several latest technologies, which include a CMOS vision sensor, high brightness LEDs, and digital image processing techniques. It belongs to a new type of simplex communication link. A digital camera is used to capture images contained in the LED beacon signal. The captured digital images are then processed by the developed algorithms and the information is extracted. The issues examined include the structure of the transmitter and the receiver, the signaling method, the transmission protocol of the LED panel, the relationship between the camera capturing rate and the LED pattern update rate, the digital camera exposure technology, and the efficiency of the image processing algorithms.

8.8 Digital Camera-Based Signal Processor

The LED transmitter panel can be divided into partitions (say, nine partitions). Within each partition, the LED is turned on and off simultaneously. However, a different pattern can be shown on the LED panel at a time. A microcontroller can be used for controlling the on-pattern of the LED panel in different periods to display a location code. Whereas different patterns on the LED panel are being flashed, the rate is sufficiently fast so that human eyes cannot detect the changes.

The receiver consists of a digital camera, a camera controller, memories for storing images, and an image processor. Figure 8.7 shows the structure of the digital camera-based LED signal receiver. The operation of the digital camera is controlled for image capturing and parameter modification. The captured images are stored in the memory unit and then processed.

The link between the LED panel and the digital camera is a simplex communication link. The protocol is designed to be simple and workable. The protocol design should prevent the need of synchronization between the transmitter and the receiver.

After the images are captured by the digital camera and stored in the memory unit, the transmitted data is extracted. Figure 8.8 shows a flowchart for the recovery of a transmitted data (a location code) from the captured images. An image processor can extract the location code from the images with the algorithms developed in [12,13].

FIGURE 8.7 The structure of the digital camera-based i.LIGHT receiver.

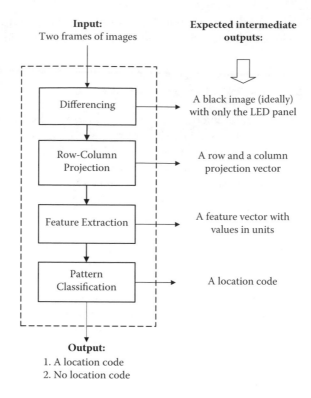

FIGURE 8.8   Flowchart for the recovery of a location code from the captured images by using an image processing algorithm.

8.9   Conclusion

Exit signs, traffic lights, traffic signal devices, and message display boards are being replaced by light emitting diodes (LEDs). This is due to their high brightness, long life expectancy, high tolerance to humidity, low power consumption, and minimal heat generation. In addition to their normal function of being an indication and illumination device, LEDs can be used as a communication device for the transmission and broadcasting of information and data. Hence, they can become an integral part of a wireless optical communication facility in a mechatronic system. For example, they can serve as a short-range beacon to support indoor and vehicle-to-roadside communications. This chapter presented some development work for this dual use of LEDs. Applications of this technology can broadcast audio information, vehicle location, navigation information, and so on, in addition to the normal functions of signaling and illumination.

Acknowledgment

This work has been supported by The University of Hong Kong CRCG grant 10206566.

References

1. Stringfellow, G.B. and Craford, M.G., High brightness light emitting diodes, *Semiconductors and Semimetals*, Vol. 48, 1–63, 1997.
2. Craford, M.G., LEDs challenge the incandescents, *IEEE Circuits and Devices Magazine*, Vol. 8, No. 5, 24–29, September 1992.
3. Werner, K., Higher visibility for LEDs, *IEEE Spectrum*, Vol. 31, No. 7, 30–39, July 1994.

4. Smyth, P.P., Eardley, P.L., Dalton, K.T., Wisley, D.R., McKee, P., and Wood, D., Optical wireless—a prognosis, *Proceedings on Wireless Data Transmission*, SPIE Vol. 2601, October 23–25, 1995, pp. 212–225.

5. Pang, G., Chan, C.H., and Kwan, T., Tricolor light emitting diode dot matrix display system with audio output, *IEEE Transactions on Industry Applications*, Vol. 37, No. 2, 534–540, March–April 2001.

6. Pang, G., Kwan, T., Chan, C.H., and Liu, H., LED traffic light as communication device, *Proceedings of IEEE International Conference on Intelligent Transport Systems*, Tokyo, Japan, October 5–8, 1999, pp. 788–793.

7. Pang, G.K.H., Kwan, T.O., Liu, H.S., and Chan, C.H., LED wireless, *IEEE Industry Application Society Magazine*, Vol. 8, No. 1, 21–28, January–February 2002.

8. Pang, G.K.H., Ho, K.L., Kwan, T.O., and Yang, E., Visible light communication for audio systems, *IEEE Transactions on Consumer Electronics*, Vol. 45, No. 4, 1112–1118, November 1999.

9. Chu, T.S. and Gans, M.J., High speed infrared local wireless communication, *IEEE Communications Magazine*, Vol. 25, No. 8, 4–10, August 1997.

10. Meyer, M., Infrared LEDs, *Compound Semiconductor*, Vol. 5, No. 9, 39–40, May–June 1996.

11. Bentzen, B. and Mitchell, P., Audible signage as a wayfinding aid: comparison of verbal landmarks and talking signs, *Journal of Visual Impairment and Blind*, Vol. 89, 494–505, 1995.

12. Liu, H.S. and Pang, G.K.H., Positioning beacon system using digital camera and LEDs, *IEEE Transactions on Vehicular Technology*, Vol. 52, No. 2, 406–419, March 2003.

13. Pang, G.K.H. and Liu, H.S., LED location beacon system based on processing of digital image, *IEEE Transactions on Intelligent Transportation Systems*, Vol. 2, No. 3, 135–150, September 2001.

9
Bluetooth Networking

V.C.M. Leung

V.W.S. Wong

Summary

Network communication is crucial for communication among mechatronic devices and in complex mechatronic systems. Bluetooth technology is particularly useful in this context. Over the past few years, intensive research on Bluetooth networking has been going on at the University of British Columbia (UBC) under the Enabling Technologies for Ubiquitous Personal Area Networking project. This chapter summarizes some of the main results of this research. In the area of Bluetooth scatternet formation and scheduling, the following have been developed: (1) the two-phase scatternet formation (TPSF) algorithm and its improved variant, TPSF+, (2) the Bluescouts scatternet formation method employing mobile agents, and (3) an adaptive scheduling algorithm (ASA) for master and slave nodes in a scatternet. In the area of Bluetooth access point design, a collision avoidance scheduling (CAS) algorithm has been developed that enables high-capacity Bluetooth access points employing multiple Bluetooth transceivers.

9.1 Introduction

Communication between mechatronic devices and within complex mechatronic systems is important in carrying out their intended tasks. Bluetooth technology plays an important role in this context. Bluetooth is a short-range radio technology that uses the frequency hopping spread spectrum (FHSS) technique with time-division duplex (TDD) in the license-free 2.4 GHz industrial, scientific, and medical (ISM) band. In Bluetooth systems, the ISM band is divided into 79 frequency sub-bands. A Bluetooth channel is defined by a pseudo-random frequency hopping (FH) sequence hopping through the 79 frequency bands at a rate of 1600 hops per second. The result is a slotted channel with the slot duration equal to 625 μs.

Due to the low cost and low power consumption of Bluetooth devices, Bluetooth is considered a promising technology for wireless personal area networks (WPANs), particularly in mechatronic applications. For example, various consumer electronics devices (e.g., cellular telephones, personal digital assistants, personal computers, keyboard, and mouse) have begun to provide Bluetooth wireless communication capability. The complete Bluetooth specification is standardized within the Bluetooth SIG (Special Interest Group). The specification considered in this chapter is version 1.2 [1].

The Bluetooth specification has defined various packet types for both data and control packets. Data packets with Forward Error Correction (FEC) capability that occupy either one, three, or five time slots are called Data-Medium Rate 1 (DM1), Data-Medium Rate 3 (DM3), and Data-Medium Rate 5 (DM5) packets, respectively. These DM packets use the (15, 10) shortened Hamming code for FEC. On the other hand, data packets without FEC capability and that occupy either one, three, or five time slots are called Data-High Rate 1 (DH1), Data-High Rate 3 (DH3), and Data-High Rate 5 (DH5) packets, respectively.

To facilitate a wider adoption of this technology, a part of the Bluetooth specification (i.e., the physical and data link layers) is also standardized within the IEEE 802.15.1a WPAN working group [2]. Due to the widespread deployment of Bluetooth radios in cellular telephones, worldwide shipment of Bluetooth devices have increased significantly over the past few years, increasing the potential for widespread application in mechatronic systems.

At the University of British Columbia, the project titled "Enabling technologies for ubiquitous personal area networking" has been funded by the Canadian Natural Sciences and Engineering Research Council (NSERC) under the Strategic Project Grant program and has been supported by several companies in the Canadian wireless industry. The expertise of its research team spans several areas including radio propagation, digital transmissions, networking and chip design, which are all relevant in the development of advanced mechatronic systems. One of the major tasks in the project is the research on Bluetooth network architecture and protocols, and some of the main results achieved in this task are reported in this chapter. Other major tasks of the project include: propagation measurement and channel modeling for Bluetooth radio transmissions, design of advanced receivers and next-generation Bluetooth signaling techniques, and the development of a system on a chip (SoC) implementation platform. A particular highlight is the development of new noncoherent sequence estimation schemes [3,4] and a new decoding scheme [5] that allows various performance-complexity trade-offs and yields a significantly better performance (up to 4 dB gain) than the current limiter-discriminator-integrator Bluetooth receivers. In our investigation of programmable baseband architectures for Bluetooth amenable to flexible SoC implementation, it has been determined that the level of flexibility is best achieved using a synthesizable embedded programmable logic core [6]; both software- [7] and firmware-based programmability for the logic core is being studied.

The rest of this chapter is organized as follows: Section 9.2 presents the two-phase scatternet formation (TPSF) algorithm and its variant (TPSF+). Section 9.3 describes the use of mobile agents in the Bluescouts scatternet formation method. Section 9.4 presents an adaptive scheduling algorithm for master and bridge nodes in Bluetooth scatternets. Section 9.5 presents a collision avoidance scheduling method for high capacity Bluetooth access points employing multiple radios. Section 9.6 concludes the chapter.

9.2 Two-Phase Scatternet Formation Algorithms

Each Bluetooth device has a unique 48-bit Bluetooth device address (BD_ADDR). Bluetooth devices are required to form a *piconet* before exchanging data. Each piconet has a *master* unit that controls the channel access and frequency hopping sequence of all other nodes within the piconet, which are referred to as the *slave* units. In a Bluetooth piconet, the master node can simultaneously control up to seven slaves that are actively communicating with the master. The master will assign a 3-bit Logical Transport Address (LT_ADDR) to each of the active slave.

Several piconets can be interconnected via *bridge* nodes to create a scatternet. Bridge nodes are capable of time-sharing between multiple piconets, receiving packets from one piconet and forwarding them to another. As a bridge node is connected to multiple piconets, it can be a master in one piconet and act as slaves in other piconets. This is called a *master/slave bridge* (or *M/S bridge*). Alternatively, a bridge node can act as a slave in each of the piconets which it is connected to. This is called a *slave/slave bridge* (or *S/S bridge*).

If a set of Bluetooth devices are within the transmission range of each other and need to exchange data, then a piconet formation algorithm is necessary to create a connected topology for the devices and to assign the role of each device (i.e., master or slave). The current Bluetooth specification [1] has standardized the procedures for piconet formation. This involves a sequence of steps including the inquiry, inquiry scan, page, and page scan phases.

Similarly, given a set of communicating Bluetooth devices that are either distributed over an area exceeding the coverage of a piconet, or whose number exceeds the eight active devices that can communicate with each other in a piconet, a scatternet formation algorithm (SFA) is necessary to create a connected topology for the devices, and to assign the role of each device (i.e., master, slave, or bridge). Although the current Bluetooth specification [1] defines what a scatternet is, a specific SFA is not specified. How to efficiently self-organize the nodes into a high-performance ad hoc network has been the subject of intensive research [8–13].

In the current Bluetooth specification, all packet transmissions within a scatternet need to be routed via the master or bridge nodes. When the traffic load is high, these master or bridge nodes may become the network bottleneck. Furthermore, in order to support dynamic joining or leaving of mobile devices within a scatternet, a number of time slots in the master nodes may need to be allocated for the dynamic topology configuration updates. This further reduces the time slots available at the master nodes for data transmissions.

In [14] a two-phase scatternet formation (TPSF) algorithm is proposed with the aim of supporting dynamic topology changes while maintaining a high aggregate throughput. In the first phase, a *control scatternet* is constructed for control purposes (i.e., to support dynamic join/leave, route discovery, and so on). The second phase is invoked whenever a node needs to initiate data communications with another node. A dedicated piconet/scatternet is constructed on demand, between the communicating nodes. As the *on-demand scatternet* can dedicate all the time slots to a single communication session, it has the capability to provide a high throughput and a small end-to-end data transfer delay. The on-demand scatternet is torn down when the data transmissions are completed.

The control scatternet formation consists of three steps. In the first step, each node performs device discovery by exchanging control information with its neighbors. The second step consists of role determination. The node with the highest number of neighbors among all its neighbors is selected as the master node. The last step is the creation of the control scatternet. The topology of the control scatternet has the following features. Each bridge node belongs to at most two piconets and acts as an S/S bridge. Once the control scatternet has been formed, the master and bridge nodes stay in active mode, while all the pure slave nodes are put in the park mode.

The on-demand scatternet formation procedure is initiated by a source node that wants to transmit data to a destination node, and consists of two steps. First, a dynamic source routing based protocol is applied to enable piconet route discovery in the control scatternet. The piconet route discovery is achieved via the exchange of Route Request (RREQ) and Route Reply (RREP) messages. Then, all the master nodes along the piconet route select the participating nodes for the on-demand scatternet. This part is achieved via the exchange of Path Request (PREQ) and Path Reply (PREP) messages. Finally, the destination node initiates the connection setup via the paging procedure.

In [15] the mobility of Bluetooth devices is considered within a limited range and TPSF+ is proposed, which enhances the second phase (i.e., the on-demand scatternet formation phase) of the original TPSF algorithm. TPSF+ has the advantage of involving a small number of nodes participating in the on-demand scatternet route discovery procedure so as to avoid unnecessary route discoveries. The on-demand scatternet route discovery is limited to several piconets instead of all the piconets within the control scatternet. Results show that the route discovery procedure for TPSF+ is more efficient than the original TPSF.

The performance gain is illustrated here by comparing the proposed SFA, TPSF+ [15], with Bluenet [13]. In the simulation model, there are 40 nodes in total, and these nodes are placed in an area of 20×20 m². For each data point, the simulation was run 100 times and each run time was 120 seconds. The nonpersistent TCP (Transmission Control Protocol) on/off traffic is used. During the "on" periods,

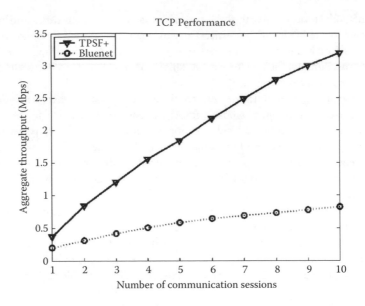

FIGURE 9.1   Aggregate throughput versus number of sessions.

packets are generated at a constant burst data rate of 1440 Kbps. During the "off" periods, no traffic is generated. Burst times and idle times follow the exponential distributions with an average "on" time of 0.5 seconds and an average "off" time of 0.5 seconds. The packet size is 1000 bytes. The performance metrics include the *aggregate throughput* and the *average end-to-end delay*. The aggregate throughput is defined as the total throughput obtained by all the communication sessions. The end-to-end delay is determined from the time when the packet is created at the source node to the time when the packet is received at the destination node.

Figures 9.1 and 9.2 show the aggregate throughput and end-to-end delay of TPSF+ in comparisons with that of Bluenet, respectively. It is apparent that TPSF+ achieves a much higher aggregate throughput and

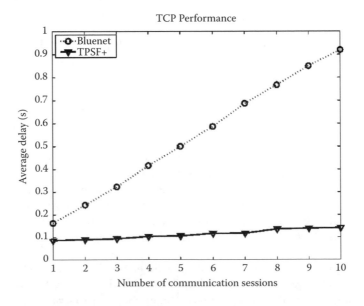

FIGURE 9.2   End-to-end delay versus number of sessions.

much lower end-to-end delay than Bluenet in a multi-hop scenario, with the performance differences increasing with the number of sessions. The performance bottleneck in Bluenet is due to the traffic load at the master and bridge nodes. TPSF+ avoids this bottleneck problem by setting up dedicated on-demand scatternet for each communication session.

9.3 Bluescouts—Mobile Agents for Scatternet Formation

The lack of a standardized procedure for the creation of scatternets in the Bluetooth specification has resulted in a number of SFAs being proposed in the literature [8–13]. We have reconsidered some of the assumptions made by existing SFAs and designed an efficient SFP that addresses the scatternet formation problem from a completely different perspective [16,17]. In particular, we have proposed a solution in which compact programs called *mobile agents* are able to both relocate themselves spatially in the network, and employ available resources at the local node in order to configure scatternets on-demand.

The rationale behind existing SFAs is not much different from earlier approaches for topology configuration in wireless ad hoc networks. In most cases, nodes rely on either an algorithm that chooses a single "super-leader," or one that designates a few nodes as "cluster-heads." These nodes then rely on the message passing and client-server communications models in order to coordinate negotiations with the rest of the available nodes and settle on a particular scatternet topology. In practice, the existing approaches to the scatternet formation have two problems. The first is the common assumption that all the nodes are present when the scatternet formation process begins. The second is the lack of supporting evidence that the topologies they individually produce are in fact the most favorable for the type of services that the network will sustain. We contend that both of these issues are potentially detrimental to the chances of any of the existing SFAs' widespread adoption.

The previous arguments motivate the need for an SFA that: (1) is able to configure the network topology in a fully asynchronous fashion so that the dependence of concurrent network nodes' participation can be dropped, and (2) supports the incorporation of custom policies into the topology configuration algorithm. We have conceived an SFA that achieves both of these objectives by decoupling the actual scatternet formation algorithm from the device inquiry/paging process that existing SFAs rely upon. Our approach relies on the spatial coordination of nodes, whose participation, as required during the reconfiguration of the scatternet topology, is controlled by mobile programs known as *mobile agents*. The result is a system that enables a dynamic and gradual organic-like growth as nodes attempt to join an existing network. To accomplish this, our protocol makes use of a platform that relies on both message passing and mobile-code mechanisms in order to coordinate the time and execution stage of the SFA [18]. We configure our mobile agents with a particular policy in order to create a tree-shaped topology and test the effectiveness of our mobile-code approach by simulations.

We first note that a piconet master is potentially busier as it operates as both medium-access and data-forwarding coordinator in contrast to their slave-node counterparts. In addition, they are evidently outnumbered by slave-nodes. Therefore, it is reasonable to expect that new nodes joining a piconet will be discovered by slaves, forcing them to assume a master role. This situation has been regarded as undesirable, as it leads to both longer scatternet diameter and data forwarding delays. Besides, it also leads to increased average path length and potentially lower network throughput due to packet collisions in the wireless medium.

To minimize these issues, a newly discovered node that has assumed a master role initiates the necessary process in order to configure itself as a slave, and thus attempts to keep the number of piconets low. This is accomplished by sending a mobile agent that carries the newly discovered node's hardware address and its current clock value to the slave node which discovered it. The slave node then forwards the mobile agent to the piconet's master, which, in turn, attempts to page the new node as instructed by the agent employing the values it brought along. This is accomplished by allowing the local Node Controller Interface of the Bluetooth protocol stack to be directly accessed by the mobile agent interpreter. The new node may be able to receive the page signal transmitted by the master, given that it enters the PAGE_SCAN state just before it dispatches the mobile agent and that they are both within radio range of each other. The new node may then switch to a slave role if the circumstances allow it to do so.

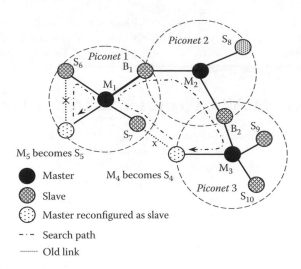

FIGURE 9.3 Scatternet reconfiguration.

Clearly, it is possible that the new node and the piconet's master may not be within radio range of each other, in which case the new node simply keeps its master role, resulting in the creation of a scatternet. New nodes attempting to join the existing scatternet perform the same procedure, except that the mobile agent will create copies of itself in order to explore the various paths that possibly lead to other master nodes. These agent clones do not operate independently from the rest, as this might incur page signal collisions perceived by the new node, if it happens to be within radio range of more than one master node while it is being paged. Instead, agents employ a depth-first spatial spreading method in order to coordinate the overall scatternet reconfiguration process as shown in Figure 9.3. The need for concurrent operation of the scatternet nodes is thus eliminated, as the agents operate in a tightly controlled asynchronous fashion.

Performance evaluations of our approach have yielded good results as evidenced by high slave-to-master ratios in the resulting piconets shown in Figure 9.4, which approaches the theoretical limit. Note that the slave-to-master ratio in a scatternet is always less than the maximum number of active slaves (7) allowed in a piconet, because each piconet has one or more bridge nodes that are shared with adjacent piconets. More importantly, the use of agents enables any other policy to be programmed into the agents in order to reconfigure the scatternet in a custom manner. Factors such as the type of device, link capacity, or even available services may be considered according to the needs of the user, making our approach very flexible.

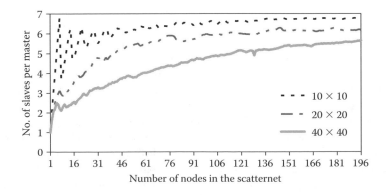

FIGURE 9.4 Slave-to-master ratio.

9.4   Adaptive Bluetooth Scheduling Algorithm

Bluetooth supports different traffic types between two Bluetooth devices. For voice traffic, the master maintains a Synchronous Connection-Oriented (SCO) link to a specific slave by reserving time slots at regular T_{SCO} intervals. In the Bluetooth specification, the T_{SCO} interval can be equal to either two, four, or six time slots. A SCO link does not support packet retransmission.

On the other hand, the Extended Synchronous Connection-Oriented (eSCO) link supports limited retransmission of packets. The master maintains an eSCO link to a specific slave by using reserved slots at regular T_{eSCO} intervals. The T_{eSCO} interval is negotiated between the master and the slave during link setup. The range of the T_{eSCO} interval is between 4 and 256 slots. Besides, the master also maintains an additional W_{eSCO} slots after the reserved slots as a retransmission window. The range for W_{eSCO} is from 0 to 256 slots.

For data traffic, the master can establish an Asynchronous Connectionless Link (ACL) to any slaves in TDD slots not reserved for SCO links. Therefore, if a SCO link is present, there will be either two or four time slots available for all the ACL connections. As a result, data traffic using ACL is a best-effort service.

Scheduling in Bluetooth scatternets can be divided into two tasks; namely, *intra-piconet scheduling* and *inter-piconet scheduling*. Intra-piconet scheduling focuses on the scheduling of packet transmission within a piconet. In a piconet, the master controls the access of all devices to the channel through a Time Division Duplex (TDD) master-slave polling scheme. For communications between a master and its slave, the master first sends a data packet to the slave in an even-numbered slot as a polling message. If the master queue for the slave is empty, it can instead send a POLL packet as a polling message. When the slave receives the polling message, it immediately replies with a data packet to the master. If the slave queue is empty, it can instead send a NULL packet as a reply message. As a slave node cannot transmit a packet without first being polled by its master, intra-piconet scheduling is controlled by the master. On the other hand, inter-piconet scheduling focuses on when a bridge node switches between piconets and how a bridge node communicates with the masters in different piconets.

An efficient intra-piconet scheduling algorithm should minimize the number of wasted slots, assign time slots dynamically based on the traffic rate, and maintain fairness among different slaves. As each master-slave polling operation includes the transmission from the master to its slave and a reply from the slave to its master, an empty queue in either side will result in a wasted slot. The master has limited knowledge on the buffer occupancy of its slaves. Each slave may have a different bandwidth requirement.

The main issue in inter-piconet scheduling is the switching of a bridge node between piconets. As each Bluetooth device has one transceiver, it can only participate in one piconet at a time. Because each master uses its own local clock, a bridge node has to re-synchronize with the new master when it switches to a new piconet. The switch between two piconets may result as a *slot loss*. Another problem occurs when two masters try to access the bridge node simultaneously. This is referred to as the *bridge node conflict*. As a bridge node can only listen to one master at a time, the other master will not be able to communicate with the bridge node and will waste slots for the polling operation. Lastly, an efficient inter-piconet scheduling scheme needs to dynamically allocate bandwidth on each link and maintain fairness among all nodes.

In [19], we proposed an Adaptive Scheduling Algorithm (ASA) that focuses on achieving max–min fairness for all nodes within the scatternet, preventing bridge node conflict, and dynamically allocating bandwidth for each link based on traffic changes. As a result, it gives improved performance in aggregate throughput and average delay.

ASA uses a time-sliding window (TSW) for traffic estimation. The TSW maintains a time-based history and can obtain a smooth average for bursty traffic. Besides estimating the packet arrival rate, each node also needs to determine the time for it to accumulate enough packet requests for data transmission. In order to maintain fairness for all nodes within the scatternet, ASA defines the *maximum usable serving slots* (*MUSS*) to limit the serving time between two nodes. The size of MUSS must be large enough for both uplink and downlink transmissions. If the scatternet supports 1-slot, 3-slot, and 5-slot packets, then the size of MUSS must be larger than or equal to ten Bluetooth time slots. The *trigger point* for a one-way transaction of a node is defined as half the size of *MUSS*. In ASA, the time for a node to reach the trigger point is defined as the *transmission request arrival time*.

For intra-piconet scheduling in ASA, each master node maintains an *active list* and a *waiting list* to schedule the serving order for all connected slave nodes. The *active list* contains all the slave nodes that have accumulated enough packets to reach the trigger point. The *waiting list* contains all the slave nodes that do not have enough packets in the queue. The master follows the order of nodes in its active list and serves them in a "round-robin" fashion. The master starts to serve the next slave node either when it has finished serving the current slave node for *MUSS* or when there is no data to send between the two nodes. A master uses the dynamic switch schedule to reserve a fixed size *switch schedule slot* (*SS_Slot*), which is equal to the size of *MUSS*.

For inter-piconet scheduling in ASA, a *switch schedule* is maintained between a master and a bridge node to organize the time for them to communicate with each other. Each time when a master meets with a bridge node, they negotiate their *next meeting time* and update their switch schedule. In ASA, a master uses the *hold mode* to allow the bridge node to switch between piconets. By assigning the hold mode, the master puts the connected node into the sleep state in which the node is not required to listen to the master for a period of time. When the time expires, the connected node switches to the active mode and actively listens to the master. Therefore, after determining the new meeting time, a master node knows how long it should put a bridge node into the hold mode. The negotiation process is initialized by the master. We compare the performance of our proposed ASA with credit-based scheduling (CBS) [20] and flexible scatternet-wide scheduling (FSS) [21] via network simulations. Figure 9.5 shows the scatternet topology. We consider bursty on–off UDP (User Datagram Protocol) traffic. The on and off periods follow exponential distributions with average lengths of 2 s and 1 s, respectively. UDP traffic is encapsulated in DH1 packets. The following bi-directional traffic flows are considered: M1-M2, M2-M4, M4-M1, M2-M3, M3-M5, and M5-M2. Moreover, we also consider traffic between each master and its slaves. Each sender in piconet of M1, M3, M4, and M5 generates packets at a rate of 34.5 Kbps, and each sender in piconet of M2 generates packets at a rate of 29 Kbps. We choose these rates by dividing the max-min bandwidth share for the link by half. We then vary the traffic rate by a factor of 0.6, 1.5, 1.8, 2, 2.4, 2.8, and 3 to observe the performance change. Figure 9.6 shows that ASA achieves a higher throughput than the other two schemes especially when the traffic rate is high. Figure 9.7 shows that ASA maintains a lower delay than the other two schemes.

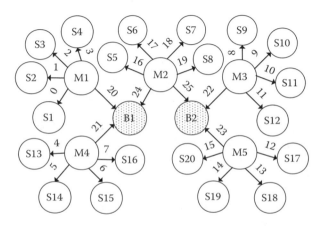

FIGURE 9.5 Simulation topology.

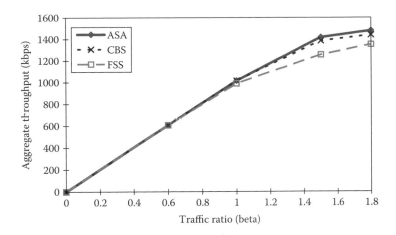

FIGURE 9.6 Aggregate throughput (on–off ratio = 2:1).

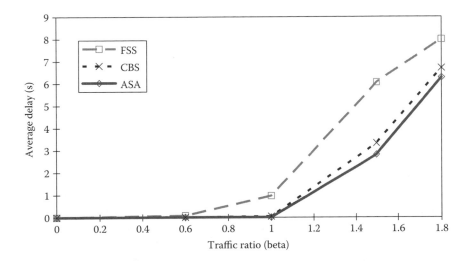

FIGURE 9.7 Average delay (on–off ratio = 2:1).

9.5 Collision Avoidance Scheduling in Bluetooth Access Points

The proliferation of Bluetooth-enabled personal-computing, communication, and mechatronic devices could fuel the demand of Bluetooth access points that enable these devices to access the Internet. Typically, a Bluetooth access point is connected to the Internet via wireline facility such as a local area network (LAN). According to the Bluetooth specification [1], each piconet supports a raw data rate of one megabits per second and a maximum of seven active slaves. Thus, a Bluetooth access point, acting as a master and equipped with one radio, only can provide a maximum of one megabits per second access bandwidth and seven simultaneous connections. Use of multiple radios in a Bluetooth access point can be an effective option to overcome such a limitation [22]. An example of a high-capacity Bluetooth access point with four Bluetooth radios which function as masters of four piconets is shown in Figure 9.8. Here "*Pi*" represents the *i*-th piconet with its master located at the access point, and "*R*" is the radius of the radio coverage, usually around 10 m, of the master nodes. However, the throughput of the Bluetooth access point in Figure 9.8 is less than four times that of an isolated piconet, due to collisions of transmissions

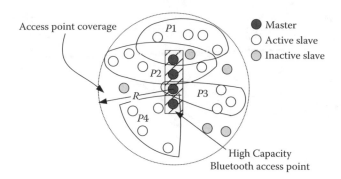

FIGURE 9.8 A Bluetooth access point with four piconets.

between piconets that hop onto the same frequency channel. The collision problem becomes more severe when the number of collocated piconets and traffic load increase [22,23].

Bluetooth employs FHSS transmissions over 79 channels each 1 MHz wide. The pseudo-random hop sequence of each piconet is determined by the master node, and all slave nodes are synchronized to it. In the United States, FCC Part 15 rules, which apply to license-free devices operating in the ISM band, do not allow the frequency-hop sequences of different piconets to be coordinated. Similar rules apply in other jurisdictions. However, as all the radios in the Bluetooth access point can share a common baseband module, it is possible to align the time slot boundaries of all the piconets by applying a common clock to all the master nodes in the Bluetooth access point. This approach effectively eliminates collisions that spill into adjacent time slots. Furthermore, as the multiple Bluetooth radios share a common antenna, when one of the radios is transmitting, other radios could become saturated if they are receiving. Therefore, not only should the piconets align their time slots, but they also should align their use of time slots for master–slave and slave–master transmissions, so that the entire Bluetooth access point is either transmitting or receiving in any given time slot. As all transmissions in a piconet are scheduled by the master node, and all master nodes in the Bluetooth access point employ a common baseband module, it follows that the baseband module can schedule transmissions in all the corresponding piconets. Furthermore, as the baseband module has knowledge of all the (independently chosen) hop sequences employed by all the piconets, one can design a collision avoidance scheduling (CAS) algorithm for deployment in the baseband module to prevent transmissions over different piconets from interfering with each other.

The CAS algorithm employs a frequency occupation table in the baseband module to keep track of the usage of the 79 channels. This is necessary as transmissions could span 1, 3, or 5 time slots during each hop. When several piconets attempt to hop to a specific channel, one of the piconets (the "winning piconet") is selected at random as the candidate for accessing the channel, whereas other piconets defer access to the next hop. The frequency occupation table is next consulted to see if the channel is already occupied by an ongoing transmission in another piconet. If so, the winning piconet defers channel access until the next hop. If no ongoing transmission is found in the channel, the winning piconet is allowed to access the channel, and the frequency occupation table is updated to reflect the number of slots in which the winning piconet will occupy the channel.

We have evaluated system performance through simulations, in which we consider SCO and ACL links with DH1, DH3, and DH5 packets, which occupy 1, 3, and 5 slots, respectively. A typical set of results is shown in Figure 9.9, which compares the system throughput of a Bluetooth access point employing CAS with that without CAS. It can be seen that CAS gives throughput improvements that increase with the number of piconets, i.e., the number of collocated Bluetooth radios at the Bluetooth access point. Whereas the performance improvements are relatively small for DH1 packets, the amount of improvements increases as longer packets are used. With 10 Bluetooth radios in the Bluetooth access point and DH5 packets, CAS gives a throughput improvement of better than 1.5 Mb/s or about 20% over that of a Bluetooth access point without CAS.

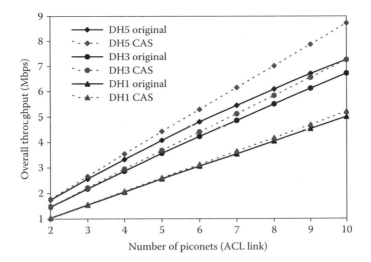

FIGURE 9.9 Throughput versus number of piconets (ACL link).

9.6 Conclusion

Bluetooth technology is applicable in communication among mechatronic devices and within a complex mechatronic system. In this chapter, we presented some highlights of the Bluetooth networking research completed at UBC under the "Enabling Technologies for Ubiquitous Personal Area Networking" project, which considered large-scale Bluetooth networking in ad hoc networking and access-point networking scenarios. For Bluetooth ad hoc networking, we have addressed the scatternet formation problem and intra-piconet and inter-piconet scheduling problem. We have presented our proposed TPSF and TPSF+ algorithms. TPSF+ can obtain more recent on-demand scatternet route information than the original TPSF in a dynamic environment with node mobility. Simulation results show that TPSF+ has a higher aggregate throughput and lower end-to-end delay when compared with Bluenet. We have also described the use of mobile agents in the Bluescouts scatternet formation method, which facilitate organic growth of a scatternet with reconfiguration for topology optimization accomplished via mobile processing. Results show that large-scale scatternets formed by Bluescouts have a slave-to-master ratio that approaches the theoretical maximum value. For Bluetooth scheduling, we have proposed an ASA for Bluetooth scatternets. ASA can dynamically allocate bandwidth to each link based on traffic demands while maintaining max–min fairness for all the nodes within the scatternet. Besides, ASA can also prevent bridge node conflicts. Simulation results show that ASA can maintain a high aggregate throughput and low delay on bursty on–off UDP traffic when compared with FSS and CBS. For Bluetooth access point networking, we have considered the frequency hop collision problem in high-capacity Bluetooth access points employing multiple Bluetooth radios, and proposed a collision-avoidance scheduling algorithm that effectively addresses this problem to give improved system throughput.

Acknowledgment

This chapter summarizes the research contributions of research associates Yoji Kawamoto of Sony Corporation during his visit to UBC, Dr. Zhifeng Jiang, Dr. Qixiang Pang, and graduate students Chu Zhang, Sergio González-Valenzuela, and Raymond Lee. Their work was partially supported by the Canadian Natural Sciences and Engineering Research Council (NSERC) under grant number STPGP 257684-02.

References

1. *Specification of the Bluetooth System*, version 2.0, November 2004, available at http://www.bluetooth.org.
2. *The IEEE 802.15 WPAN Task Group 1a*, http://www.ieee802.org/15/pub/TG1a.html.
3. Jain, M., Lampe, L., and Schober, R., Sequence detection for Bluetooth systems, in *Proceedings of IEEE Globecom'04*, Dallas, TX, November 2004.
4. Lampe, L., Schober, R., and Jain, M., Noncoherent sequence detection receiver for Bluetooth systems, *IEEE Journal on Selected Areas in Communications*, Vol. 23, No. 9, 1718–1727, September 2005.
5. Lampe, L., Jain, M., and Schober, R., Improved decoding for Bluetooth systems, *IEEE Transactions on Communications*, Vol. 53, No. 1, 1–4, January 2005.
6. Wu, J.C.H., Aken' Ova, V., Wilton, S.J.E., and Saleh, R., SoC implementation issues for synthesizable embedded programmable logic cores, in *Proceedings of IEEE Custom Integrated Circuits Conference*, Santa Clara, CA, September 2003, pp. 45–48.
7. Wilton, S.J.E., Kafafi, N., Wu, J.C.H., Bozman, K., Aken'Ova, V., and Saleh, R., Design considerations for soft embedded programmable logic cores, *Solid State Circuits Journal*, Vol. 40, No. 2, 485–497, February 2005.
8. Law, C., Mehta, A.K., and Siu, K.-Y., Performance of a new Bluetooth scatternet formation protocol, in *Proceedings of ACM Symposium on Mobile Ad Hoc Networking and Computing*, Long Beach, CA, October 2001, pp. 183–192.
9. Tan, G., Miu, A., Guttag, J., and Balakrishnan, H., An efficient scatternet formation algorithm for dynamic environments, in *Proceedings of IASTED Communications and Computer Networks (CCN) Conference*, Cambridge, MA, November 2002.
10. Zaruba, G.V., Basagni, S., and Chlamtac, I., Bluetrees—scatternet formation to enable Bluetooth-based ad hoc networks, in *Proceedings of the IEEE International Conference on Communications (ICC'01)*, Helsinki, Finland, June 2001, pp. 273–277.
11. Salonidis, T., Distributed topology construction of Bluetooth personal area networks, in *Proceedings IEEE INFOCOM*, Anchorage, AK, April 2001, pp. 1577–1586.
12. Basagni, S. and Petrioli, C., A scatternet formation protocol for ad hoc networks of Bluetooth devices in *Proceedings of IEEE Vehicular Technology Conference (VTC-Spring)*, Birmingham, AL, May 2002, pp. 424–428.
13. Wang, Z., Thomas, R., and Haas, Z., Bluenet—a new scatternet formation scheme, in *Proceedings of the 35th Annual Hawaii International Conference on System Sciences (HICSS'02)*, Honolulu, Hawaii, January 2002.
14. Kawamoto, Y., Wong, V., and Leung, V., A two-phase scatternet formation protocol for Bluetooth wireless personal area networks, in *Proceedings of IEEE Wireless Communications and Networking Conference (WCNC)*, New Orleans, LA, March 2003.
15. Zhang, C., Wong, V.W.S., and Leung, V.C.M., TPSF+: a new two-phase scatternet formation algorithm for Bluetooth ad hoc networks, in *Proceedings of IEEE Globecom'04*, Dallas, TX, November–December 2004, pp. 3599–3603.
16. González-Valenzuela, S., Vuong, S.T., and Leung, V.C.M., BlueScouts—a scatternet formation protocol based on mobile agents, in *Proceedings of IEEE ASWN'04*, Boston, MA, August 2004, pp. 109–118.
17. González-Valenzuela, S., Vuong, S.T., and Leung, V.C.M., Programmable agents for efficient topology formation of Bluetooth scatternets, *International Journal of Wireless and Mobile Computing* (in press).
18. Sapaty, P., *Mobile Processing in Distributed and Open Environments*, John Wiley & Sons, New York, 2000.
19. Lee, R. and Wong, V., An adaptive scheduling algorithm for Bluetooth ad hoc networks, in *Proceedings of IEEE International Conference on Communications (ICC'05)*, Seoul, Korea, May 2005, pp. 3532–3537.

20. Baatz, S., Frank, M., Kuhl, C., Martini, P., and Scholz, C., Adaptive scatternet support for Bluetooth using sniff mode, in *Proceedings of IEEE Conference on Local Computer Networks*, Tampa, FL, November 2001, pp. 112–120.

21. Zhang, W. and Cao, G., A flexible scatternet-wide scheduling algorithm for Bluetooth networks, in *Proceedings of IEEE IPCCC*, Phoenix, AZ, April 2002, pp. 291–298.

22. Lim, Y., Kim, J., Min, S.L., and Ma, J.S., Performance evaluation of the Bluetooth-based public Internet access point, in *Proceedings of International Conference on Information Networking 2001*, Beppu, Japan, February 2001, pp. 643–648.

23. El-Hoiydi, A., Interference between Bluetooth networks—upper bound on the packet error rate, *IEEE Communications Letters*, Vol. 5, No. 6, 245–247, June 2001.

10

Brain–Machine Interfacing and Motor Prosthetics

A.E. Brockwell

M. Velliste

Summary

Interaction of humans and mechatronic systems can be facilitated through a suitable human—machine interface. Going a significant step further, the machine interface may be linked directly to the brain so that a human brain can directly communicate with a mechatronic device through wireless means. Such a communication link has obvious advantages, particularly for disabled users and operators of mechatronic devices. In recent years, neuroscientists have made significant progress in understanding the relationship between neuronal firing patterns and limb motion in primates. The results have made it possible to begin work on a new generation of prosthetic devices. These devices will apply state-of-the-art robotics and statistical signal processing techniques to data collected from electrodes implanted directly in the brain. Neural signals will be recorded, transmitted through wireless communication to a CPU contained within the robotic arm (or leg), and processed to translate intent into actuator command. Such brain-controlled prosthetic arms and other mechatronic devices could enable paralyzed people or amputees to recover the lost ability to interact with the physical environment; for example, to feed themselves. A prototype of such a device has been developed in the laboratory, and successfully operated by monkeys. In this chapter, we discuss the state of this work (as of late 2005). We also describe some of the remaining hurdles to be overcome before such devices become practical and safe for human use.

10.1 Introduction

Artificial limbs, or "prostheses," were described as early as 484 B.C. by Herodotus (for one translation, see [1]), who tells of a Persian soldier who cuts off his foot to escape from stocks, and subsequently used a wooden replacement. Apart from this description, and a prosthetic leg dated approximately 300 B.C.

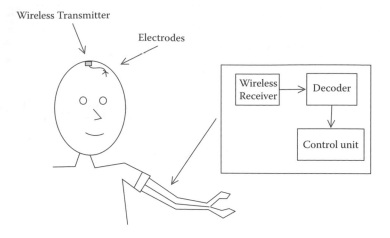

FIGURE 10.1　A potential design for a cortical neural prosthetic device.

found in a tomb in Italy, relatively little is known about early prosthetic limbs. By the 16th century, iron legs were being created for soldiers. In particular, Ambroise Paré, a French army surgeon, designed an iron leg with an articulated knee joint. Since then, a number of improvements in material technology and methods of attachment have been made, but control of limb prostheses is still somewhat limited. In the best cases, one has limited control of the device through attachments to other intact parts of the body.

Prosthetic devices present a practical example of brain–machine cooperation. Advanced prosthetic devices are indeed mechatronic systems. Interaction of humans and mechatronic systems can be facilitated through a suitable human–machine interface. Going a significant step further, the machine interface may be linked directly to the brain so that a human brain can directly communicate with a mechatronic device through wireless means. Such a communication link has obvious advantages, particularly for disabled users and operators of mechatronic devices.

With recent developments in neuroscience, bioengineering, robotics, and statistical signal processing, it is now possible to build a new generation of prosthetic devices for arm and/or leg amputees, controlled directly from the brain's cortex. Such devices are referred to as *cortical neural prostheses* (CNPs, see, for example, [2]). A basic design for a CNP is shown in Figure 10.1. The idea would be to use electrodes implanted directly into a part of the brain that codes for volitional movement, such as the motor cortex. These would be connected to a low-power preliminary signal-processing device in the skull with wireless transmission (potentially Bluetooth; see Chapters 7 and 9) capability. This device would perform "spike detection," that is, it would look for the action potentials in the continuously recorded voltage signals associated with each neuron being monitored. The spike event times for each neuron, in turn, would be transmitted to a wireless receiver in the robotic arm. An additional signal processor in the arm would translate the sequence of spike events into "intended motion," and the results would then be fed into a control unit that would move the robotic arm accordingly.

Before such a device can be put into practical use by humans, there are a number of challenges to be overcome. In the remaining sections of this chapter, we describe experimental results obtained to date in humans, monkeys, and rats, and outline some of these challenges.

10.2　Achievements to Date

In recent years, humans have been implanted with intracortical microelectrodes and have used the recorded signals to control a 1- or 2-dimensional (1-D or 2-D) computer cursor (see [3–6]). However, due to the experimental nature of the implants and the risks associated with surgery, most of the research on CNPs to date has been done in monkeys and rats. A noninvasive alternative to implanted electrodes is to use electroencephalogram (EEG) electrodes on the surface of the scalp. Humans have carried out 2-D cursor control using an EEG-based brain-computer interface (BCI) [7]. A disadvantage of this approach, however, is that the electrical signals recorded at the scalp are far removed from the underlying individual cell activity. Thus, it is not possible to extract accurate information about the subject's natural movement intent. In EEG-based experiments, subjects have controlled cursor movement using cognitive tasks unrelated to the movement task, such as mental arithmetic, relaxation, and visualization. However, this requires high mental effort from the subjects.

On the other hand, individual cell spiking activity, when simultaneously recorded from a large population of cells using intracortical microelectrode arrays, contains a direct representation of a subject's desired arm movement parameters (direction, speed, target position, wrist orientation, hand posture, grip force, etc.). It has been shown that direction and speed (i.e., velocity) of natural arm movements can be predicted from motor cortical spiking activity [8,9]. Subsequently, Schwartz [10] has demonstrated that not only the velocity of movement, but the entire trajectory of the hand can be reconstructed from motor cortical activity. Since then, correlation of cortical activity with other behavioral parameters has been shown, and real-time cortical control has been achieved in several laboratories. Apart from the human studies cited above, spike recordings have been used in rats for 1-D control of a lever [11] or an audio cursor [12]. Monkeys have used their cortical spiking activity to control a 2-D cursor [13,14], a 2-D cursor combined with virtual grip force control [15], or a 3-D virtual reality cursor [16,17].

There have been a number of experiments carried out involving real-time control of a real robotic arm. The first of these was a simple 1-D device [11]. In two recent studies, a robotic arm was introduced into the control loop [15,17], but the subjects still controlled the device and received visual feedback through a cursor interface. Researchers [18] have also trained a monkey to use a robotic arm under 3-D cortical control to retrieve pieces of fruit placed on the end of the arm. The arm used in that preliminary study had the joint configuration of a typical industrial robot arm and therefore did not look like a natural arm. In another laboratory, work is currently under way to have monkeys control an anthropomorphic (humanlike) arm to feed by themselves [19]. The arm has 4 degrees of freedom: 3 at the shoulder and one at the elbow, just like a human arm. Monkeys have successfully used the arm under 3-D cortical control to retrieve food rewards from four different target positions in 3-D space.

10.3 Information Encoding in the Motor Cortex

In this section we will describe the encoding of brain signals, which is crucial for the development and operation of a brain–machine communication link.

10.3.1 Signals

The signals used by a CNP are recorded using microelectrodes inserted into the brain's cortex (a typical implant contains 64 or more individual recording channels). The electrode tips are small enough to pick up the action potentials of individual neurons, but background activity arising from synchronous firing of many neurons is also picked up. This background activity, termed local field potentials (LFPs), can be spectrally separated from the spikes. Frequencies below about 200 Hz are generally considered LFPs, whereas frequencies between 400 Hz and 6 kHz make up the spike signals. Spike signals are presumed to contain more detailed information than LFPs. Figure 10.2 shows an example of simultaneously recorded traces from three different electrodes after band-pass filtering to obtain the spike signals. The important features are the intermittent bursts in the signal, referred to as *spikes* or *action potentials*. The wave shape and amplitude of a particular spike carries no information other than enabling the discrimination of spikes from different neurons recorded on the same channel. Each discriminated waveform is termed a *unit*.

Behaviorally relevant information is carried in the timing and frequency of spikes. In particular, motor cortical cells are directionally tuned, i.e., their firing rate is maximal in a *preferred direction* (PD) of movement, and the firing rate varies roughly proportionally to the angle of deviation from the PD. This

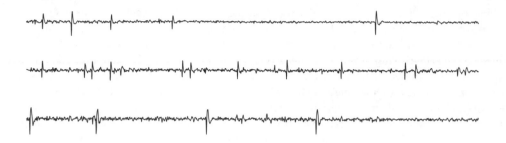

FIGURE 10.2 Voltage traces recorded simultaneously from three different electrodes (data from the Motor Lab, University of Pittsburgh).

smooth broad tuning, referred to as *cosine tuning*, can be seen in Figure 10.3, where a monkey used a set of 32 simultaneously recorded cortical units to control a cursor in 3-D virtual reality. The spike events of one of these cells are shown during 10 repetitions of movements to 8 different targets. Each tick represents the occurrence of a spike event, with time on the *x*-axis. The 10 rows of ticks in each pane represent the firing of the same cell during 10 repeated trials to each target (targets were presented in random order), whereas the vertical black line represents the beginning of each trial. The target panes are arranged spatially the way the monkey sees the targets. Targets 1–4 are away from the monkey, and targets 5–8 are toward the monkey. The cell can be seen to fire most during movements to target 6 (down, to the right, and toward the monkey).

10.3.2 Decoding

Based on the neurophysiological findings just described, the neuronal signals in the motor cortex are assumed to represent the position and velocity of the arm's endpoint, i.e., the hand. The *decoding* problem can be stated formally as follows: In successive 10-msec time bins, we measure the number of spikes on each of P neurons. Let $N_t^{(j)}$ denote the number of spikes for the j-th neuron, in time bin t. Also, we denote the average hand velocity and position, in time bin t, by v_t and x_t, respectively. Both v_t and x_t are 3-D vectors. Based on the directional tuning results already obtained, a reasonable statistical model

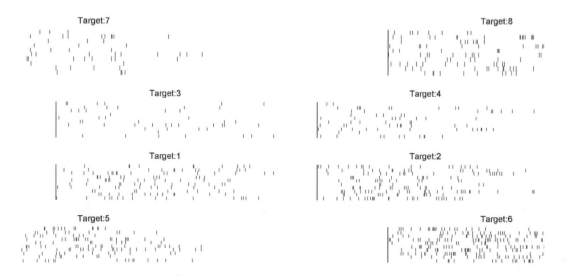

FIGURE 10.3 Spike raster plot showing the directional tuning of a single cell during repeated brain-controlled movements to 8 different targets in 3-D virtual reality (data from the Motor Lab, University of Pittsburgh).

for the relationship between x_t, v_t and $N_t^{(j)}$ is

$$N_t^{(j)} \sim Poisson(\delta\lambda(x_{t-lag_j}, v_{t-lag_j}, \theta_j)) \tag{10.1}$$

where δ is the length of a time bin, θ_j is a vector of parameters associated with the j-th neuron, and $\lambda(x, v, \theta)$ is a *tuning-function*, which determines the firing rate (spikes per second) as a function of the hand position and velocity, as well as neuron parameters. The term lag_j represents a neuron-specific delay between neuron behavior and arm motion. The parameter vector θ_j usually specifies a base firing rate, as well as a 3-D preferred direction. In particular, we can take θ_j to contain four parameters $\beta_0, \ldots, \beta_3$ and set

$$\lambda(x, v, \theta) = \exp(\beta_0 + (\beta_1, \beta_2, \beta_3) \cdot v)$$

The decoding problem is simply to estimate x_t and v_t, given current and past observations of spike counts, where N_t denotes the vector of measurements $(N_t^{(1)}, \ldots, N_t^{(P)})$. As a further challenge, the estimates must be obtained in real time if a CNP is to be controlled.

One method used by neuroscientists is referred to as *population vector algorithm* (PVA) [8,9]. The idea is fairly simple and proceeds roughly as follows: During a training period, the preferred direction vector $(\beta_1, \ldots, \beta_3)$ is determined for each neuron. Then, for decoding during the t-th time bin, one simply records the number of spikes from each neuron, smoothes appropriately to get an estimate of the rate, and then creates a vector by linearly combining the P preferred direction vectors with weights proportional to the corresponding firing rates. Other decoding methods used by various investigators include optimal linear estimators [20] and filtering approaches based on the Kalman filter.

10.4 Outstanding Challenges

Conceptually, the CNP design is straightforward. One records spiking activity of multiple neurons in the motor cortex, uses an appropriate method to carry out decoding, thereby obtaining intended arm velocity and position, and feeds this information into the robotic control (mechatronic) device. However, there are a number of limitations of current technology that need to be addressed before the CNP becomes practical for humans.

10.4.1 Electrode Implantation

The electrodes implanted in the motor cortex are critical to the success of the CNP. In particular, they need to (1) be able to measure spiking activity of multiple isolated neurons, and (2) be able to remain in place in the brain tissue without deteriorating, causing damage to the surrounding tissue, or being rejected by the immune system. Broadly speaking, there are two categories of implantable electrodes: (older) microwire electrodes and (newer) silicon microfabricated electrode arrays.

Microwire probes are simply very narrow wires, made of platinum, gold, tungsten, iridium or stainless steel, which are coated with insulating material but exposed at the tip. These probes are inserted slowly into the brain, where they are left while signals are recorded. A key problem is the tissue reaction. In the short term, there is an inflammatory reaction, whereas in the longer term the wires generally become encapsulated by glial (non-neuron) cells. This encapsulation generally inhibits the quality of the measured signal. Typical reports from laboratories indicate that because of this reaction, electrodes become significantly less effective after a matter of months. Clearly, this would not be satisfactory for a CNP device for a human.

Silicon machined probes, such as the "Utah" [21] and "Michigan" [22] probes, are considered to be the next generation of probes, although they still suffer from the same problems with tissue reaction as do microwires. Because they are machined, they can have more complex designs; in particular, various design strategies can be adopted to make them more "tissue friendly." Researchers are currently considering modifications of the electrode surface by adding structures that can be bound to bioactive compounds. These bioactive compounds could be chosen to attract growing neurites [23], reduce the inflammatory response, and so on. It is hoped that such approaches could eventually lead to safe chronic human implants.

10.4.2 Robotics

Ideally, robotic prosthetic limbs are mechatronic devices that will need to be light, versatile enough to provide natural limb-like functionality, contain their own power source, and be attached securely to the body. Robotics and mechatronics researchers continue to make progress on the first three of these problems, but attachment to the human body remains a difficult problem. It is currently possible to attach a robotic prosthetic arm to the body using a system of straps, using suction attachments, or by surgical attachment. The first two approaches, while noninvasive, could be problematic if significant loads or stresses are to be placed on the limb. Surgical attachment has the potential to provide a sturdy attachment, and techniques like those used for standard hip replacement could be used, but surgery always carries extra risk. It is worth noting that for wheelchair-bound subjects, it may be most effective simply to mount the CNP directly on the wheelchair. This would provide a sturdy attachment and also allow the CNP to be powered directly from the wheelchair's power supply.

10.4.3 Signal Processing (Decoding)

The decoding module, which translates measured motor cortical spiking activity into intended limb motion, is obviously a key component of the CNP. Although the PVA method described earlier has been proven to work, it is clearly desirable to make the best possible use of the available information. Improvements in decoding efficiency will allow either equivalent performance with fewer neurons (hence smaller implanted electrode arrays) or improved performance (e.g., smoother and more accurate movements) with the same number of neurons. With more sophisticated methods, it may also be possible to extract more aspects of arm and hand movement, such as the amount of elbow raising, turning of the wrist, hand opening/closing, and so on.

In Reference [24], it was demonstrated that by combining equation (10.1) with an equation describing the dynamics of hand position and velocity, one can pose the decoding problem as a nonlinear filtering problem and use associated techniques to reduce decoding error (the same kind of approach was also used by Reference [25] to decode hippocampus signals from rats). In particular, we can use the equations

$$v_{t+1} = v_t + \varepsilon_{t+1} \tag{10.2}$$

$$x_{t+1} = x_t + \delta v_t \tag{10.3}$$

where $\{\varepsilon_t\}$ is a sequence of independent and identically distributed 3-D normal random variables with zero mean and some specified covariance matrix. These equations describe the evolution of hand position/velocity in the absence of any other information. The first equation simply says that velocity drifts around randomly without changing too rapidly, whereas the second one simply says that velocity must be the derivative of position (more sophisticated models could of course be used).

Equations (10.1), (10.2), and (10.3) make up what is often referred to as a *state-space model*, where $\{x_t\}$ and $\{v_t\}$ are unobserved processes, and $\{(N_t^{(j)}, j = 1, 2, \quad , P), t = 1, 2, \quad \}$ is an observed sequence of multivariate spike counts. In this context, the problem of finding optimal estimates of each unobserved pair (v_t, x_t), given observations up to time t, is simply the well-studied *filtering problem* in engineering. If all parts of the equations were linear, and the distributions were Gaussian, then the Kalman filter [26] would give an exact solution. In this case, the system is nonlinear and non-Gaussian, which means we cannot use the standard Kalman filter. However, it is possible to use a more recently developed method sometimes referred to as the *particle filter* [27,28]. This method is computationally more intensive than the standard Kalman filter, but is applicable to nonlinear non-Gaussian systems.

The use of nonlinear non-Gaussian models in this context is particularly important. It has been recognized by researchers for some time that the model (10.1), (10.2), and (10.3), and its linear Gaussian version do not fully reflect the true flow of information in the motor cortex. There are a number of important complicating factors. For example, there is empirical evidence to suggest (among other things) that

- The tuning function λ may depend on position, velocity, and other important quantities of interest, such as wrist angle, gripping position, and so on.
- Even after accounting for velocity and position, the spiking activity of neurons may be correlated.
- The lag associated with each neuron may vary slowly over time, and may also be task-dependent.

Furthermore, monkeys typically go through periods in which they choose not to pay attention to the task at hand, and during these periods, the behavior of a neuron may change dramatically. It seems reasonable to expect the same from humans.

Building models that provide accurate probabilistic descriptions of the relationship between spiking activity and intended limb motion remains an active area of research, but as the models improve, as long as appropriate decoding methods such as particle filtering (which is guaranteed to give minimum mean-squared error estimates if the model is "correct") are used, we can expect improved decoding performance.

10.5 Discussion

Cortical signals hold the promise of providing enough information to control mechatronic devices such as robotic prosthetic limbs with the full range of flexibility of the natural limb. The quality of a CNP device will ultimately be determined by a number of factors, including quality of implanted electrodes, robustness and utility of the motorized robotic limb, strength and comfort of the CNP attachment to the body, and quality of the decoding module. In our own work, we have concentrated particularly on the decoding problem. As neuroscientists gain more understanding of the complex relationship between intended motion and neural spiking activity, it becomes possible to extract more information from the neural signals. At the same time, this extraction also becomes a more difficult and computationally intensive exercise in nonlinear filtering. However, with developments in filtering algorithms, it is possible to overcome these difficulties. With additional improvements in mechatronic technology in the other aforementioned areas in the coming years, it is conceivable that the promise of CNPs may be realized.

Acknowledgment

The authors are grateful to Tracy Cui and Chance Spalding for their comments on this chapter, and to Andrew Schwartz for helpful discussions. This work was supported in part by research grants NSF IIS-0083148, NIH-NINDS-N01-NS-2-2346, and NIH-R01 MH064537-04AZ.

References

1. Herodotus, A., *The Histories*, Penguin Classics (Reissue edition), Middlesex, England, 2003.
2. Schwartz, A.B., Cortical neural prosthetics, *Annual Review of Neuroscience*, Vol. 27, 487–507, 2004.
3. Kennedy, P.R., Bakay, R.A., Moore, M.M., and Adams, K., Direct control of a computer from the human central nervous system, *IEEE Transactions on Rehabilitation Engineering*, No. 8, 198–202, 2000.
4. Kennedy, P.R., Kirby, M.T., Moore, M.M., King, B., and Mallory, A., Computer control using human intracortical local field potentials, *IEEE Transactions on Neural Systems and Rehabilitation Engineering*, Vol. 12, 339–344, 2004.
5. Serruya, M.D., Caplan, A.H., Saleh, M., Morris, D.S., and Donoghue, J.P., The BrainGate™ pilot trial: building and testing a novel direct neural output for patients with severe motor impairment, in *Abstract Viewer/Itinerary Planner*, Washington, D.C., Society for Neuroscience, 2004.
6. Mukand, J., Williams, S., Polykoff, G., and Apple, D.F., Poster board 76: Feasibility study of the brainbatet neural interface system for individuals with quadriplegia, *American Journal of Physical Medicine and Rehabilitation*, Vol. 84, No. 3, 222, 2005.
7. Wolpaw, J.R. and McFarland, D.J., Control of a two-dimensional movement signal by a noninvasive brain-computer interface in humans, *Proceedings of National Academic Science*, Washington, D.C., Vol. 101, 17849–17854, 2004.

8. Georgopoulos, A.P., Kettner, R.E., and Schwartz, A.B., Primate motor cortex and free arm movements to visual targets in three-dimensional space, II. Coding of the direction of movement by a neuronal population, *Neuroscience*, Vol. 8, 2928–2937, 1988.

9. Georgopoulos, A.P., Schwartz, A.B., and Kettner, R.E., Neuronal population of movement direction, *Science*, Vol. 243, 234–236, 1989.

10. Schwartz, A.B., Direct cortical representation of drawing, *Science*, Vol. 265, 540–542, 1994.

11. Chapin, J.K., Moxon, K.A., Markowitz, R.S., and Nicolelis, M.A., Real-time control of a robot arm using simultaneously recorded neurons in the motor cortex, *Nature Neuroscience*, Vol. 2, 664–670, 1999.

12. Gage, G.J., Ludwig, K.A., Otto, K.J., Ionides, E.L., and Kipke, D.R., Naïve coadaptive cortical control, *Journal of Neural Engineering*, Vol. 2, 52–63, 2005.

13. Serruya, M.D., Hatsopoulos, N.G., Paninski, L., Fellows, M.R., and Donoghue, J.P., Instant neural control of a movement signal, *Nature*, Vol. 416, 141–142, 2002.

14. Musallam, S., Corneil, B.D., Greger, B., Scherberger, H., and Andersen, R.A., Cognitive control signals for neural prosthetics, *Science*, Vol. 305, 258–262, 2004.

15. Carmena, J.M., Lebedev, M.A., Cist, R.E., ODoherty, J.E., Santucci, D.M., Dimitrov, D.F., Patil, P.G., Henriquez, C.S., and Nicolelis, M.A.L., Learning to control a brainmachine interface for reaching and grasping by primates, *PLoS Biology*, Vol. 1, 193–208, 2003.

16. Taylor, D.M., Helms Tillery, S.I., and Schwartz, A.B., Direct cortical control of 3D neuroprosthetic devices, *Science*, Vol. 296, 1829–1832, 2002.

17. Taylor, D.M., Helms Tillery, S.I., and Schwartz, A.B., Information conveyed through brain-control: cursor versus robot, *IEEE Transactions on Neural Systems and Rehabilitation Engineering*, Vol. 11, 195–199, 2003.

18. Helms-Tillery, S.I, Taylor, D.M., and Schwartz, A.B., The general utility of a neuroprosthetic device under direct cortical control, *Proceedings of IEEE EMBS 25th International Conference*, Cancun, Mexico, Vol. 3, 2003, pp. 2043–2046.

19. Spalding, M.C., Velliste, M., Jarosiewicz, B., Kirkwood, G.C., and Schwartz, A.B., Direct brain control of an anthropomorphic robotic arm during a feeding task, in *Abstract Viewer/Itinerary Planner*, Washington, D.C., Society for Neuroscience, 2004.

20. Salinas, E. and Abbott, L.F., Vector reconstruction form firing rates, *Journal of Computational Neuroscience*, Vol. 1, 89–107, 1994.

21. Rousche, P.J. and Normann, R.A., Chronic recording capability of the Utah intracortical electrode array in cat sensory cortex, *Journal of Neuroscience Methods*, Vol. 82, 1–15, 1998.

22. Kipke, D.R., Vetter, R.J., Willams, J.C., and Hetke, J.F., Silicon-substrate intracortical microelectrode arrays for long-term recording of neuronal spike activity in cerebral cortex, *IEEE Transactions on Neural Systems and Rehabilitation Engineering*, Vol. 11, No. 2, 2003.

23. Cui, X., Lee, V.A., Raphael, Y., Wiler, J.A., and Hetke, J.F., Surface modification of neural recording electrodes with conducting polymer/biomolecule blends, *Journal of Biomedical Materials Research*, Vol. 56, No. 2, 261–272, 2001.

24. Brockwell, A., Rojas, A., and Kass, R., Recursive Bayesian decoding of motor cortical signals by particle filtering, *Journal of Neurophysiology*, Vol. 91, 1899–1907, 2004.

25. Brown, E.N., Frank, L.M., Tang, D., Quirk, M.C., and Wilson, M.A., A statistical paradigm for neural spike train decoding applied to position prediction from ensemble firing patterns of rat hippocampal place cells, *Neuroscience*, Vol. 18, 7411–7425, 1998.

26. Kalman, R.E., A new approach to linear prediction and filtering problems, *Journal of Basic Engineering (ASME)*, Vol. 82D, 35–45, 1960.

27. Kitagawa, G., Monte Carlo filter and smoother for non-Gaussian nonlinear state space models, *Journal of Computational and Graphical Statistics*, Vol. 5, No. 1, 1–25, 1996.

28. Doucet, A., de Freitas, N., and Gordon, N. (Eds.), *Sequential Monte Carlo Methods in Practice*, Springer, New York, 2001.

III

Control Technologies

11

Adaptive Control of a Gantry System

C.S. Teo

K.K. Tan

S. Huang

S.Y. Lim

Summary

Proper control is crucial in the operation of mechatronic devices. The gantry positioning system is a mechatronic device that relies on feedback control. This chapter addresses the control of an H-type gantry positioning stage. The stage is posed as a three-degree-of-freedom system. Based on this structure, a mathematical model is obtained using the Lagrangian equations. An adaptive control method is formulated for the positioning of the stage based on this model, with minimal *a priori* information assumed. A stability analysis is provided for the proposed control scheme. Real-time experimental results are presented to illustrate the practical application of the scheme and to verify the adequateness of the gantry model.

11.1 Introduction

Mechatronics and robotics go hand in hand because robots are mechatronic devices. Among the various configurations of long travel and high-precision Cartesian robotic systems, one of the most popular is the H-type, which is more commonly known as the moving gantry stage. In this configuration, two motors, which are mounted on two parallel slides, move a gantry simultaneously in tandem. This gantry configuration has been in use for large overhead traveling cranes in ports, rolling mills, and flying shear. When precision of positioning is of the primary concern, direct-drive linear motors are usually used, and fitted with aerostatic bearings for optimum performance. When used particularly with permanent magnet linear motors (PMLM), the moving gantry stage can be designed to provide high-speed, high-accuracy X, Y, and Z motion to facilitate such mechatronic applications as automated processes in flat panel display, printed circuit board manufacturing, precision metrology, and circuit assembly, where high part placement accuracy for overhead access is necessary. The stage possesses a high power density due to the dual drives, and it can yield high-speed motion with no significant lateral offset when the two drives are appropriately coordinated and synchronized in motion. In certain applications such as in wafer steppers, the dual drives can also be used to produce a small "theta" rotary motion without any additional rotary actuators. The application domain of the moving gantry stage is rapidly expanding because of the increasingly stringent requirements arising from developments in mechatronics, precision engineering, and nanotechnology ([1,2]; also see Chapter 5 and Chapter 6 of this book). To date, there is not one precision machine manufacturer who does not provide for a moving gantry stage in one form or other.

The main challenges to be addressed to harness the full potential of this configuration of Cartesian stages lie mainly in the control system. Various research papers on independent axis control have been published, including Reference [3], but these mostly ignore the coupling effect between the axes. In certain instances, efficient synchronization among the axes is crucially important to minimize the positional offsets between the two axes. This offset arises due to different drive and motor characteristics, nonuniform load distribution of the gantry and attached end-effectors, and time-varying thermo-mechanical properties. Researches such as [4] and [5] seek to achieve this objective. In Reference [4], Tan et al. used an observer-augmented scheme to more effectively deal with the interaxis offset phenomenon through a disturbance observer. However, these are all essentially nonparametric schemes, or partially parametric ones based on linear-dominant linear models for the respective servo dynamics without explicitly modeling the cross-axis effects. In particular, disturbances in the form of dynamic load changes, which can be fairly asymmetric in nature, have to be adequately addressed. To the best of our knowledge, there seem to be little effort devoted to address the motion control aspects of a moving gantry robotic system on a highly precise scale even though these are fast becoming important enabling technologies to facilitate the fulfillment of high accuracy mechatronic processes. Even though adaptive control approaches such as in Reference [6] have long been in existence, and the use of the Lagrangian equation in the model gantry system is not uncommon [7], the merger of these two approaches provide appealing solutions to the mentioned issues.

In this chapter we present the development of an adaptive control scheme based on a full model of the gantry stage. The adaptive control scheme is designed based on the analytical model, which is able to adaptively estimate the model parameters without assuming much *a priori* information. A stability analysis is provided to show the error convergence under the scheme. Full experimental results on a real gantry stage are presented as well to illustrate the applicability of the method.

11.2   Dynamic Model of the Gantry Stage

Prior to addressing the control of the gantry stage, a brief description of the typical gantry setup and its constituent subassemblies is given in order to better align the terms used in the subsequent development. The targeted gantry stage to be controlled is shown in Figure 11.1. It may be structurally simplified and modeled as the three-degree-of-freedom system shown in Figure 11.2.

FIGURE 11.1   Targeted H-type gantry positioning stage.

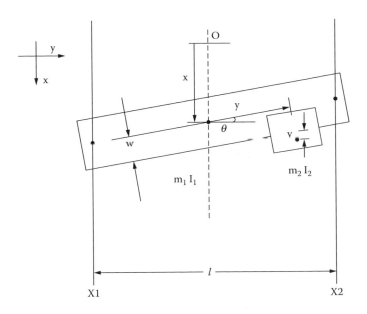

FIGURE 11.2 Schematic representation of gantry positioning stage.

Based on the Lagrangian equation approach, the model may be obtained as

$$DX + CX + BF = BU,$$ (11.1)

with $X = [x \quad \theta \quad y]^T$ where $x = x_1 + \dfrac{x_2 - x_1}{2}$, and the parameters D, C, B, F, and U are obtained as

$$D = \begin{bmatrix} m_1 + m_2 & -m_2(d\sin\theta + y\cos\theta) & -m_2\sin\theta \\ -m_2(d\sin\theta + y\cos\theta) & m_1\left(\frac{1}{2}\right)^2 + m_2\left(\left(\frac{1}{2} + y\right)^2 + y^2 + d^2\right) & m_2 d \\ -m_2\sin\theta & m_2 d & m_2 \end{bmatrix}, \quad (11.2)$$

$$C = m_2 \begin{bmatrix} 0 & y\theta\sin\theta - (d\theta + y)\cos\theta & -\theta\cos\theta \\ y\theta\sin\theta - (d\theta + y)\cos\theta & (y\sin\theta - d\cos\theta)x - \left(\frac{l}{2} + 2y\right)y & \left(\frac{l}{2} + 2y\right)\theta - x\cos\theta \\ -\theta\cos\theta & \left(\frac{l}{2} + 2y\right)\theta - x\cos\theta & 0 \end{bmatrix}, \quad (11.3)$$

$$B = \begin{bmatrix} 1 & 1 & 0 \\ l\cos\theta & -l\cos\theta & 0 \\ 0 & 0 & 1 \end{bmatrix}, \quad (11.4)$$

$$F = [F_{x1} \quad F_{x2} \quad F_y]^T, \quad (11.5)$$

$$U = [u_{x1} \quad u_{x2} \quad u_y]^T. \quad (11.6)$$

F_{x1}, F_{x2}, and F_y are the frictional forces, and u_{x1}, u_{x2}, and u_y are the generated mechanical forces along x_1, x_2, and y, respectively. The frictional forces are assumed to be adequately described by the Tustin model

$$F_z = d_z z + f_z \operatorname{sgn}(z). \tag{11.7}$$

The Tustin model has proven to be useful, and it has been validated adequately in many successful applications, including Reference [8]. Interested readers may see Reference [9] for a detailed derivation of this model.

11.3 Adaptive Control of Gantry Stage

For the actual physical system, it is a challenging and difficult task to obtain the exact values of the parameters m_1, m_2, d_i, and f_i ($i = x_1$, x_2, y) of the model accurately. To this end, an adaptive controller is designed based on the model that does not require accurate initial estimates of the model parameters.

Define the filtered error $s = \Lambda e + e$, where $e = X_d - X$, and X_d, representing the desired trajectories, is twice differentiable. Equation 11.1 can be expressed as

$$Ds = -Vs + m_1 D_0 (\Lambda e + X_d) + m_2 \left[\frac{1}{2} Vs + D_1 (\Lambda e + X_d) + C_0 X \right] + \sum_{i=1}^{3} (d_i BF_{0i} + f_i BF_{1i}) - BU, \tag{11.8}$$

where the parameters D, C, and F are further broken down as follows:

$$D = m_1 D_0 + m_2 D_1, \tag{11.9}$$

$$C = m_2 C_0, \tag{11.10}$$

$$F = \sum_{i=1}^{3} (d_i F_{0i} + f_i F_{1i}), \tag{11.11}$$

with

$$D_0 = \begin{bmatrix} 1 & 0 & 0 \\ 0 & (^1/_2)^2 & 0 \\ 0 & 0 & 0 \end{bmatrix}, \tag{11.12}$$

$$D_1 = \begin{bmatrix} 1 & -d\sin\theta - y\cos\theta & -\sin\theta \\ -d\sin\theta - y\cos\theta & (^1/_2 + y)^2 + y^2 + d^2 & d \\ -\sin\theta & d & 1 \end{bmatrix}, \tag{11.13}$$

$$C_0 = \begin{bmatrix} 0 & 0 & 0 \\ 0 & \left(-\left(\dfrac{l}{2} + 2y\right)\right)y + (y\sin\theta - d\cos\theta)x & \left(\left(\dfrac{l}{2} + 2y\right)\theta - x\cos\theta\right) \\ 0 & \left(\left(\dfrac{l}{2} + 2y\right)\theta - x\cos\theta\right) & 0 \end{bmatrix}, \tag{11.14}$$

$$F_{01} = [x_1 \quad 0 \quad 0]^T,$$ (11.15)

$$F_{02} = [0 \quad x_2 \quad 0]^T,$$ (11.16)

$$F_{03} = [0 \quad 0 \quad y]^T,$$ (11.17)

$$F_{11} = [\text{sgn}(x_1) \quad 0 \quad 0]^T,$$ (11.18)

$$F_{12} = [0 \quad \text{sgn}(x_2) \quad 0]^T,$$ (11.19)

$$F_{13} = [0 \quad 0 \quad \text{sgn}(y)]^T.$$ (11.20)

The additional representative vector V is expressed as

$$V = \begin{bmatrix} 0 & -d\theta\cos\theta - y\cos\theta + y\theta\sin\theta & -\theta\cos\theta \\ -d\theta\cos\theta - y\cos\theta + y\theta\sin\theta & 2\left(\frac{1}{2} + y\right)y + 2yy & 0 \\ -\theta\cos\theta & 0 & 0 \end{bmatrix}.$$ (11.21)

We propose an adaptive controller given by

$$U = B^{-1}Ks + \hat{m}_1 B^{-1} D_0(\Lambda e + X_d) + \hat{m}_2 B^{-1}\left[\frac{1}{2}Vs + D_1(\Lambda e + X_d) + C_0 X\right] + \sum_{i=1}^{3}(\hat{d}_i BF_{0i} + \hat{f}_i BF_{1i}),$$ (11.22)

with the following adaptation rules:

$$\hat{m}_1 = \gamma_1 s^T D_0(\Lambda e + X_d),$$ (11.23)

$$\hat{m}_2 = \gamma_2 s^T\left[\frac{1}{2}Vs + D_1(\Lambda e + X_d) + C_0 X\right],$$ (11.24)

$$\hat{d}_i = \gamma_{3i} s^T BF_{0i},$$ (11.25)

$$\hat{f}_i = \gamma_{4i} s^T BF_{1i},$$ (11.26)

where K is positive definite, and $\hat{m}_1, \hat{m}_2, \hat{d}_i,$ and $\hat{f}_i$ are estimates of $m_1, m_2, d_i,$ and f_i, respectively.

11.4   Stability Analysis

In this section, we show that the proposed adaptive controller can guarantee the stability of the closed-loop system, and the filtered error s will approach zero as $t \to \infty$.

Define the Lyapunov function

$$v = s^T D s + \frac{1}{\gamma_1} m_1^2 + \frac{1}{\gamma_2} m_2^2 + \sum_{i=1}^{3} \left(\frac{1}{\gamma_{3i}} d_i^2 + \frac{1}{\gamma_{4i}} f_i^2 \right). \tag{11.27}$$

The derivative of v is

$$v = -2s^T K s + 2m_1 s^T D_0 (\Lambda e + X_d) + 2m_2 s^T [V_0 s + D_1 (\Lambda e + X_d) + C_0 X]$$

$$+ 2 \sum_{i=1}^{3} s^T (d_i BF_{0i} + f_i BF_{1i}) - 2\frac{1}{\gamma_1} m_1 \hat{m}_1 - 2\frac{1}{\gamma_2} m_2 \hat{m}_2 - 2 \sum_{i=1}^{3} \left(\frac{1}{\gamma_{3i}} d_i \hat{d}_i + \frac{1}{\gamma_{4i}} f_i \hat{f}_i \right). \tag{11.28}$$

Incorporating the adaptive laws (11.23)–(11.26), v becomes

$$v = -2s^T K s. \tag{11.29}$$

This implies that s, $\hat{m}_1$, $\hat{m}_2$, $\hat{d}_i$, $\hat{f}_i$ are bounded. From the definition of filtered error, because Λ is positive definite and s is bounded, it follows that e is bounded as well. This also implies that e is bounded and, in turn, that X, X are bounded. In addition, from (11.8) we can conclude that s is bounded. Furthermore, (11.29) and the definition of v jointly imply that

$$\lim_{t \to \infty} K s^2 = V(0) - \lim_{t \to \infty} V(\infty). \tag{11.30}$$

Finally, applying Barbalat's lemma, we obtain $\lim_{t \to \infty} s(t) = 0$.

11.5   Experimental Results

Two sets of experiments were carried out, one based on three independent proportional-integral-derivative (PID) controller and another using the said adaptive controller. For user-defined parameters, the PID controllers' proportional, integral, and derivative gains are tuned as 100, 50, and 1, respectively, for the x-axis; and 50, 1, and 0, respectively, for the y-axis. For the adaptive controller, the parameters are configured as follows: $\gamma_1 = 100$, $\gamma_2 = \gamma_{31} = \gamma_{32} = \gamma_{33} = \gamma_{41} = \gamma_{42} = \gamma_{43} = 1$, $K = \text{diag}(100\ 10\ 10)$, and Λ equates the identity matrix, i.e., $\Lambda = \text{diag}(1\ 1\ 1)$.

11.5.1   Hardware

The stage used for the experimental setup is the gantry stage as described and shown earlier in Figure 11.1. The motor specifications are listed in Table 11.1 below.

The length l of the gantry and the distance d from the slider mass center to the y-axis are found to be 0.415 m and 0.015 m, respectively.

TABLE 11.1 Specifications of Gantry Motors

Item	Units	*X*-Axis Servo Motor	*Y*-Axis Servo Motor
		SEM MT22G2-10	Yaskawa SGML-01AF12
Power	W	350	100
Max Torque	N.m	0.70	0.318
Max Velocity	rpm	5000	3000
Encoder Resolution	μm	10	10

11.5.2   Desired Trajectories

The desired trajectories (position, velocity, and acceleration) are as shown in Figure 11.3. The trajectory would span a distance of 0.01 m, periodically in 4 s. The maximum velocity and acceleration attained are 0.094 m/s and 0.145 m/s², respectively.

11.5.3   Results

The experimental results are shown in Figures 11.4 and 11.5. The adaptive controller is able to yield an individual axis error of under 0.38 mm at steady state as compared to the PID performance of 0.96 mm for both x_1 and x_2 axes, whereas the *y*-axis error is kept under 2 mm for both controllers. In addition, the adaptive controller is able to minimize the interaxis offset error (by manipulation of the parameter *K*; see (11.22)), whereas the decoupled PID controller was only able to track individual trajectories independently. This performance is reflected by the resultant interaxis offset error of 0.32 mm using the adaptive controller, as compared to 0.81 mm for the PID controller. It is seen that for a short time duration from $t = 0$ to $t = 3$, the PID outperforms the adaptive controller. This is expected as the learning parameters have been initialized to zero with no *a priori* knowledge assumed of the system masses and friction. Subsequently, after some parameter adaptation, the proposed approach quickly yields significantly improved performance over the PID control.

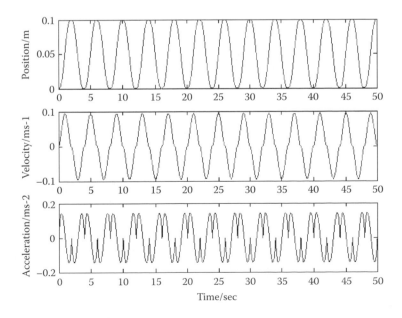

FIGURE 11.3   Desired trajectories for position, velocity, and acceleration.

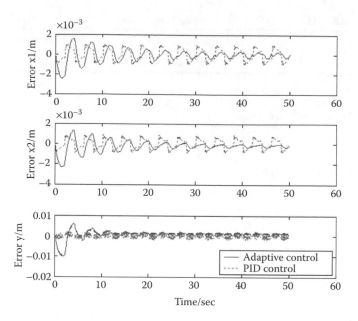

FIGURE 11.4   Tracking error for individual axis.

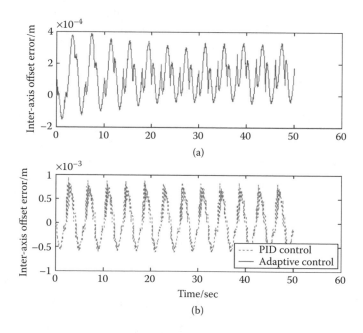

FIGURE 11.5   Inter-axis offset error between X_1 and X_2.

11.6 Conclusion

Using an appropriate dynamic model of a typical H-type gantry stage based on the Lagrangian equation, an adaptive controller was developed to minimize the tracking error as well as interaxis offset error. Minimal *a priori* information of the model was assumed to ensure robustness. The stability of the control scheme has been proven via a Lyapunov-based analysis. Furthermore, the experimental results have shown the superior performances of the adaptive controller over PID control.

References

1. Goto, S. and Nakamura, N., Accurate contour control of mechatronics servo system using Gaussian Networks, *IEEE Transactions on Industrial Electronics*, Vol. 43, No. 4, 469–476, 1996.
2. Syh-Shiuh Yeh, and Pau-Lo Hsu, Analysis and design of the integrated controller for precise motion systems, *IEEE Transactions on Control Systems Technology*, Vol. 7, No. 6, 706–717, 1999.
3. Yang, X. and Taylor, D.G., H control design for positioning performance of gantry robots, *Proceedings of the American Control Conference*, Chicago, IL, 2000, pp. 3038–3042.
4. Tan, K.K., Lim, S.Y., Huang, S., Dou, H., and Giam, T.S., Coordinated motion control of moving gantry stages for precision applications based on an observer-augmented composite controller, *IEEE Transactions on Control Systems Technology*, Vol. 12, No. 6, 984–991, 2002.
5. Kim, S., Chu, B., Hong, D., Park, H.K., Park, J.M., and Cho, T.Y., Synchronizing dual-drive gantry of chip mounter with LQR approach, *IEEE International Conference on Advanced Intelligent Mechatronics*, Port Island, Japan, 2003, pp. 838–843.
6. Butler, H., Honderd, G., and Van Amerongen, J., Model Reference Adaptive Control of a Gantry Crane Scale Model, *IEEE Control Systems Magazine*, Vol. 11, No. 1, 57–62, 1991.
7. Fang, Y., Dixon, W.E., Dawson, D.M., and Zergeroglu, E., Nonlinear coupling control laws for an underactuated overhead crane system, *IEEE Transactions on Mechatronics*, Vol. 8, No. 3, 418–423, 2003.
8. Tan, K.K., Lee, T.H., Huang, S., and Jiang, X., Friction modeling and adaptive compensation using a relay feedback approach, *IEEE Transactions on Industrial Electronics*, Vol. 48, No. 1, 169–176, 2001.
9. Teo, C.S., Tan, K.K., Huang, S., Lim, S.Y., and Tay, E.B., Dynamic modeling and adaptive control of a H-type Gantry stage, *Mechatronics—The Science of Intelligent Machines*, accepted for publication.

12

Virtual Computer Numerical Control System

K. Erkorkmaz

Y. Altintas

C.-H. Yeung

Summary

Computer numerical control (CNC) machines used in automated manufacturing are indeed mechatronic systems. Control plays a crucial role in their operation. This chapter presents a comprehensive virtual simulation model of a realistic and modular CNC system. The Virtual CNC architecture represents an actual CNC, but with modular feed drives, sensors, motors, and amplifiers. The CNC software library includes a variety of trajectory interpolation and axis control laws. Constant, trapezoidal, and cubic acceleration profiles can be selected as a trajectory generation module. The control laws can be selected ranging from a simple proportional-integral-derivative (PID) to complex pole placement, generalized predictive control, or sliding mode control with friction compensation. When the Virtual CNC is assembled, its performance can be tested using frequency- and time-domain response analyses, which are automated. The Virtual CNC includes both analytical tuning methods for linear controllers, as well as fuzzy logic–based expert auto-tuning system for adaptive sliding mode control. Tool-path and feed-rate optimization strategies have also been implemented for accurate high-speed cornering applications. The chapter includes detailed experimental verification of the developed algorithms.

12.1 Introduction

The objective of virtual manufacturing technology is to design a completely digital factory where the part is modeled, machined with optimized process parameters, and resulting errors are predicted with corrective actions being taken in a computer simulation environment. The field of mechatronics is intimately connected with the governing technologies. Machine tools are mechatronic systems. In particular, CNC machines prevalent in automated manufacturing systems are complex and sophisticated mechatronic systems. This chapter presents a virtual model of a CNC system for machine tools.

The CNC system is a mechatronic system that consists of mechanical feed drives, motors, amplifiers, position-velocity-acceleration sensors, servo motors, controllers, and real-time computer algorithms that generate time-stamped position commands through trajectory generation and close the axis servo loops [1]. The Virtual CNC (VCNC) requires a realistic mathematical model of each CNC component and its logical interconnection. There has been significant research reported in modeling various trajectory generation algorithms [2], control laws [3–7], and physical components of the drives such as motors, amplifiers, ball-screw, and linear drives with various friction and backlash characteristics [8–11]. The position errors originating from the CNC are dependent on the robustness and tracking ability of the axis control laws. Tuning plays an important role in the controller design. Over the last two decades, the use of fuzzy logic control has been widely proposed in literature for the auto-tuning of control parameters [12]. This concept has been successfully applied in a hierarchical control structure for deployable orbiting manipulators by Goulet et al. [13]. Fuzzy logic has also been adopted in the present chapter for automatic tuning of sliding mode axis controllers.

Maintaining the contouring accuracy at high feed rates is vital to preserving the tolerance integrity of produced parts. Contour errors originate from servo tracking errors in individual axes, which become prevalent when trying to track reference trajectories with high frequency content and high torque demands beyond the actuators' limits. The look-ahead functionality in CNC systems has been developed to alleviate this problem to a certain extent by ensuring that acceleration commands in the interpolated trajectory never exceed the drives' dynamic capability [14]. This follows similar research conducted in time-optimal trajectory generation for robots [15] and machine tools [16]. However, limiting the magnitude of acceleration or torque is not sufficient for guaranteeing the tracking and contouring accuracy. Additional factors such as jerkiness (rate of change of acceleration) of the commanded trajectory, friction, backlash, and structural resonances in the drive system play an important role in determining the overall contouring performance.

In this chapter, a contouring strategy is presented that takes advantage of the VCNC's ability to accurately predict contour errors in the presence of the above-mentioned factors, and applies corrective measures by modifying the tool path and adjusting the feed rate, so that sharp corners can be traveled in minimum time without violating the specified tool-path tolerance. The contour error estimate is obtained using the technique developed by Erkorkmaz and Altintas [17]. One possible solution for tracking cornered tool paths is to replace the sharp corner with a double clothoid curve, as proposed in References [18,19]. Another innovative idea for accurately turning sharp corners has been proposed by Weck and Ye [20], which consists of low pass filtering the motion commands before applying them to a high-bandwidth servo controller with feedforward action. This has a similar effect to geometrically smoothening out the sharp corners. In the present chapter, the smooth transition at sharp corners is achieved with a quintic spline [21], which is interpolated with minimum feed fluctuation using the recursive strategy developed in Reference [22].

The remainder of the chapter is organized as follows. The architecture of the VCNC is presented in Section 12.2, which includes the actuators, trajectory generation, and control laws. This is followed by application examples in Section 12.3, which consists of experimental verification of the VCNC's ability to accurately predict contouring errors, automatic tuning of sliding mode axis control using fuzzy logic, and modification of corner tool-path and feed-rate profiles to enable sharp corners to be tracked in minimum time within given tolerances. Experimental validation is provided in all cases. The chapter is concluded in Section 12.4.

12.2 VCNC Structure

The architecture of the developed VCNC system is illustrated in Figure 12.1, which resembles the real, reconfigurable, and open CNC developed in the Manufacturing Automation Laboratory, University of British Columbia (UBC) [23]. The VCNC accepts reference tool paths generated on CAD/CAM systems in

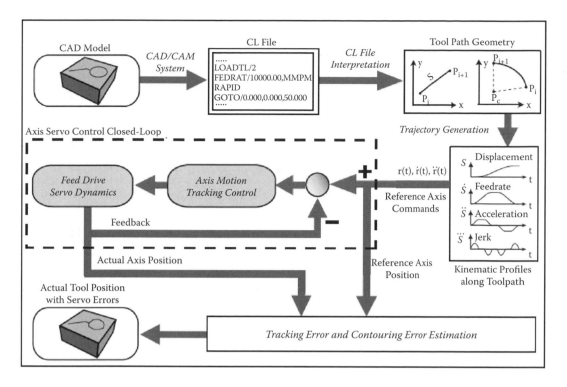

FIGURE 12.1 Architecture of virtual CNC system.

the form of industry-standard Cutter-Location (CL) format. Each block in the CL file contains NC block numbers; tool paths in the form of linear, circular, and spline segments; the cutter dimensions; tool center coordinates; and feed speed for machining a particular part using a CNC machine tool. Each tool path segment, such as linear, circular, or spline path, is passed through the trajectory generation algorithm, which creates displacement, feed, acceleration, and jerk expressions at divided segments [1].

Trajectory generation sets up the real-time interpolation parameters such as discrete displacement along the path and its frequency. The interpolator generates discrete displacement commands for each axis at every control interval. The axis commands are passed on to the control law, which shapes the overall response of the feed drive transfer function, consisting of digital to analog (D/A) converter, amplifier, servo motor, inertia, viscous damping, guideway friction, and lead-screw backlash. The axis can be configured to have acceleration, velocity, and position sensors with defined accuracy and noise parameters. The position error of each axis is evaluated in the feedback loop and combined to predict the contouring error at each control interval. The user can select any control law, lead-screw, or linear-drive parameters, as well as amplifier, motor, friction field, and sensor so that most machine tools can be reconfigured automatically by the user who can add new modules or modify the existing algorithms that are created in MATLAB environment. The VCNC system consists of three main modules: feed drive, trajectory generation, and axis-tracking control modules.

12.2.1 Dynamic Model of Feed Drive

The parametric model of the feed drive is shown in Figure 12.2, which consists of digital-to-analog-conversion (DAC) quantization, amplifier saturation and gain, motor torque gain, nonlinear friction field, equivalent inertia, viscous damping, ball-screw and gear transmission ratios, backlash (if applicable), and feedback measurement noise characteristics. The control signal u_a [V] generated by the axis controller is applied to the current amplifier that has a gain of K_a [A/V]. In the motor armature, the motor dynamic torque T_m [N.m] is produced, which is assumed to be linearly proportional to the motor current with the

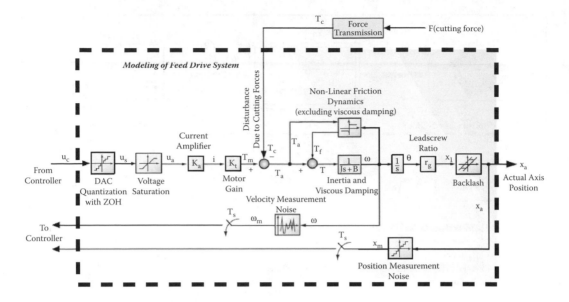

FIGURE 12.2 Overall block diagram of feed drive model.

torque constant K_t [N.m/A]. The total dynamic torque delivered by the motor is spent in overcoming the external disturbance torque T_c [N.m] due to cutting forces, and the nonlinear friction torque T_f [N.m] due to static and Coulomb friction. The remaining torque T [N.m] is used in accelerating the inertia (J [kg.m^2]) of the rigid body motion of the table and overcoming the system's viscous damping (B [kg.m^2/s]). In return, angular velocity ω [rad/s] of the motor shaft is produced, which is transferred to angular position θ [rad] through integration. The angular position of the motor shaft is transmitted as a linear displacement of the table by the ball-screw and nut mechanism, which has a transmission gain of r_g [mm/rad]. This factor also includes the effect of any gear ratio that might be used between the motor and ball screw. If a linear drive system is used, $r_g = 1$, and J and B are substituted by mass and rectilinear damping constant. The nonlinear guideway friction is defined as:

$$T_f(k) = \begin{cases} 0, & \omega(k) = 0 \text{ and } T_a(k) = 0 \\[2mm] T_a(k), & \omega(k) = 0 \text{ and } T_{stat}^- < T_a(k) < T_{stat}^+ \\[2mm] T_{stat}^+, & \omega(k) = 0 \text{ and } T_a(k) \geq T_{stat}^+ > 0 \\[2mm] T_{stat}^-, & \omega(k) = 0 \text{ and } T_a(k) \leq T_{stat}^- < 0 \\[2mm] T_{stribeck}^+(\omega(k)), & \omega(k) > 0 \\[2mm] T_{stribeck}^-(\omega(k)), & \omega(k) < 0 \end{cases} \qquad (12.1)$$

where
$$T_a(k) = T_m(k) - T_c(k) \Leftarrow T_m(k) = K_a K_t \cdot u_a(k)$$

The friction torque, while the axis is in motion, is defined by the Stribeck curve [8]. Because the effect of viscous damping B is included in the rigid body axis dynamics, the Stribeck curve can be represented as

$$T_{stribeck}^{+/-}(\omega(k)) = T_{stat}^{+/-} \cdot e^{-\omega(k)/\omega_1^{+/-}} + T_{coul}^{+/-} \cdot (1 - e^{-\omega(k)/\omega_2^{+/-}}) \qquad (12.2)$$

Here, T_{stat} and T_{coul} are the static friction and the Coulomb friction torque, respectively. Also, ω_1 determines the spacing between the boundary lubrication and partial fluid lubrication zones, and ω_2 determines the spacing between the partial lubrication and full fluid lubrication regions. The superscript "+/−" corresponds to positive and negative directions of motion. The spacing or the dead zone between the screw and nut creates backlash, which is modeled by a disengaged zone, engagement in the positive direction, and engagement in the negative direction. The analytical expression for the backlash model is summarized as

$$
x_a(k) = \begin{cases} x_a(k-1); & d_e^- \leq x_l(k) \leq d_e^+ \\ x_l(k) - D_b/2; & x_l(k) \geq d_e^+ \Rightarrow set \quad d_e^- = x_l(k) - D_b \quad \text{and} \quad d_e^+ = x_l(k) \\ x_l(k) + D_b/2; & x_l(k) \leq d_e^- \Rightarrow set \quad d_e^- = x_l(k) \quad \text{and} \quad d_e^+ = x_l(k) + D_b \end{cases}
$$

$$(12.3)$$

where D_b denotes the backlash of the feed drive mechanism. Also, x_a and x_l are the actual axis position due to backlash and ideal axis position, respectively, and d_e^- and d_e^+ are the positions of negative and positive ends of the dead-band, respectively. The axis position is assumed to be at the middle of the dead-zone at initial conditions ($d_e^- = -D_b/2$, $d_e^+ = +D_b/2$).

12.2.2 Trajectory Generation Mechanism

The trajectory generation mechanism implemented in the VCNC is illustrated in Figure 12.3. After interpreting the CL file, the start and end coordinates of the tool path, the types of tool movement, and the feed rate are recognized and stored in a buffer. By executing the buffer block by block, the descriptions for each tool-path segment are obtained and then passed to the trajectory generation process sequentially. The trajectory generation algorithm identifies the distance to be traveled, and divides it into acceleration, constant velocity, and deceleration subsegments depending on the kinematic profiles selected in the CNC design. The VCNC system presented here gives a choice of constant, trapezoidal, or cubic acceleration profiles. Step changes in the constant acceleration profile leads to infinite jerk and severe acceleration discontinuities, which contain high frequencies in the generated position commands. Such high frequency content may excite structural vibrations and cause severe tracking errors in high-speed machines. Cubic acceleration leads to much smoother position commands with continuity in velocity, acceleration, and jerk, which is better suited for high-speed and high-precision drives, but is at the expense of increased computational load on the CNC computer. Details of the trajectory generation algorithms can be found in Reference [24] and are not repeated here.

12.2.3 Axis Control Module

There are a significant number of control laws that can be implemented in CNC systems. Typically, any axis control law has two components: the feedforward part that processes the reference position commands, and the feedback part that shapes the measured states such as position, velocity, and acceleration to stabilize the closed-loop dynamics, as shown in Figure 12.4. The presented VCNC system has a number of user-reconfigurable control laws, which have all been experimentally proven on our open CNC system [23]. The conventional control laws include cascaded P position and PI velocity (P-PI) control, PID control, and lead-lag control [1]. More sophisticated control laws include pole placement control (PPC) [25], zero-phase tracking error control (ZPTEC) [7], generalized predictive control (GPC) [4], and sliding mode control (SMC) [3]. Whereas the conventional control laws can be found in standard control texts, details of the more advanced control laws can be found in the referenced publications.

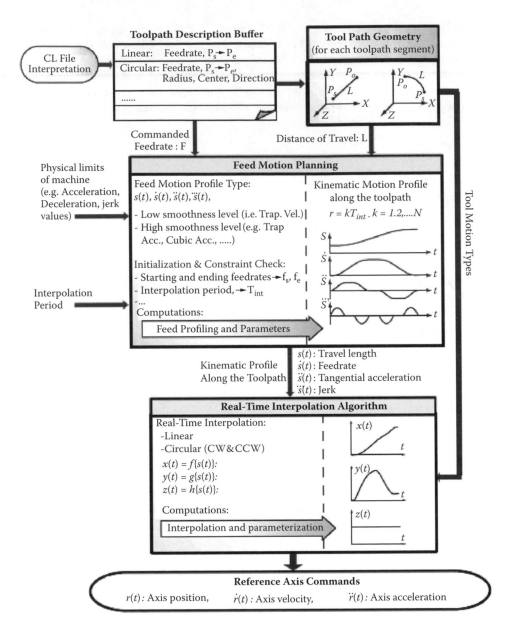

FIGURE 12.3 Trajectory generation mechanism.

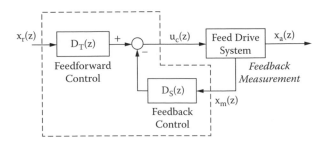

FIGURE 12.4 Axis control law in a standard form.

12.3   Application Examples

In the following, application examples are presented for the VCNC, which target the optimization of a machine tool's contouring performance in a virtual environment without having to conduct time-consuming trials on the actual machine. These examples include prediction of the high-speed contouring accuracy, fuzzy logic–based tuning of sliding mode control law, and tool trajectory modification to improve the cornering performance.

12.3.1   Prediction of High-Speed Contouring Accuracy

The VCNC system is experimentally verified on a three-axis vertical machining center driven by ball-screw drives. Various tracking control schemes have been tried out, such as P-PI, PID, SMC, GPC, and PPC with feed-forward friction compensation. PID and SMC results are reported here as examples. The machine tool is controlled by an in-house-developed open CNC system [23], which allows modular integration of any trajectory generation, control law, and compensation strategies. Detailed drive and controller parameters are provided in Reference [26]. Standard circular and diamond toolpath contouring tests were conducted: The diamond had a side length of 50 mm, and the circular path radius was 50 mm. The reference trajectories have been generated to achieve a feedrate of 200 mm/s with maximum acceleration and jerk values of 2000 mm/s² and 50000 mm/s³, respectively. The selected acceleration profile was trapezoidal, resulting in piecewise constant jerk commands along the toolpath. The experiments were conducted under air-cutting conditions to avoid structural deformations of the machine under cutting load. The measured and predicted tracking and contour error profiles are shown in Figures 12.5–12.8. The most critical parts on the diamond paths are located at the corners where transients in reference trajectory occur. The most significant deviations on the circular paths are located at the quadrants where the direction of the axis motion changes and the errors are mostly due to friction, which holds the slide until sufficient torque is accumulated to overcome the static friction disturbance. The experimental

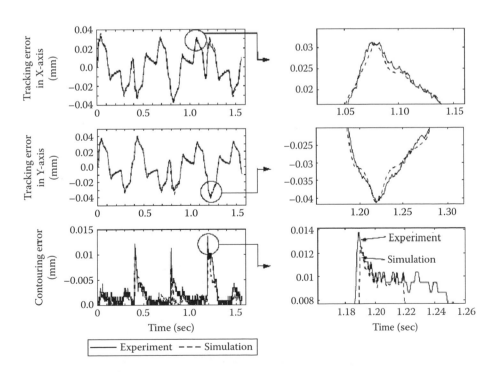

FIGURE 12.5   Diamond-shaped contouring tests with PID control.

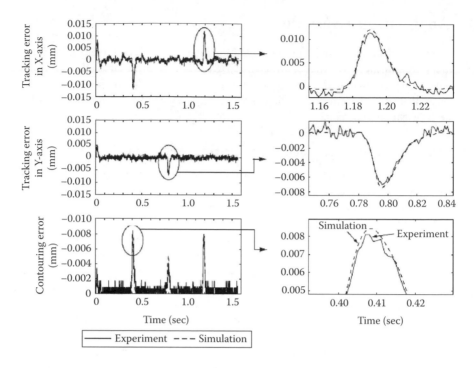

FIGURE 12.6   Diamond-shaped contouring tests with sliding mode control (SMC).

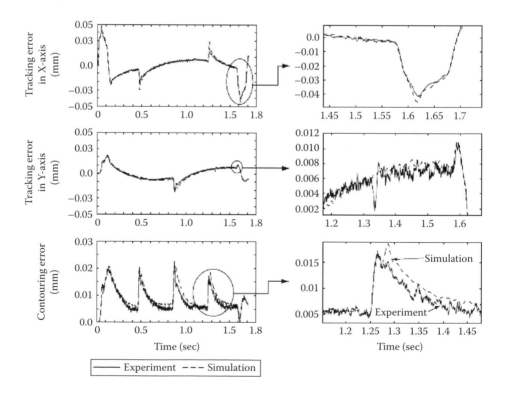

FIGURE 12.7   Circle-shaped contouring tests with PID control.

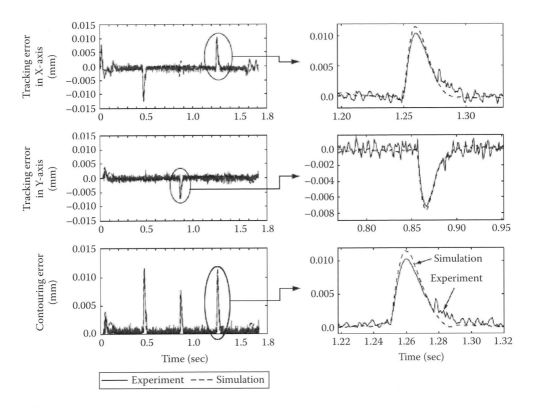

FIGURE 12.8 Circle-shaped contouring tests with sliding mode control (SMC).

results clearly indicate the accurate prediction capability of the VCNC system, also indicating that the SMC yields significantly less contouring and tracking errors than the standard PID controller.

12.3.2 Auto-Tuning of Axis Servo Controller Using Fuzzy Logic

It is desirable to tune the feed drives in a simulated virtual environment provided that the mathematical model of the system is sufficiently accurate. The virtual tuning allows modification to the machine-tool drive mechanism as well as proper selection of motors and sensors during the initial design. A fuzzy logic-based tuning method for SMC is presented here. The adaptive SMC law is given as [3]:

$$u_c^{SMC}(k) = J_e[\lambda(x_r(k) - x_m(k)) + \ddot{x}_r(k)] + B_e \dot{x}_m(k) + K_S \cdot S(k) + \hat{d}(k)$$

$$\text{where} \begin{cases} S(k) = \lambda[x_r(k) - x_m(k)] + [\dot{x}_r(k) - \dot{x}_m(k)] \\ \hat{d}(k) \approx \rho \cdot \left\{ \lambda \cdot \left[\frac{T_s \cdot z}{z-1} \cdot \dot{x}_r(k) - \frac{T_s \cdot z}{z-1} \cdot \dot{x}_m(k) \right] + [\dot{x}_r(k) - \dot{x}_m(k)] \right\} \\ J_e = \frac{J}{K_a K_t r_g}; \quad B_e = \frac{B}{K_a K_t r_g}; \end{cases} \quad (12.4)$$

Here, J_e and B_e are the equivalent moment of inertia and viscous damping constant, respectively, as reflected on the motor shaft; S is a stable sliding surface function; and $\hat{d}$ is the axis disturbance estimated

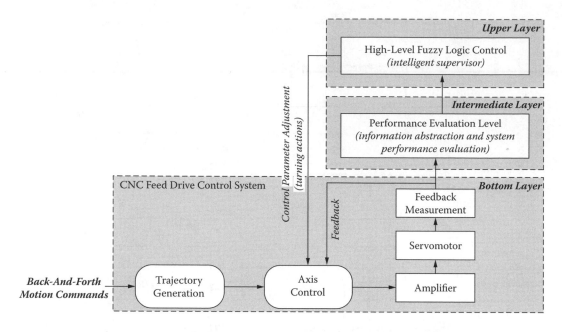

FIGURE 12.9 Hierarchical auto-tuning strategy.

using a simple observer for adaptation. The control parameters that need to be tuned are sliding surface bandwidth λ [rad/s], feedback gain K_S [V/(mm/s)], and disturbance adaptation gain ρ [V/mm]. Furthermore, λ is assumed to be fixed and is determined according to the achievable bandwidth of the drive, and the two control parameters (K_S and ρ) are considered in the auto-tuning process. Considering Figure 12.9, the auto-tuning process has a three-level hierarchical structure: the bottom layer for SMC that needs to be tuned, the intermediate layer for system performance evaluation, and the upper layer for decision making using fuzzy logic. A smooth back-and-forth motion command is applied to the closed loop. The system performance is observed and characterized in the intermediate layer, where the system-response descriptors such as oscillation of the control signal, stability of the loop, and tracking error are evaluated. If the system response is not found to be satisfactory, then new control parameters are assigned to the SMC through fuzzy-logic tuning in the upper layer. The control parameters are iteratively adjusted until the overall system performance becomes satisfactory. Full details of the tuning algorithm are found in Reference [26]. Results of the fuzzy tuning procedure along with experimental validation are shown in Figure 12.10.

12.3.3 Accurate Contouring at Sharp Corners

Two tool-path modification strategies have been adopted for smoothing the motion around sharp corners. The under-corner approach is suitable for machining corners in a shorter time while remaining within a specified tolerance. This technique can be applied with high bandwidth servo controllers that are capable of accurately tracking the reference toolpath, such as adaptive sliding mode control. The over-corner approach is based on smoothing the tool path outside the corner while remaining within the given tolerance. This approach is suitable for correcting the under-cut problem, caused by the large phase lag in low bandwidth controllers such as P, P-PI, or PID. The cornering feedrate is adjusted iteratively such that contour error violation does not occur. Both techniques have been experimentally validated, as shown in Figures 12.11 and 12.12. Full details of these algorithms are provided in Reference [27].

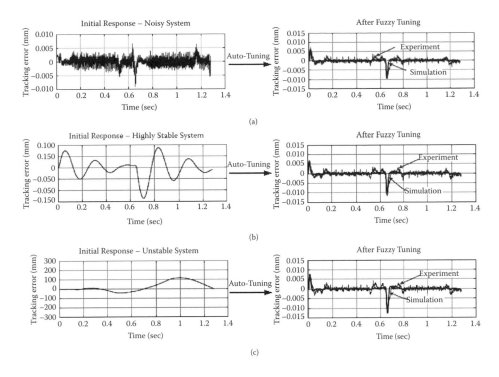

FIGURE 12.10 System performance before and after fuzzy tuning: (a) initially noisy system, (b) initially highly stable and sluggish system, (c) initially unstable system.

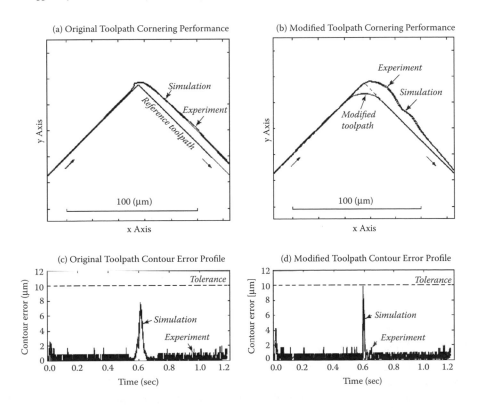

FIGURE 12.11 Contouring performance for 90° under-corner spline with sliding mode control.

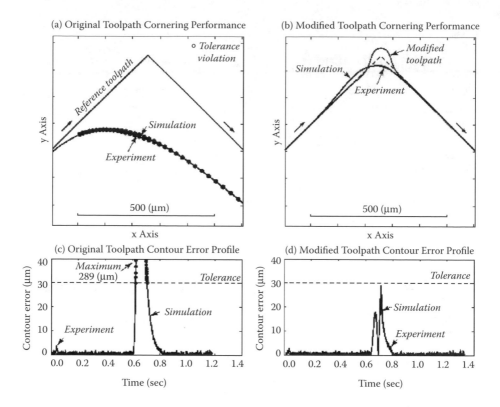

FIGURE 12.12 Contouring performance for 90° over-corner spline with P-PI servo control.

12.4 Conclusion

CNC machines are mechatronic systems. Control plays a crucial role in their operation. A comprehensive virtual model of a modular CNC was presented in this chapter. The VCNC allows integration of trajectory planning and interpolation routines, mathematical models of ball-screw and linear drives, friction, feedback sensors, amplifiers, D/A converters, and flexible motion control laws. The system allows the designer to try out various feed-drive design alternatives, control laws, and sensors with different resolutions. Accurate modeling of drives allows realistic prediction of the machine's high-speed contouring capability. Auto-tuning of sophisticated axis control laws allows tracking performance to be optimized prior to implementation on the real machine. Modification of sharp corners with smooth spline segments allows desired contouring tolerances to be maintained while traveling sharp corners in minimum time. These features have all been verified experimentally on a three-axis machining center.

Acknowledgment

This research is sponsored by NSERC and Pratt & Whitney Canada under research chair and strategic grant agreements.

References

1. Altintas, Y., *Manufacturing Automation: Metal Cutting Mechanics, Machine Tool Vibrations, and CNC Design*, Cambridge University Press, Cambridge, England, 2000.
2. Erkorkmaz, K. and Altintas, Y., High speed CNC system design: Part I—Jerk Limited Trajectory Generation and Quintic Spline Interpolation, *International Journal Machine Tools and Manufacture*, Vol. 41, No. 9, 1323–1345, 2001.

3. Altintas, Y., Erkorkmaz, K., and Zhu, W.-H., Sliding mode controller design for high speed drives, *Annals of CIRP*, Vol. 49, No. 1, 265–270, 2000.

4. Boucher, P., Dumur, D., and Rahmani, K.F., Generalized predictive cascade control (GPCC) for machine tool drives, *Annals of CIRP*, Vol. 39, No. 1, 357–360, 1990.

5. Erkorkmaz, K. and Altintas, Y., High speed CNC system design: Part III—high speed tracking and contouring control of feed drives, *International Journal of Machine Tools and Manufacture*, Vol. 41, No. 11, 1637–1658, 2001.

6. Pritschow, G., On the influence of the velocity gain factor on the path deviation, *Annals of the CIRP*, Vol. 45, No. 1, 367–371, 1996.

7. Tomizuka, M., Zero phase error tracking algorithm for digital control, *ASME Journal of Dynamic Systems, Measurement, and Control*, Vol. 109, 65–68, 1987.

8. Armstrong, H.B., Dupont, P., and Canudas De Wit, C., A survey of models, analysis tools and compensation methods for the control of machines with friction, *Automatica*, Vol. 30, No. 7, 1083–1138, 1994.

9. Erkorkmaz, K. and Altintas, Y., High speed CNC system design: Part II—modeling and identification of feed drives, *International Journal of Machine Tools and Manufacture*, Vol. 41, No. 10, 1487–1509, 2001.

10. Erkorkmaz, K., Optimal Trajectory Generation and Precision Tracking Control for Multi-Axis Machines, Ph.D. thesis, University of British Columbia, Department of Mechanical Engineering, Vancouver, BC, Canada, 2004.

11. Kao, J.Y., Yeh, Z.M., Tarng, Y.S., and Lin, Y.S., A study of backlash on the motion accuracy of CNC lathes, *International Journal of Machine Tools and Manufacture*, Vol. 36, No. 5, 539–550, 1996.

12. de Silva, C.W., *Intelligent Control–Fuzzy Logic Applications*, CRC Press, Boca Raton, FL, 1995.

13. Goulet, J.-F., de Silva, C.W., and Modi, V.J., Hierarchical knowledge-based control of a deployable orbiting manipulator, *Acta Astronautica*, Vol. 50, No. 3, 139–148, 2002.

14. Weck, M., Meylahn, A., and Hardebusch, C., Innovative algorithms for spline-based CNC controller, *Production Engineering Research and Development in Germany; Annals of the German Academic Society for Production Engineering*, Vol. 6, No. 1, 83–86, 1999.

15. Bobrow, J.E., Dubowsky, S., and Gibson, J.S., Time-optimal control of robotic manipulators along specified paths, *International Journal of Robotics Research*, Vol. 4, No. 3, 3–17, 1985.

16. Butler, J., Haack, B., and Tomizuka, M., Reference generation for high speed coordinated motion of a two axis system, *Symposium on Robotics, ASME Winter Annual Meeting*, Chicago, IL, DSC-11, 1988, pp. 457–470.

17. Erkorkmaz, K. and Altintas, Y., High speed contouring control algorithm for CNC machine tools, *Proceedings of ASME Dynamic Systems and Control Division*, IMECE'98, DSC-64, 1998, pp. 463–469.

18. Jouaneh, M.K., Wang, Z., and Dornfeld, D.A., Trajectory planning for coordinated motion of a robot and a positioning table: Part 1—path specification, *IEEE Transactions on Robotics and Automation*, Vol. 6, No. 6, 735–745, 1990.

19. Jouaneh, M.K., Dornfeld, D.A., and Tomizuka, M., Trajectory planning for coordinated motion of a robot and a positioning table: Part 2—optimal trajectory specification, *IEEE Transactions on Robotics and Automation*, Vol. 6, No. 6, 746–759, 1990.

20. Weck, M. and Ye, G., Sharp corner tracking using the IKF control strategy, *Annals of the CIRP*, Vol. 39, No. 1, 437–441, 1990.

21. Wang, F.-C. and Yang, D.C.H., Nearly arc-length parameterized quintic-spline interpolation for precision machining, *Computer Aided Design*, Vol. 25, No. 5, 281–288, 1993.

22. Erkorkmaz, K. and Altintas, Y., Quintic spline interpolation with minimal feed fluctuation, *ASME Journal of Manufacturing Science and Engineering*, Vol. 127, No. 2, 339–349, 2005.

23. Altintas, Y. and Erol, N.A., Open architecture modular tool kit for motion and machining process control, *Annals of the CIRP*, Vol. 47, No. 1, 295–300, 1998.

24. Yeung, C.H., A Three-Axis Virtual Computer Numerical Control (CNC) System, M.A.Sc. thesis, University of British Columbia, Department of Mechanical Engineering, Vancouver, BC, Canada, 2004.

25. Astrom, K.J. and Wittenmark, B., *Computer-Controlled Systems: Theory and Design*, 3rd ed., Prentice-Hall, Upper Saddle River, NJ, 1997.

26. Yeung, C.-H., Altintas, Y., and Erkorkmaz, K., Virtual CNC system—Part I: system architecture, *International Journal of Machine Tools and Manufacture*, Vol. 46, No. 10, 1107–1123, 2006.

27. Erkorkmaz, K., Yeung, C.-H., and Altintas, Y., Virtual CNC system—Part II: high speed contouring application, *International Journal of Machine Tools and Manufacture*, Vol. 46, No. 10, 1124–1138, 2006.

13

Model Predictive Control of a Flexible Robot

T. Fan

C.W. de Silva

Summary

Robots are mechatronic devices whose control is directly related to proper operation. Structural flexibility adds an extra complexity to the underlying control problem. In this chapter, we present a model predictive control (MPC) strategy for motion control of a flexible-link robotic manipulator. In Section 13.2, the dynamic model for a prototype flexible-link manipulator system (FLMS) is developed. Section 13.3 presents the new MPC algorithm, its underlying strategy, and issues related to the application of MPC in the motion control of flexible-link robot manipulators. Section 13.4 describes the experimental test bed of a flexible-link robot manipulator designed and developed in our laboratory. This robot is employed to implement and investigate the MPC scheme developed here. Section 13.5 gives the simulation results for a single flexible-link robot manipulator under control of the MPC scheme. Concluding remarks are given in Section 13.6.

13.1 Introduction

A robotic manipulator is a mechatronic system [1] usually consisting of a mechanical subsystem with components such as links, joints, and end-effector; an actuation subsystem formed by such components as DC motors and PWM amplifiers; a measurement subsystem constituting such sensors as encoders; a data acquisition subsystem, for example, a motion control interface card; a control subsystem, which has motion controllers, and so on. Structural flexibility in links and joints of a robot complicates the control problem associated with this mechatronic system. In this chapter, we focus on the dynamic modeling and controller design aspects of a flexible robotic manipulator system.

Robot manipulators have a wide range of applications, from industrial automation and medical operations to exploring hazardous environments such as space, underwater, and nuclear power plants. In all these applications, completion of a generic task requires accurate control of the movements of the end-effectors of the utilized manipulators. In general, the control of robotic manipulators consists of motion control and contact force control. For unconstrained (free) motion, there is no physical interaction between the end-effector and the environment; hence, only the motion control is required. For constrained motion (for example, if contact forces arise between the end-effector and the environment), both motion and contact forces need to be controlled. Only the unconstrained motion control of the robotic manipulator is addressed in this chapter.

Structural flexibility refers to the deflection of a structure under applied or inertial (acceleration) forces/torques. Here, the structural flexibility is addressed from the control point of view. A robotic manipulator is considered flexible if the flexural effects are so significant that they cannot be neglected during the controller design stage for the system to meet the performance specifications. Flexibilities in a robotic manipulator may result from joint flexibilities [2] and link flexibilities. Joint flexibility arises primarily because of the elastic behavior of the joint transmission elements such as gears (e.g., harmonic drives [1]) and shafts of the actuators. Link flexibility is a consequence of lightweight constructional features of large-dimension manipulators that are designed to operate at high speed with low inertia or handle heavy payloads.

Research on the dynamic modeling and control of flexible manipulators has received increased attention due to the advantages of flexible manipulators over rigid ones [3], which include higher payload to robot weight ratio and low energy consumption. Achieving accurate, high-speed manipulation with a lightweight structure is clearly a desirable objective in robotic tasks. Due to the flexible nature of a robotic system, the dynamics can be highly nonlinear, coupled, and complex. Additional deflection variables have to be included as extra degrees of freedom [4] to account for the effects of the link deflections on the kinematics, dynamics, and control of such robots. Several factors contribute to the complexity and difficulty of the associated control problem. The model complexity, uncertainties in modeling, dynamic interactions, nonlinear dynamics, and non-minimum-phase characteristics are among the major factors.

The motion control algorithms for conventional robotic manipulators generally are based on the rigid model of the system, the flexibility of the system being considered negligible. There are two primary solutions used by conventional robotic manipulators to avoid the flexibility problems of the system. The first is to make the manipulator rigid by increasing the stiffness of the system. This will reduce the system vibration, and good positional accuracy can be achieved. High stiffness usually is achieved by using heavy and bulky structural components. Most designs of conventional industrial robotic manipulators are based on this concept. This will reduce the efficiency and operating speed/bandwidth, increase the cost and energy consumption, and limit the performance of the manipulator. The second solution is to reduce the speed of the manipulator. By moving slowly enough, it may be possible to ensure that the flexible modes of the system are not excited; then the system can be controlled as a rigid system. This can lead to slow performance and extend the task completion time beyond acceptable limits. To overcome such performance limitations of existing robotic systems, it is important to directly address the flexibility issue of robotic manipulators. With the rapid advances in hardware and software, the implementations of high-performance, advanced motion control methods have become possible even for complex plants and mechatronic systems. It is expected that the MPC method developed in the present work will become an important step in this direction.

The development of the present MPC algorithm concerns two major aspects: dynamic model development and real-time MPC controller design. The nonlinear dynamic model of the system is derived from the Euler–Lagrange equations of motion. More realistic boundary conditions that represent the balance of moments and shear forces at the end of each link are used in the dynamic model of the system. Local linearization is used to derive a linear model of the system at an operating point. The model is relinearized for large operating-point changes. The parameters of the linear model are identified online using a recursive least squares (RLS) technique. The MPC controller design is based on this linear model. A computationally efficient MPC algorithm is developed here to facilitate real-time implementation of the overall adaptive scheme. The input constraints are not managed through optimization but by a local anti-windup scheme. This will reduce the optimization of a quadratic programming (QP) problem to a simple least-squares (LS) problem. The computer simulations are given, which show that the anti-windup scheme provides quite similar performance to a constrained QP MPC. The terminal constraints with weighting factors are included in the cost function. This ensures the nominal stability of the MPC controller based on Lyapunov's second method. Physical implementation of the developed MPC algorithm in a prototype FLMS is explored. The performance of the MPC scheme is evaluated using computer simulations of the prototype FLMS. The results show that the MPC can effectively control the motion of a flexible-link robot manipulator.

13.2 Dynamic Modeling

In this section, a model for the flexible-link robotic manipulator, which is used for simulation and controller design, is developed. First, a flexible-link model is developed based on the Euler–Bernoulli beam theory. Then, the dynamic model of the manipulator is derived using Euler–Lagrange equations.

13.2.1 Flexible-Link Model

For the flexible link we use the Euler–Bernoulli model, in which rotary inertia and shear deformation effects are ignored [4]. Consider the flexible link shown in Figure 13.1, which undergoes transverse vibrations.

The displacement of the segment with respect to the x-axis is $w(x,t)$. Also, $V(x,t)$ is the shear force and $M(x,t)$ is the bending moment. The link mass per unit length is ρ, and the flexural rigidity of the link is EI, which are assumed constant along the length of the link. The force balance (D'Alembert's principle) and moment balance (in the absence of rotary inertia) of the segment of the link lead to the Euler–Bernoulli beam Equation [4]:

$$EI\frac{\partial^4 w(x,t)}{\partial x^4} + \rho\frac{\partial^2 w(x,t)}{\partial t^2} = 0 \tag{13.1}$$

The solution of the Euler–Bernoulli beam equation can be found using the method of separation variables. Separating the shape and time functions according to $w(x,t) = \phi(x)\delta(t)$, the beam Equation (13.1) becomes

$$\frac{EI}{\rho\phi(x)}\frac{d^4\phi(x)}{dx^4} = \frac{-1}{\delta(t)}\frac{d^2\delta(t)}{dt^2} = \omega^2 \tag{13.2}$$

where ω has to be a constant because it is equal to both a spatial function and a temporal function. The solution of the temporal function of Equation (13.2) is

$$\delta(t) = A\cos(\omega t) + B\sin(\omega t) \tag{13.3}$$

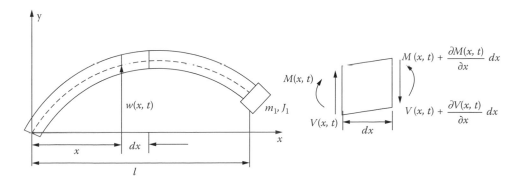

FIGURE 13.1 A flexible link in bending.

where A and B are constants determined by initial conditions. The solution of the spatial function of equation (13.2) is

$$\phi(x) = C_1 \cosh(\beta x) + C_2 \sinh(\beta x) + C_3 \cos(\beta x) + C_4 \sin(\beta x) \tag{13.4}$$

where C_1, C_2, C_3, and C_4 are constants, and $\beta^4 = \frac{\rho \omega^2}{EI}$. The function $\phi(x)$ is the mode shape function (eigenfunction) of the beam, and ω is the natural frequency (undamped) of vibration. The unknown constants C_1 through C_4, and β in Equation (13.4) can be determined from the boundary conditions of the beam. The natural frequencies are expressed as

$$\omega = \beta^2 \sqrt{\frac{EI}{\rho}} \tag{13.5}$$

It is usually assumed that the boundary conditions for a flexible manipulator link are clamped-free; i.e., the flexible link is fixed (clamped) at the $x = 0$ end, having zero deflection and no change in slope, and is free of shear forces or bending moments at the $x = l$ end. In the present work, however, we use more accurate boundary conditions representing the balance of bending moment and shear force at $x = l$ of the flexible link (where there is a clamped mass). Specifically, the following boundary conditions are used to obtain the mode shape functions:

At $x = 0$,

$$w(0,t) = 0, \quad \frac{\partial w(0,t)}{\partial x} = 0$$

At $x = l$,

Bending movement:

$$EI \frac{\partial^2 w(x,t)}{\partial x^2}\bigg|_{x=l} = -J_L \frac{d^2}{dt^2}\left(\frac{\partial w(x,t)}{\partial x}\bigg|_{x=l}\right)$$

Shear force:

$$EI \frac{\partial^3 w(x,t)}{\partial x^3}\bigg|_{x=l} = m_L \frac{d^2}{dt^2}(w(x,t)|_{x=l})$$

where J_L and m_L are the equivalent moment of inertia and mass at the end of the link, respectively. These boundary conditions follow from the fact that J_L and m_L generate a concentrated inertia force $m_L \frac{\partial^2 w(x,t)}{\partial t^2}$ and an inertial rotary (bending) moment $-J_L \frac{d^2}{dt^2}(\frac{\partial w(x,t)}{\partial x})$ at $x = l$. Application of these boundary conditions to Equation (13.4) yields the frequency equation:

$$1 + \cos(\beta l)\cosh(\beta l) - \frac{m_L \beta}{\rho}(\sin(\beta l)\cosh(\beta l) - \cos(\beta l)\sinh(\beta l))$$

$$-\frac{J_L \beta^3}{\rho}(\sin(\beta l)\cosh(\beta l) + \cos(\beta l)\sinh(\beta l)) + \frac{m_L J_L \beta^4}{\rho^2}(1 - \cos(\beta l)\cosh(\beta l)) = 0 \tag{13.6}$$

This frequency equation has an infinite number of solutions, $\{\beta_i, i = 1, \cdots, \infty\}$. Each solution β_i is related to one natural frequency ω_i through Equation (13.5). The corresponding mode shape function for the i-th mode is

$$\phi_i(x) = a_i[\cosh(\beta l) - \cos(\beta l) - b_i(\sinh(\beta l) - \sin(\beta l))] \qquad (13.7)$$

where

$$b_i = \frac{\cosh(\beta l) + \cos(\beta l) - \dfrac{J_L \beta^3}{\rho}(\sinh(\beta l) + \sin(\beta l))}{\sinh(\beta l) + \sin(\beta l) - \dfrac{J_L \beta^3}{\rho}(\cosh(\beta l) - \cos(\beta l))} \qquad (13.8)$$

The constant a_i normalizes the mode shape functions, such that

$$\int_0^l \phi_i^2(x)dx = l \qquad (13.9)$$

The orthogonal condition for the clamped-mass beam is

$$\int_0^l \phi_i(x)\phi_j(x)\rho dx + m_p\phi_i(l)\phi_j(l) + J_p\phi_i'(l)\phi_j'(l) = 0 \quad (i \neq j) \qquad (13.10)$$

These mode shape functions are used as admissible functions for the method of assumed modes, which is employed in the derivation of the equations of motion in Subsection 13.2.2. The overall response of the clamped-mass beam is given by the modal summation

$$w(x,t) = \sum_{i=1}^{\infty} \phi_i(x)\delta_i(t) \qquad (13.11)$$

13.2.2 Derivation of Equations of Motion

Consider a planar n-link flexible manipulator with revolute joints. Figure 13.2 shows a two-link example. The links are subjected to bending deformations only in the plane of motion, and torsional effects are neglected. A link is modeled as an elastic beam that is clamped to a rotary actuator at the joint end and having a mass with rotary inertia at the other end (clamped mass-boundary conditions). A finite-dimensional model of link flexibility can be obtained by the assumed-modes technique [4]. The deflection of each link is described by a weighted summation of the mode shape functions; thus

$$w_i(x_i,t) = \sum_{j=1}^{m_i} \phi_{ij}(x_i)\delta_{ij}(t) \qquad (13.12)$$

Here, x_i denotes the location along the neutral axis of the link i, ϕ_{ij} is the clamped-mass mode shape function presented in Subsection 13.2.1, and δ_{ij} is the modal coordinate. The geometry of the system is described in terms of the coordinates shown in Figure 13.2.

Frame F^0 is the fixed, world coordinate frame with the joint of link 1 located at its origin, frame F^1 is fixed at the joint end of link 1, and frame F^2 is fixed at the joint end of link 2. The geometric relations

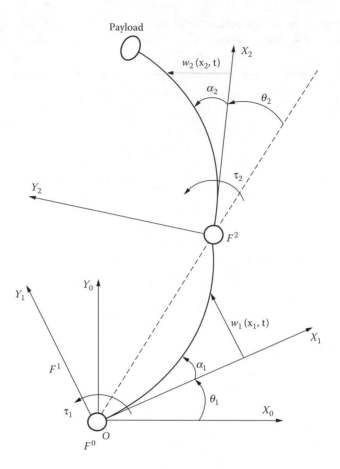

FIGURE 13.2   Schematic diagram of the flexible manipulator model.

among the coordinate frames can be described using homogeneous transformation matrices [5]. The transformation between frames F^1 and F^0 is

$$\mathbf{H}_{0,1}(t) = \begin{bmatrix} c_1 & -s_1 & 0 & 0 \\ s_1 & c_1 & 0 & 0 \\ 0 & 0 & 1 & 0 \\ 0 & 0 & 0 & 1 \end{bmatrix} \tag{13.13}$$

where $s_1 = \sin\theta_1(t)$ and $c_1 = \cos\theta_1(t)$. The rigid body components of rotation and translation between frames F^2 and F^1 is

$$\hat{\mathbf{H}}_{1,2}(t) = \begin{bmatrix} c_2 & -s_2 & 0 & 0 \\ s_2 & c_2 & 0 & 0 \\ 0 & 0 & 1 & 0 \\ 0 & 0 & 0 & 1 \end{bmatrix} \tag{13.14}$$

where $s_2 = \sin\theta_2(t)$ and $c_2 = \cos\theta_2(t)$. Additional rotation and translation due to the structural deformation of link 1 must be considered. Based on ideas introduced by Book [1], the structural deformation of link 1 relative to the rigid link is described by the transformation matrix:

$$D_{1,2}(t) = \begin{bmatrix} c_{\alpha 1} & -s_{\alpha 1} & 0 & l_1 \\ s_{\alpha 1} & c_{\alpha 1} & 0 & w_1(l_1,t) \\ 0 & 0 & 1 & 0 \\ 0 & 0 & 0 & 1 \end{bmatrix} \tag{13.15}$$

where α_1 is the angle of rotation of link 1 due to bending, l_1 is the length of link 1, $s_{\alpha 1} = \sin\alpha_1(t)$, and $c_{\alpha 1} = \cos\alpha_1(t)$. The deflections of the flexible links are assumed to be small compared to their lengths, so the high-order terms of deformation can be neglected. The changes of link lengths due to deformation are also neglected. We have

$$\alpha_i = \left.\frac{\partial w_i(x_i,t)}{\partial x_i}\right|_{x_i=l_i} \tag{13.16}$$

The homogeneous transformation between frames F^2 and F^1 is

$$H_{1,2}(t) = D_{1,2}(t)\hat{H}_{1,2}(t) = \begin{bmatrix} c_{\alpha 1,2} & -s_{\alpha 1,2} & 0 & l_1 \\ s_{\alpha 1,2} & c_{\alpha 1,2} & 0 & w_1(l_1,t) \\ 0 & 0 & 1 & 0 \\ 0 & 0 & 0 & 1 \end{bmatrix} \tag{13.17}$$

where $s_{\alpha_1,2} = \sin(\alpha_1(t)+\theta_2(t))$ and $c_{\alpha_1,2} = \cos(\alpha_1(t)+\theta_2(t))$. The homogeneous transformation between frame F^2 and the world coordinate frame F^0 is

$$H_{0,2}(t) = H_{0,1}(t)H_{1,2}(t) = \begin{bmatrix} c_{1,\alpha 1,2} & -s_{1,\alpha 1,2} & 0 & c_1 l_1 - s_1 w_1(l_1,t) \\ s_{1,\alpha 1,2} & c_{1,\alpha 1,2} & 0 & s_1 l_1 + c_1 w_1(l_1,t) \\ 0 & 0 & 1 & 0 \\ 0 & 0 & 0 & 1 \end{bmatrix} \tag{13.18}$$

where $s_{1,\alpha_1,2} = \sin(\theta_1(t)+\alpha_1(t)+\theta_2(t))$ and $c_{1,\alpha_1,2} = \cos(\theta_1(t)+\alpha_1(t)+\theta_2(t))$.

The absolute position and linear velocity for any point on link 1 and link 2 can be calculated based on the preceding transformations. For a point $\mathbf{r}_1$ on link 1 in frame F^1,

$$\mathbf{r}_1^1(t) = [x_1 \quad w_1(x_1,t) \quad 0 \quad 1]^T \tag{13.19}$$

The absolute position of $\mathbf{r}_1$ in frame F^0 is

$$\mathbf{r}_1^0(t) = H_0^1(t)\mathbf{r}_1^1(t) = [c_1 x_1 - s_1 w_1(x_1,t), \quad s_1 x_1 + c_1 w_1(x_1,t), \quad 0 \quad 1]^T \tag{13.20}$$

The absolute velocity of this point in frame F^0 is

$$\mathbf{v}_1^0(t) = \frac{d\mathbf{r}_1^0(t)}{dt} = \begin{bmatrix} -[x_1 s_1 + c_1 w_1(x_1,t)]\theta_1 - w_1(x_1,t)s_1 \\ [x_1 c_1 - s_1 w_1(x_1,t)]\theta_1 + w_1(x_1,t)c_1 \\ 0 \\ 0 \end{bmatrix} \tag{13.21}$$

For a point $\mathbf{r}_2$ on link 2 in frame F^2,

$$\mathbf{r}_2^2(t) = \begin{bmatrix} x_2 & w_2(x_2,t) & 0 & 1 \end{bmatrix}^T \tag{13.22}$$

The absolute position of $\mathbf{r}_2$ in frame F^0 is

$$\mathbf{r}_2^0(t) = \mathbf{H}_0^2 \mathbf{r}_2^2(t) = \begin{bmatrix} l_1 c_1 - w_1(l_1,t)s_1 + x_2 c_{1,\alpha_1,2} - w_2(x_2,t)s_{1,\alpha_1,2} \\ l_1 s_1 + w_1(l_1,t)c_1 + x_2 s_{1,\alpha_1,2} + w_2(x_2,t)c_{1,\alpha_1,2} \\ 0 \\ 1 \end{bmatrix} \tag{13.23}$$

The absolute velocity of this point in frame F^0 is

$$\mathbf{v}_2^0(t) = \frac{d\mathbf{r}_2^0(t)}{dt} = \begin{bmatrix} -(l_1 s_1 + w_1(l_1,t)c_1)\theta_1 - w_1(l_1,t)s_1 - x_2 s_{1,\alpha_1,2}(\theta_1 + \alpha_1 + \theta_2) \\ -w_2(x_2,t)c_{1,\alpha_1,2}(\theta_1 + \alpha_1 + \theta_2) - w_2(x_2,t)s_{1,\alpha_1,2} \\ (l_1 c_1 - w_1(l_1,t)s_1)\theta_1 + w_1(l_1,t)c_1 + x_2 c_{1,\alpha_1,2}(\theta_1 + \alpha_1 + \theta_2) \\ -w_2(x_2,t)s_{1,\alpha_1,2}(\theta_1 + \alpha_1 + \theta_2) + w_2(x_2,t)c_{1,\alpha_1,2} \\ 0 \\ 0 \end{bmatrix} \tag{13.24}$$

Next, the position vectors, $\mathbf{r}_0^1(t)$ and $\mathbf{r}_0^2(t)$, and the velocity vectors, $\mathbf{v}_0^1(t)$ and $\mathbf{v}_0^2(t)$, are used to form the kinetic energy and potential energy for the flexible-link manipulator. The kinetic energy of the hub i is

$$T_{hi} = \frac{1}{2} m_{hi} (\mathbf{v}_i^0)^T \mathbf{v}_i^0 \Big|_{x_i=0} + \frac{1}{2} J_{hi} \theta_i^0$$

where m_{hi} is the mass, J_{hi} is the moment of inertia, and θ_i^0 is the absolute angular velocity of the hub. Also,

$$\theta_i^0 = \sum_{j=1}^{i} (\theta_j + \alpha_{j-1}) \tag{13.25}$$

The kinetic energy of link i is

$$T_{li} = \frac{1}{2} \int_0^{l_i} \rho_i(x_i)(\mathbf{v}_i^0)^T \mathbf{v}_i^0 dx_i \tag{13.26}$$

and the kinetic energy of a payload with mass m_p and moment of inertia J_p located at the end-effector is

$$T_p = \frac{1}{2} m_p \left(\dot{\mathbf{v}}_n^0 \right)^T \dot{\mathbf{v}}_n^0 \Big|_{x_n = l_n} + \frac{1}{2} J_p \left(\dot{\theta}_n^0 + \dot{\alpha}_n \right)^2 \tag{13.27}$$

The total kinetic energy of the system is

$$T = \sum_{i=1}^{n} T_{hi} + \sum_{i=1}^{n} T_{li} + T_p \tag{13.28}$$

The total potential energy of the system is given by

$$V = \sum_{i=1}^{n} \frac{1}{2} \int_0^{l_i} (EI)_i \left[\frac{\partial^2 w_i(x_i, t)}{\partial x_i^2} \right]^2 dx_i \tag{13.29}$$

Equations (13.28) and (13.29) are used to form the Lagrangian $L = T - V$ for the system. On the basis of the assumed-modes method, the dynamic model is obtained by satisfying the Lagrangian equations:

$$\frac{d}{dt} \frac{\partial L}{\partial \dot{q}_i} - \frac{\partial L}{\partial q_i} = f_i, \quad i = 1, \dots, n. \tag{13.30}$$

where $\{q_i(t)\}$ are generalized coordinates and $\{f_i(t)\}$ are the corresponding generalized forces. The dynamics of a planar n-link flexible manipulator can be obtained as

$$\mathbf{M}(\theta, \delta) \begin{bmatrix} \mathbf{q} \\ \delta \end{bmatrix} + \begin{bmatrix} \mathbf{h}_1(\theta, \delta, \theta, \delta) \\ \mathbf{h}_2(\theta, \delta, \theta, \delta) + \mathbf{K}\delta \end{bmatrix} = \begin{bmatrix} \mathbf{u} \\ 0 \end{bmatrix} \tag{13.31}$$

where θ is the $n \times 1$ vector of joint variables, δ is the $m \times 1$ vector of deflection variables, $\mathbf{h}_1$ and $\mathbf{h}_2$ are the terms due to Coriolis and centrifugal forces, and $\mathbf{K}$ is the stiffness matrix.

13.3 Model Predictive Control

In this section, the MPC method is presented and its use in the control of a flexible-link manipulator is described.

13.3.1 The Basic Concepts of MPC

MPC is a model-based control technique. It uses an explicit internal model to generate predictions of future plant behavior and hence the corresponding control inputs. The various MPC algorithms mainly differ in the model they use to represent the plant, noise representation, and the cost function to be minimized. They all share the essential features of predictive control: an explicit internal model (used to predict the plant output in a future time interval), determination of the control signal by optimizing the predicted plant behavior, and the concept of receding horizon (at each instant the horizon is moved forward into the future, and the first value of the control sequence calculated at each step is applied to the plant).

As noted, in the receding horizon framework, only the first computed control action is implemented. At the next sampling time, the optimization is carried out again with new measurements from the plant.

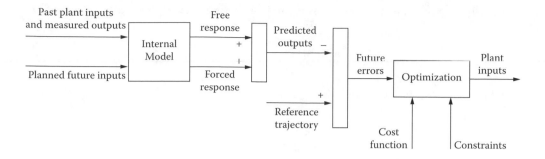

FIGURE 13.3 The basic structure of Model Predictive Control (MPC).

Thus, both the control horizon and the prediction horizon move or recede further by one step as time moves ahead by one step. The online optimization can be typically reduced to either a linear program or a quadratic program. The purpose of taking new measurements at each time step is to compensate for unmeasured disturbances and model inaccuracy, both of which cause the system output to be different from the one predicted by the model. This also introduces a feedback element into the controller. If the exact measurements of the states are not available, a state observer should be designed to estimate the states at each sampling time. The basic structure shown in figure 13.3 is used to implement the MPC. An internal model is used to predict the future plant outputs based on the past and current values of the inputs and outputs and on the proposed optimal future control actions. These actions are calculated by the optimizer while taking into account the cost functions and the constraints. It is seen that the plant model plays a decisive role in the controller. The chosen model must be capable of capturing the process dynamics so as to accurately predict the future outputs, while being sufficiently simple to implement and analyze.

13.3.2   Design of Online System Identifier

After determining the first principles of nonlinear model of the flexible manipulator, the next step is to design the online system identification module. Local linearization is used to obtain a linear model of the system. This linear model is an approximation of the nonlinear system in the neighborhood of an operating point. Because multilink robotic manipulators are highly nonlinear systems, one has to relinearize the model when the operating point has changed significantly. For example, the model may be relinearized at every $10°$ movement of the revolute joints in the prototype system. This will make the linear model sufficiently close to the nonlinear model. Also, if the manipulator is moving at high speed, high frequency modes may not be negligible. Then, because the assumed-modes method is used for the nonlinear model, the model structure will have to be modified accordingly. The parameters of this linear model will be identified online using the RLS technique. The MPC controller is designed based on this model. Because the model parameters are identified online, the MPC controller has an adaptive feature. It is expected to have a good stability performance under large payload changes, parameter uncertainties, and model errors.

13.3.3   Design of Model Predictive Controller

The MPC controller is designed online based on the linear model from the real-time system identifier. The reasons for choosing MPC for low-level direct control of the flexible manipulator are the following:

- It is suitable for multivariable control problems.
- It can take account of actuator limitations and systematically handle the constraints.
- The design objectives can be specified in a flexible manner.
- It can be used to control systems with time delays, nonminimum-phase characteristics, and instabilities.
- It takes full advantage of the power available in modern computer hardware and software.

The MPC controller developed here uses the following quadratic cost function:

$$J(H_p, H_u, \lambda) = \sum_{i=1}^{H_p} \left\| \hat{\mathbf{y}}(k+i\,|\,k) - \mathbf{r}(k+i\,|\,k) \right\|_Q^2 + \sum_{i=1}^{H_u} \left\| \Delta\hat{\mathbf{u}}(k+i-1\,|\,k) \right\|_R^2 + \lambda\hat{\mathbf{x}}(k+H_p\,|\,k) \qquad (13.32)$$

where H_p is the prediction horizon, H_u is the control horizon, $\mathbf{Q}$ and $\mathbf{R}$ are weighting matrices, and λ is the parameter for terminal constraint. The cost function J penalizes deviations of the predicted controlled output $\hat{y}(k+i\,|\,k)$ from a reference trajectory $r(k+i\,|\,k)$. The following input constraints are used:

$$\begin{aligned} |\mathbf{u}(t)| &\le \mathbf{u}_{max} &\quad \forall t \\ |\Delta\mathbf{u}(t)| &\le \Delta\mathbf{u}_{max} &\quad \forall t \end{aligned} \qquad (13.33)$$

Here, $\mathbf{u}_{max}$ is a vector that includes the maximum control inputs for each joint and $\Delta\mathbf{u}_{max}$ is a vector that includes the maximum rate of change (slew rate) of the control inputs for each joint. Because the internal model is linear, the QP optimization problem for MPC design is convex. The QP optimization has to be done at every time step to generate the control inputs. For fast moving robot manipulators, the efficiency of the QP solver is very important. In the absence of constraints, the optimal solution can be computed analytically [6].

To ensure real-time implementation of the overall adaptive scheme, a simple unconstrained MPC is used for control of the flexible-link robotic manipulator. Input constraints are not managed through optimization but by using a local anti-windup scheme. Figure 13.4 shows the structure of unconstrained MPC controller. Here, $T(k)$ is the reference trajectory, $\varepsilon(k)$ is the tracking error, $z(k)$ is the controlled output, $\mathbf{y}(k)$ is the measured output, $\mathbf{K}_{MPC}$ is the controller gain matrix, and ζ and Ψ are the internal model matrices. Input constraints are handled by a simple anti-windup scheme. Saturation constraints are enforced by simply "clipping" the manipulated variable changes so that they satisfy all the constraints. For the input constraints given in equation (13.33), the MPC algorithm computes an input move $\Delta\mathbf{u}(k)$ without taking any constraints into account. Then the input move that is actually applied to the plant is

$$\Delta^{\cdot}\mathbf{u}(k) = \begin{cases} \min(\Delta\mathbf{u}(k), \Delta\mathbf{u}_{max}(k), \mathbf{u}_{max} - \mathbf{u}^{\cdot}(k-1)) & \text{if } \Delta\mathbf{u}(k) > 0 \\ \max(\Delta\mathbf{u}(k), \Delta\mathbf{u}_{max}(k), \mathbf{u}_{max} - \mathbf{u}^{\cdot}(k-1)) & \text{if } \Delta\mathbf{u}(k) < 0 \end{cases} \qquad (13.34)$$

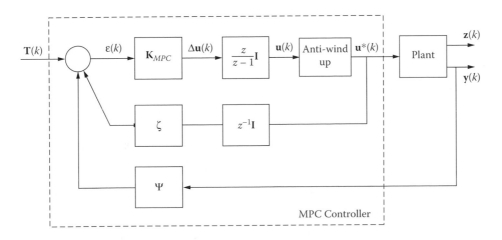

FIGURE 13.4 Unconstrained MPC controller structure.

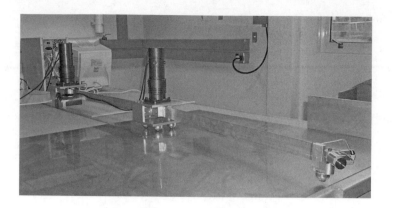

FIGURE 13.5 Prototype two-link flexible robot manipulator.

The same input move must be applied to the internal model. This not only helps to keep the predictions accurate, but also avoids problems analogous to "integrator windup." This approach will not give the true optimal solution in general, but it can provide very similar results. The main advantages of this control algorithm are its simplicity and computational efficiency.

Closed-loop stability is important for the MPC controller developed here. Based on Lyapunov's theorem, it may be shown that nominal stability can be achieved by imposing terminal constraints [6]. Based on the idea of using a terminal constraint set to guarantee closed-loop stability, a terminal constraint, $\lambda \hat{x}(k + H_p|k)$, is included in the cost function as given in equation (13.32), and a multistage MPC algorithm is developed as follows:

- Define a terminal constraint set $\lambda \hat{x}(k + H_p \,|\, k)$ which contains the origin, rather than a single point.
- First stage: choose the initial terminal constraint set such that MPC can drive the state into this set with $\lambda \ll 1$.
- Gradually increase λ until MPC achieves guaranteed system stability for initial conditions within the set with $\lambda \gg 1$, assuring that state approaches the origin.

13.4 Experimental Test Bed

A prototype flexible-link robot manipulator system has been designed and developed in our laboratory at the University of British Columbia. This robotic system forms the test bed for the simulation and experimental studies of the MPC scheme developed in the present work. A view of the prototype manipulator is shown in Figure 13.5. This planar manipulator is composed of two revolute joints having a vertical axis of rotation, two tip-mounted load cells, and an ultrasound tip displacement sensor. Two harmonic-drive DC servomotors form the actuators. A graphical user interface software package with pull-down menu and dialog boxes is developed using Visual C++'s Microsoft Foundation Class (MFC).

13.5 Simulation and Results

The prototype two-link flexible robot manipulator shown in Figure 13.5 has two flexible links; however, only the outboard link is simulated here. A dynamic model of the outboard link has been developed, which is essentially a single flexible-link manipulator. The MPC controller is designed for this manipulator, and the performance of the controller is evaluated using computer simulations. By modeling the link with a single structural mode and neglecting damping and gravity, the equations of motion are obtained as

$$I_{01}\theta(t) + I_{111}\delta(t) = \tau(t)$$

$$I_{111}\theta(t) + I_{211}\delta(t) + I_{311}\delta(t) = 0$$

$$(13.35)$$

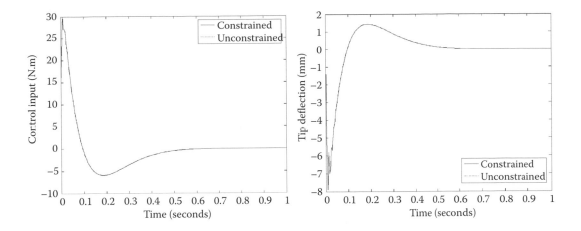

FIGURE 13.6 Case 1: (a) control inputs and (b) tip position responses.

where I_{01} is the moment of inertia about the joint axis, and I_{111}, I_{211}, and I_{311} are mode shape function integrals. The parameter values used are $I_{01} = 0.2817\ kg.m^2$, $I_{111} = 0.4807\ kg.m^2$, $I_{211} = 0.8451\ kg.m^2$, $I_{311} = 7096\ N.m^2$.

The mode shape function value at the tip is $\phi_1(l) = 2\ m$. Control torque τ is the input to the system, and system outputs are the joint angle θ and the tip deflection $w(l,t)$. The joint angle is measured using an optical encoder, and the structural defection of the tip is measured by a tip position sensor. The tip position is given by

$$\theta_{tip}(t) = \theta(t) + \frac{\phi_1(l)\delta(t)}{l} = \theta(t) + 2\delta(t) \tag{13.36}$$

The MPC design is based on this model. The tuning parameters are prediction horizon, control horizon, tracking error weights for joint angle and tip deflection, and the control move penalty weight. Figure 13.6 shows the system response in case 1 using MPC with the following parameters: Prediction horizon = 100; control horizon = 10; output weighting for the joint angle is 2 and for the tip deflection, 300; and weighting on the control input is 0.1. Assume the input constraint to be $|\tau_{max}| \le 35$ N.m. In this case, the input constraint is inactive. The simulation results verify that the constrained and unconstrained solutions are the same when the constraints are "loose," i.e., inactive. Figure 13.7 shows the system

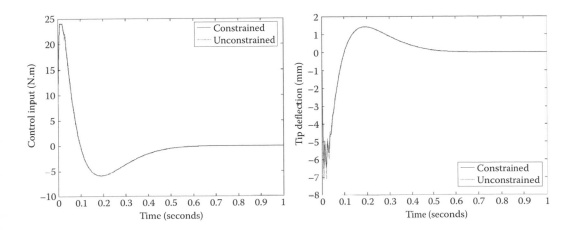

FIGURE 13.7 Case 2: (a) control inputs and (b) tip position responses.

response of case 2. In case 2, we use the same tuning parameters as in case 1. The only difference is the input constraint: $|\tau_{max}| \leq 24$ N.m. In this case, the input constraint is active. The simulation results show that the unconstrained anti-windup MPC control technique developed here is quite effective in controlling the tip position of a flexible-link manipulator.

13.6   Conclusion

Robots are mechatronic systems, and their control is important for proper operation. Structural flexibility in its links can complicate the control problem of a robot. In this chapter a model-based predictive control strategy was developed for the end point motion control of a flexible-link robot manipulator. Model development and analysis, design of a model predictive controller, computer simulation, and experimental investigation formed the focus of the work presented in the chapter. Simulation results demonstrated the effectiveness of the developed MPC method compared to conventional QP MPC. The approach may be extended to other mechatronic systems with structural flexibility.

References

1. De Silva, C.W., *Mechatronics — An Integrated Approach*, CRC Press, Taylor and Francis, Boca Raton, FL, 2005.
2. Spong, M.W., Modeling and control of elastic joint robots, *ASME Journal of Dynamic Systems, Measurement, and Control*, Vol. 109, No. 3, 310–319, 1987.
3. Book, W.J., Recursive Lagrangian dynamics of flexible manipulator arms, *International Journal of Robotics Research*, Vol. 3, No. 3, 87–101, 1984.
4. De Silva, C.W., *Vibration — Fundamentals and Practice*, 2nd ed., CRC Press, Taylor and Francis, Boca Raton, FL, 2006.
5. Spong, M.W. and Vidyasagar, M., *Robot Dynamics and Control*, John Wiley & Sons, New York, 1989.
6. Maciejowski, J.M., *Predictive Control: With Constraints*, Pearson Education Limited, Harlow, England, 2002.

14

Computer Vision in Multi-Robot Cooperative Control

Y. Wang

C.W. de Silva

Summary

Mobile robots are sophisticated mechatronic systems. Sensing, communication, and control technologies of mechatronic systems are applicable in multi-robot cooperative tasks. In this chapter, a fast computer vision algorithm is presented for a multi-robot cooperative control system. This algorithm identifies the current poses (positions and orientations) of a robot and the manipulated object (a rectangle box) from a color image, in real time. Two main challenges are faced in the multi-robot task considered here. The first one concerns the response speed of the vision subsystem. Ordinary computer vision algorithms available in the literature usually are rather slow and unacceptable in the present real-time system. The second challenge comes from uneven lighting, which makes it very difficult for the vision subsystem to trace the same color blob in different positions. A fast computer vision algorithm is presented to cope with these challenges. First, an image in the RGB (red—green—blue) color space is converted into the HSI (hue—saturation—intensity) color space. Then the saturation and the intensity components of the image are removed, and only the hue component is retained. This operation effectively removes the disturbances caused by uneven lighting conditions in the environment when the robot and the object move from one position to another. Second, filtering and template matching technologies are employed to remove the disturbances from the background and other objects in the image. Finally, coordinate transformations are used to reconstruct the poses of the robot and the object when they are moving. The experiment results are presented to show the feasibility and the effectiveness of the algorithm.

14.1 Introduction

Multi-robot systems are complex mechatronic systems. They have become a promising area of research in robotics and mechatronics [1,7]. In such a system, several autonomous robots cooperatively work to complete a common task. Because of its robustness, flexible configuration, potential high efficiency and low cost, and close similarity to human-society intelligence, multi-robot systems have attracted researchers in the robotics and mechatronics communities.

Sensing, communication, and control technologies of mechatronic systems are applicable in multi-robot cooperative tasks. For example, in a multi-robot system, it is important for the individual robots to know the latest poses (positions/orientations) of other robots and potential obstacles in the environment so as to make rational decisions. There are many approaches to measure the poses of robots and obstacles; for example, the use of sonar or laser distance finders. However, most multi-robot systems employ digital cameras to capture the poses of the robots and obstacles. The main reasons are as follows: (1) a digital image provides a rich source of information on multiple moving objects (robots and obstacles) in the environment at the same time; (2) the vast progress in computer vision research in recent years has made it possible to build fast and accurate vision subsystems at low cost; and (3) the use of cameras to observe the world is a ìnaturalî method for robots to understand their environment, similar to how humans use their eyes to observe the world.

Some work has been done to monitor multiple moving objects in a dynamic environment by using a computer vision system. Stone and Veloso [2] studied a multi-robot soccer system. In their project, they used a global camera to monitor the positions of the robots and objects in the game. Veeraraghavan et al. [3] developed a computer vision algorithm to track the vehicle motion at a traffic intersection. In their work, a multi-level approach using a Kalman filter was presented for tracking the vehicles and pedestrians at an intersection. The approach combined low-level image-based blob tracking with high-level Kalman filtering for position and shape estimation. Maurin et al. [4] presented a vision-based system for monitoring crowded urban scenes. Their approach combined an effective detection scheme based on optical flow and background removal that could monitor many moving objects simultaneously. Kalman filtering integrated with statistical methods were used in their approach. Chen et al. [5] presented a framework for spatiotemporal vehicle tracking using unsupervised-learning-based segmentation and object tracking. In their work, a method of adaptive background learning and subtraction was applied to two real-life traffic video sequences for obtaining accurate spatiotemporal information on vehicle objects.

In this chapter, a fast computer vision algorithm, used in a multi-robot transportation system developed by us, is presented. In the next section, some background knowledge of color image processing is introduced, which is followed by Section 14.3, where the multi-robot transportation system in our laboratory is described. In Section 14.4, the developed computer vision algorithm and the result obtained through that experiment are presented.

14.2   Color Image Processing

Today, color charge-coupled device (CCD) cameras have been widely employed in robotic vision systems because of their relatively low cost. Color images contain more information than gray images, and the color information is useful when multiple objects have to be distinguished and tracked. In this section, some background knowledge of color image processing is introduced.

14.2.1   Color Models

A color model is employed to describe and specify colors. Many color models are available. However, the most widely used models are RGB, CMY (cyan, magenta, and yellow), and HSI.

In the RGB model, a color is generated by combining three primary colors: red, green, and blue. Figure 14.1 presents this model, where red, green, and blue are the axes of a 3-D Cartesian coordinate system. In Figure 14.1, all values of R, G, and B are assumed to be in the range of 0–1. The color black is at the origin of the coordinate frame, and the color white is at locations 1,1, and 1. The gray scale varies along the line joining the points of black and white. All colors in this model are located in the color cube in Figure 14.1.

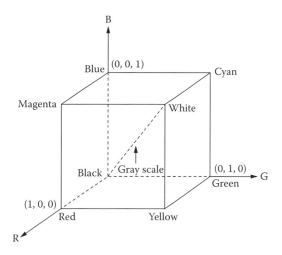

FIGURE 14.1 The RGB color model. (From Gonzalez, R.C. and Woods, R.E., *Digital Image Processing*, Prentice Hall, Upper Saddle River, NJ, 2002. With permission.)

The RGB model is adopted in most color cameras and monitors. However, in the field of color printers and photocopiers, another color model, the CMK color model, is widely used. The CMK model is similar to the RGB model except that it uses three secondary colors of light as its coordinate axes: cyan, magenta, and yellow. Most color printers and photocopiers require CMY data input. The conversion of color values from the RGB model to the CMY model is given by Gonzalez and Woods [6]:

$$\begin{bmatrix} C \\ M \\ Y \end{bmatrix} = \begin{bmatrix} 1 \\ 1 \\ 1 \end{bmatrix} - \begin{bmatrix} R \\ G \\ B \end{bmatrix} \tag{14.1}$$

However, both RGB and CMY models have the same disadvantage: Although they are good tools to generate colors, they are not good in describing colors. For example, when we observe a color, we usually do not describe it with its percentage of red, green, and blue. That is, the RGB and CMY models do not use ways similar to humans to describe colors. The most important advantage of the HSI model is that it describes and interprets colors in a way that corresponds closely with how humans see color. In addition, the HSI model decouples the gray information and the color-carrying information in a same image so that most gray-based image processing approaches can be employed in processing a color image.

In the HSI color model, hue is an attribute that describes a pure color such as red, yellow, or purple, whereas saturation represents the amount of white light added to a pure color. In addition, intensity represents the brightness of a color. When a human observes a color, he or she usually describes and interprets it with its hue, saturation, and brightness. Therefore, the HSI model is a good tool to describe colors in a way similar to how humans describe color. In a mechatronic computer-vision application, it is necessary to convert color values between the HSI model and the RGB model. The following equations indicate how to convert colors from RGB to HSI [6]:

$$H = \begin{cases} \theta & \text{if } B \leq G \\ 360 - \theta & \text{if } B > G \end{cases} \tag{14.2}$$

with

$$\theta = \cos^{-1}\left\{\frac{0.5[(R-G)+(R-B)]}{[(R-G)^2+(R-B)(G-B)]^{1/2}}\right\}$$

$$S = 1 - \frac{3}{(R+G+B)}[\min(R,G,B)] \tag{14.3}$$

$$I = \frac{1}{3}(R + G + B) \tag{14.4}$$

The following equations indicate how to convert colors from HSI to RGB [6]:
If $(0° \le H < 120°)$:

$$B = I(1-S) \tag{14.5}$$

$$R = I\left(1 + \frac{S\cos H}{\cos(60° - H)}\right) \tag{14.6}$$

$$G = 3I - (R+B) \tag{14.7}$$

If $(120° \le H < 240°)$:

$$H = H - 120° \tag{14.8}$$

$$R = I(1-S) \tag{14.9}$$

$$G = I\left[1 + \frac{S\cos H}{\cos(60° - H)}\right] \tag{14.10}$$

$$B = 3I - (R+G) \tag{14.11}$$

If $(240° \le H \le 360°)$:

$$H = H - 240° \tag{14.12}$$

$$G = I(1-S) \tag{14.13}$$

$$B = I\left[1 + \frac{S\cos H}{\cos(60° - H)}\right] \tag{14.14}$$

$$R = 3I - (G+B) \tag{14.15}$$

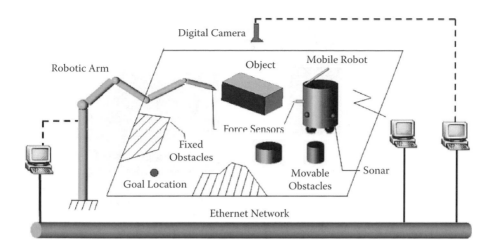

FIGURE 14.2 A schematic representation of the first version of the developed system.

14.3 Multi-Robot Transportation System

The objective of our multi-robot transportation project in the Industrial Automation Laboratory is to develop a physical mechatronic system where a group of intelligent robots work cooperatively to transport an object to a goal location and orientation in an unknown dynamic environment. Obstacles may be present and even appear randomly during the transportation process. Robot control, multi-agent technologies, and machine learning are integrated into the developed physical platform to cope with the main challenges of the problem. A schematic representation of the first version of the developed system is shown in Figure 14.2. The latest system in the laboratory uses state-of-the-art mobile robots together with fixed-base articulated robots (one state-of-the-art commercial robot and two prototypes developed in our laboratory).

In this first prototype, there is one fixed robot and one mobile robot, which cooperatively transport an object to a goal location. During the course of the transportation, the robots have to negotiate the cooperation strategy and decide on the optimal location and magnitude of the individual forces that would be applied by them so that the object would be transported quickly and effectively while avoiding any obstacles that might be present in the path. In some special cases, they also need to consider whether to move an obstacle out of the way rather than negotiating around it. Other considerations such as the level of energy utilization and damage mitigation for the transported object may have to be taken into account as well. A global camera is used to monitor and measure the current location and orientation (pose) of the object. The environment is dynamic and unpredictable with some movable obstacles that may appear randomly and some fixed obstacles. The robots, the camera, and the sensors are separately linked to their host computers, which are connected through a local network, to implement complex controls and machine intelligence.

A robotic arm and a mobile robot are used in the system shown in Figure 14.2. The system provides an opportunity to observe the cooperation between two robots with different dynamics and obtain useful information on the system behavior. Moreover, with the ability of exact localization of the robotic arm and the capability of broad ground coverage of the mobile robot, it is possible to integrate different advantages of the two robots in a complementary manner so as to improve the effectiveness of the overall system.

14.4 Computer Vision Subsystem

In Figure 14.2, a global camera is used to monitor the current world state. Based on this, a computer vision subsystem is developed. It is important for the agents to know the current positions and orientations of the robots and the object (box) for them to make proper decisions to complete their common

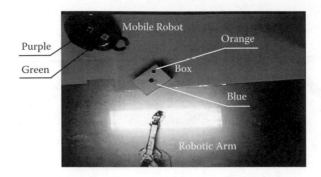

FIGURE 14.3 An original image captured by the CCD camera. (Note the two color blobs on the mobile robot and the box.)

task. Because the pose (position and orientation) of the robotic arm can be accurately computed from its encoder data, the CCD camera only needs to capture the poses of the box and the mobile robot. A C++ computer vision subsystem has been developed to accomplish this task.

To speed up the image processing task and reduce the computational load of the vision subsystem, the mobile robot and the box are marked with color blobs (a purple and a green blob on the mobile robot; an orange and a blue blob on the box) to describe their current poses, as shown in Figure 14.3.

The poses of the mobile robot and the box can be determined by quickly localizing the four color blobs from an image, followed by the necessary coordinate transformation operations. Search for the particular colors in an RGB image is not a trivial task, however. There exist two main challenges in this regard. The first one concerns the response speed of the vision subsystem. Conventional computer-vision algorithms are too slow and unacceptable for the present real-time system. The second challenge comes from object illumination. In the experimental process, because of uneven lighting in the workspace, it was found that the same color blob had very different color values when it moved into different positions. Because the mobile robot and the box are always in motion when the system is running, it is difficult to trace the same color blob in different positions under uneven lighting conditions. These problems have been resolved, as described next, using Figures 14.4 through 14.7.

The first step of the developed approach is to convert the original image from the RGB color space to the HSI color space [6] so as to remove the disturbances caused by uneven lighting conditions. When an image is captured into the computer by the frame gabber, it is represented in the RGB space, which uses red, green, and blue to describe the image color. However, although RGB representation is effective for generating colors, it is not a good approach to describe and interpret colors in a way similar to how the human eye observes an image. On the other hand, the HSI color space uses three independent components (hue, saturation, and intensity) to describe colors, and this approach is similar to the human perception of an image. In the present application, the original image is first converted from the RGB space into the HSI space. Then the saturation and intensity components in the image are removed to obtain a monochromatic image, which only includes the hue component of the original image. The result of this processing step is shown in Figure 14.4. By comparing Figure 14.3 with 14.4, it is noted that the influence of the uneven lighting is effectively removed, and only the hue of the image is retained.

The second step is to set up off-line the statistical data of sample hue values from Figure 14.4 so that the color blobs of interest can be separated from the background. In Figure 14.3, there are four types of sample hue (purple, green, orange, and blue), which have to be detected from an image. First, some hue values of the pixels are extracted from the region of interest. For example, the hue values of 4*4 = 16 pixels are extracted around the center of the purple blob on the mobile robot in Figure 14.3. Then the average h_p and the standard deviation σ_p are computed from the 16 sample data values. Here, the subscript p denotes the color purple. Next, $h_p \pm 1.2{}^*\sigma_p$ is considered as the range of the purple hue, which is of interest here. Through similar operations, the statistical results of $h_g \pm 1.2{}^*\sigma_g$, $h_o \pm 1.2{}^*\sigma_o$, and $h_b \pm 1.2{}^*\sigma_b$ are obtained as well.

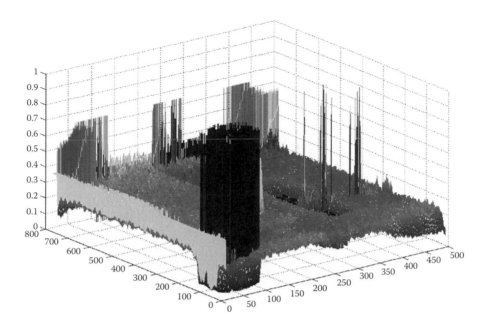

FIGURE 14.4 The hue component of the original color image in the HSI color space.

Next, the hue component image in Figure 14.4 is filtered, point by point. Only the points with hue values falling in the rage $h_i \pm 1.2^*\sigma_i$ $(i = p, g, o, b)$ are retained. The filtering result obtained in this manner is shown in Figure 14.5.

From Figure 14.5 it is clear that the background of the image is virtually removed, and only the four types of sample hue (purple, green, orange, and blue) are retained, which are distinguished with the distinct values 0.3, 0.5, 0.7, and 0.9. However, because the hue of the mobile robot is quite similar to the

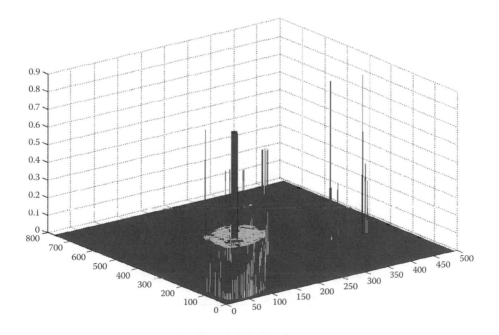

FIGURE 14.5 The hue component image after the background is filtered.

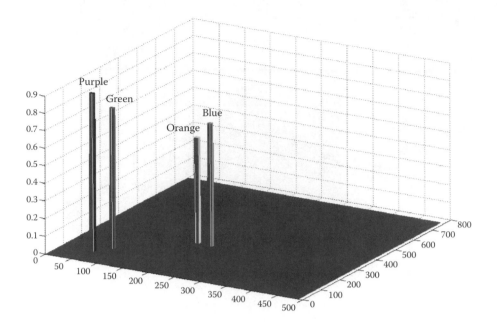

FIGURE 14.6   The result after filtering figure 14.5 using template matching.

orange hue and there still exist small regions of disturbance, the positions of the color blobs of interest are not very clear in Figure 14.5.

The third step is to use the classical template matching method [6] and some known space information (for example, the distance between the blue and the purple blobs is always greater than the distance between the blue and the orange blobs) to remove the small unwanted regions and the mobile robot hue in Figure 14.5. The processing result is shown in Figure 14.6.

From Figure 14.6 it is observed that the positions of the four color blobs are accurately established. Through coordinate transformation operations, the positions and the orientations of the robots and the box are reconstructed, as shown in Figure 14.7.

Because the image processing scheme outlined here does not require too many complex operations and some operations may be carried out off-line, the total time of acquiring and processing an image is found to be less than one second, which meets the requirement of the transportation task considered in the present application.

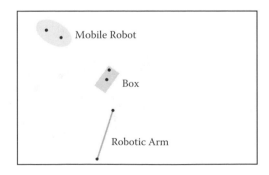

FIGURE 14.7   The reconstructed result of the poses of the robots and the box.

14.5 Conclusion

Robots are sophisticated mechatronic systems. The issues of sensing, communication, and control have to be carefully considered in developing cooperative multi-robot systems for carrying out practical tasks. In this chapter, a fast computer vision algorithm was presented for use in a multi-robot transportation system in our laboratory. In order to remove the disturbances caused by uneven lighting conditions, the original RGB image was converted into the HSI color space, and only the hue component information is retained. Next, template matching and filtering methods were used to remove the background and reconstruct the current world state. Because the algorithm did not involve a high computing requirements, the real-time performance was achieved.

References

1. Cao, Y.U., Fukunaga, A.S., and Kahng, A.B., Cooperative mobile robotics: antecedents and directions, *Autonomous Robots*, Vol. 4, No. 1, 7–27, 1997.
2. Stone, P. and Veloso, M., Layered approach to learning client behaviors in the ROBOCUP soccer server, *Applied Artificial Intelligence*, Vol. 12, No. 2–3, 165–188, 1998.
3. Veeraraghavan, H., Masoud, O., and Papanikolopoulos, N.P., Computer vision algorithms for intersection monitoring, *IEEE Transactions on Intelligent Transportation Systems*, Vol. 4, No. 2, 78–89, 2003.
4. Maurin, B., Masoud, O., and Papanikolopoulos, N.P., Tracking all traffic: computer vision algorithms for monitoring vehicles, individuals, and crowds, *IEEE Robotics and Automation Magazine*, Vol. 12, No. 1, 29–36, 2005.
5. Chen, S., Shyu, M., Peeta, S., and Zhang, C., Spatiotemporal vehicle tracking, *IEEE Robotics and Automation Magazine*, Vol. 12, No. 1, 50–58, 2005.
6. Gonzalez, R.C. and Woods, R.E., *Digital Image Processing*, Prentice Hall, Upper Saddle River, NJ, 2002.
7. De Silva, C.W., *Mechatronics—An Integrated Approach*, Taylor and Francis, CRC Press, Boca Raton, FL, 2005.

15

Robotic Tasks Using Discrete Event Controller

N.W. Koh

M.H. Ang Jr.

S.Y. Lim

Summary

Robots are practical mechatronic systems that execute their tasks with the assistance of a control system. In particular, task execution can be facilitated by a supervisory controller. This chapter presents a supervisory controller that can be used as an aid after task planning for the execution of robotic tasks and also to carry out a validation procedure for verifying whether an operation, in terms of resources, can be successfully executed. By decomposing a task into its elemental entities, each entity (or subtask) is assigned an action primitive and is allotted to a resource. The subtasks can then be carried out by calling the respective primitives, which are monitored for completion by a checker module, from a library. Supervisory control of the robotic system is based on an improved matrix model approach, catering to the generality required in robotic systems. Because the model deals only with logical matrix multiplications, computational effort is significantly reduced.

15.1 Introduction

Mechatronic systems are often treated as discrete-event systems where the tasks can be described as a sequence of events. Event-driven mechatronic systems are growing in popularity and complexity. Complex robotic applications may fall into this category. From the stance of task planning, robotic tasks, modeled as discrete event systems, have proven to be a favored method [1–3]. Task planning describes the ability to decompose a task into a set of primitives [4,5] based on a cognitive form of goal-means decomposition. From this, the aim is to recompose this set into a logical sequence with its corresponding resources (task-primitive composition), which, when executed, enables the task to be carried out autonomously or with user intervention. As the complexity of a task increases, i.e., with multiple resources and robots, a proper methodology for actually representing the task is required. If operations and resources are not correctly sequenced, the task will most probably enter into a state of deadlock, livelock, or blocking [6,7]. Therefore, it is important that the user sequences the operations and resources properly and uses the controller to check the feasibility of that sequence in terms of resource allocation. One of the more extensively used tools for the analysis, modeling, and control of a robotic or mechatronic task is Petri nets [8–11].

The purpose of the matrix-based controller is to act as a supervisor to allow for task execution, given the insertion of task primitives, in a cognitively logical series, and the allocation of resources required by that primitive. Control of a robotic system based on the matrix model approach significantly reduces computational effort because the model deals only with logical sparse matrix multiplications.

15.2　Matrix-Based Discrete Event Controller

With the use of manufacturing, engineering, and mechatronic concepts, a powerful rule-based matrix model was developed [12] and has been implemented in flexible manufacturing systems [13,14] and, more recently, in wireless sensor networks [15]. The pursuit of the work presented in this chapter is to employ the discrete event controller to showcase its potential as a tool for the execution of complex robotic tasks, taking into account the resources required to execute it and to use the controller as the basis for a robot programming tool providing for a validity check in the planned task. Preliminary work on the development and implementation of the matrix model from the modified to the improved state is found in References 16–18.

With the matrix model, assembly/job sequencing, addition of resources, analyses for deadlock and its avoidance, and a dispatching design can be carried out allowing a thorough analysis with a convenient solution to the simulation of the system in the form of a Petri net [19].

A task can be viewed as consisting of a number of jobs, which when sequenced, should depict the task in its entirety. Each job is assigned a primitive that corresponds directly to a particular resource. A job is triggered to start via the use of input signals and a set of conditions. The task is complete once an output signal is present. This description is mathematically represented using a matrix formulation described in the next section.

15.2.1　Discrete Event Model State Equation

A task can be described by

$$\bar{x} = Q_v \otimes \bar{v}_c \vee Q_r \otimes \bar{r}_c \vee Q_u \otimes \bar{u} \vee Q_D \otimes \bar{u}_D \tag{15.1}$$

Table 15.1 shows the variable definitions for equation 15.1. The condition x represents the state of the system, and the equation shows how it evolves over time. As the equations used in the matrix model are

TABLE 15.1　Variable Definitions for Equation 15.1

Variables	Dimensions	Definitions
x	$n_x \times 1$	n_x number of conditions
v	$n_v \times 1$	n_v number of jobs
r	$n_r \times 1$	n_r number of resources
u	$n_u \times 1$	n_u input signals
u_D	$n_D \times 1$	n_D number of dispatch controls
Q_v	$n_x \times j$	Job sequencing matrix
Q_r	$n_x \times k$	Resource requirements matrix
Q_u	$n_x \times l$	Process input matrix
Q_D	$n_x \times m$	Dispatching matrix

logical equations, standard matrix multiplication and addition are replaced by AND/OR algebra with all vectors and matrices being binary. Here, $\otimes$ represents an AND/OR operation, and $\vee$ an OR operation. The over-bar is a logical negation and is defined as follows: For any component $z(n)$ of a natural number vector z,

$$\bar{z} = 0 \quad \text{if} \quad z(n) > 0 , \bar{z} = 1 \quad \text{if} \quad z(n) \le 0 \tag{15.2}$$

The relevant matrices are explained as follows:

- Q_v determines which relevant job v_c should be completed before condition x is satisfied. When $v_c = 1$, a job is said to be complete.
- Q_r determines which relevant resource r_c should be present before condition x is satisfied. When $r_c = 1$, a resource is said to be currently available.
- Q_u determines which relevant input signal u should be present before condition x is satisfied. When $u = 1$, an input signal is said to be present.
- Q_D determines which relevant virtual resource u_D should be present before condition x is satisfied. When $u_D = 1$, a virtual resource is said to be present. This matrix is used to determine the priority of the operations when resources are shared.

Each condition is detailed as

$$\bar{x}_s = \bigvee_{\alpha=1}^{n_v} [Q_v(s,\alpha) \wedge \bar{v}_{c_\alpha}] \vee \bigvee_{\beta=1}^{n_r} [Q_r(s,\beta) \wedge \bar{r}_{c_\beta}] \vee \bigvee_{\gamma=1}^{n_u} [Q_u(s,\gamma) \wedge \bar{u}_\gamma] \vee \bigvee_{\delta=1}^{n_D} [Q_D(s,\delta) \wedge \bar{u}_{D_\delta}] \tag{15.3}$$

where $s = 1,\ldots, n_x$, $\bigvee$ represents a logical OR summation and $\wedge$ a logical AND. The state of each condition is dependent on the input received from the environment in terms of job completion, resource availability, input signals, and dispatch control. Upon successful satisfaction of conditions, a job and a resource will be triggered to start (Job start equation) and be released (Resource release equation), respectively.

15.2.2 Job Start Equation

The start of a job is indicated by the following equation:

$$\bar{v}_s = S_v \otimes \bar{x} \vee U_{v_c} \otimes \bar{v}_c \tag{15.4}$$

where S_v is a rectangular job start matrix, and U_{v_c} is any user-defined matrix of dimensions $n_v \times n_v$. Job v_s, $s = 1,\ldots,n_v$ starts when $v_s = 1$. Equation 15.4 can be read as follows: Job v_s will start if the relevant conditions (determined by S_v) are satisfied and the relevant job/jobs (determined by U_{v_c}) is/are complete. This general representation can be used for concurrent and dependent operations.

Each job starting is detailed as

$$\bar{v}_s = \bigvee_{\alpha=1}^{n_x} [S_v(s,\alpha) \wedge \bar{x}_\alpha] \vee \bigvee_{\beta=1}^{n_v} [U_{v_c}(s,\beta) \wedge \bar{v}_{c_\beta}] \quad U_{v_c}(\beta,\beta) \neq 1 \tag{15.5}$$

15.2.3 Resource Release Equation

The equation indicating the release of a resource is

$$\bar{r}_s = S_r \otimes \bar{x} \vee U_{r_c} \otimes \bar{r}_c \tag{15.6}$$

where S_r is a rectangular resource release matrix, and U_{r_c} is any user-defined matrix of dimensions $n_r \times n_r$. Resource r_s, $s = 1, \ldots, n_r$ is released when $r_s = 1$. Equation 15.6 can be read as follows: Resource r_s can be released if the relevant conditions (determined by S_r) are satisfied and the relevant user-defined resource/ resources (determined by U_{r_c}) is/are available.

Each resource released is detailed as

$$\bar{r}_s = \bigvee_{\alpha=1}^{n_x} [S_r(s,\alpha) \wedge \bar{x}_\alpha] \vee \bigvee_{\beta=1}^{n_r} [U_{r_c}(s,\beta) \wedge \bar{r}_{c_\beta}] \qquad [U_{r_c}(\beta,\beta) \neq 1] \qquad (15.7)$$

15.2.4 Process Output Equation

Note that S_y determines the set of conditions that need to be satisfied before a product is generated, according to the equation

$$\bar{y} = S_y \otimes \bar{x} \qquad (15.8)$$

Each process (z) output is detailed as

$$\bar{y}_z = \bigvee_{\alpha=1}^{n_x} [S_y(z,\alpha) \wedge \bar{x}_\alpha] \qquad (15.9)$$

When several resources have to be shared or requested simultaneously by several operations, matrix Q_D is deployed to coordinate the dispatching of the resources. Vector u_D would then be divided virtually into the number of times each resource is shared. As an example, say, resource R_1 is shared three times, and resource R_2 twice; then, vector $u_D = [u_{D1}\ u_{D2}\ u_{D3}\ u_{D4}\ u_{D5}]$, where $[u_{D1}\ u_{D2}\ u_{D3}]$ represents three virtual R_1s, and $[u_{D4}\ u_{D5}]$ two virtual R_2s. The term "virtual" is used here because only one set of these resources is available but is divided to satisfy the dispatching matrix to resolve any conflicts. Conflicts occur when two operations request the same resource simultaneously. This conflict resolution logic is a form of priority distinction. For a case where there are no shared resources, this matrix can be removed from the equation while still producing a dynamically consistent result.

A theoretical insight into the improved matrix-based supervisory controller is discussed in Koh et al., 2005 [18]. Figure 15.1 depicts the matrix model. Based on the system's current state and the environs, the Q matrices are used to obtain the next set of conditions, which in turn will be used by the S matrices to release the resources required for the starting of a job. Once a job is complete, the Q matrices are once again used to obtain the next set of conditions; i.e., the conditions change with the environment. The process is iterative.

The matrix-based supervisory controller allows an effective and simple online control by applying a set of discrete-event control signals. Together with the Petri net marking transition equation, the matrix formulation provides a complete dynamical description that can be used for analysis and computer simulation.

15.2.5 Petri Net Transition Equation

Given the matrix model, an *activity completion matrix Q* and an *activity start matrix S* [10] can be defined as

$$\begin{aligned} Q &= [Q_u \quad Q_v \quad Q_r \quad Q_D \quad Q_y] \\ S &= \begin{bmatrix} S_u^T & S_v^T & S_r^T & S_D^T & S_Y^T \end{bmatrix} \end{aligned} \qquad (15.10)$$

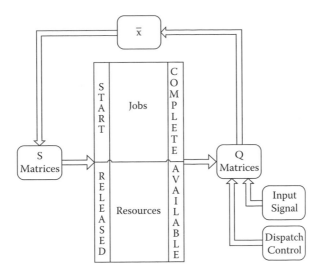

FIGURE 15.1 Depiction of the matrix-based discrete event controller.

where

- S_u is a vector with zero entries and dimensions $n_u \times n_x$ if Q_u has dimensions $n_x \times n_u$.
- S_D is a matrix with zero entries and dimensions $n_D \times n_x$ if Q_D has dimensions $n_x \times n_D$.
- Q_y and S_y are introduced to complete the matrices and are vectors of zeros.

This result identifies Q as the input incidence matrix, and S^T as the output incidence matrix, of a Petri net. The Petri net incidence matrix is then given by

$$M = S^T - Q \tag{15.11}$$

In terms of the Petri net incidence matrix, a Petri net marking transition equation can be written as

$$m_{k+1} = m_k + M^T x_k \tag{15.12}$$

where n_D is the marking vector and n_r the iteration number.

An allowable firing vector must be used in conjunction with Equation 15.12 to determine the next marking vector based on the previous one. Equation (15.13) is introduced to determine the allowable firing vector and may be written as

$$x_k = \overline{Q} \oplus \overline{m_k} = [\overline{Q}] \oplus \overline{[PI \quad v \quad r \quad u_D \quad PO]^T} \tag{15.13}$$

where $\oplus$ is an OR/AND operation, and *PI* and *PO* are the input and output signals, respectively.

15.2.5.1 Necessary Reachability Condition [19]

With the Petri net transition equation, a set of iterations depicting a piecemeal process of a task can be obtained. This provides a feasibility check to ensure if the task, given the operations and resource assignments and the planning process, is executable (or not).

Suppose that a destination marking m_d is reachable from m_0 through a firing sequence $x_1, x_2, \ldots, x_d$. The corresponding Petri net transition equation can be written as

$$m_d = m_0 + M^T \cdot \sum x_k \quad k = 1 \ldots d \tag{15.14}$$

which can be rewritten as

$$M^T x = \Delta m \qquad (15.15)$$

where $\Delta m = m_d - m_0$ and $x = \Sigma x_k$. Also, x is an $n \times 1$ column vector of nonnegative integers and is called the firing count/logical rule vector. The i-th entry denotes the number of times transition i must fire to transform m_0 to m_d. It is well known that a set of linear algebraic equations has a solution x if Δm is orthogonal to every solution y of its homogeneous system

$$My = 0 \qquad (15.16)$$

Let r be the rank of M and partition M in the following form:

$$M = \begin{bmatrix} M_{11} & M_{12} \\ M_{21} & M_{22} \end{bmatrix} \qquad (15.17)$$

with dimensions

- $M_{11} \rightarrow (m - r) \times r$
- $M_{21} \rightarrow (m - r) \times (n - r)$
- $M_{12} \rightarrow r \times r$
- $M_{22} \rightarrow r \times (n - r)$

where M_{12} is a nonsingular square matrix of order r. A set of $(m - r)$ linearly independent solutions y for Equation 15.16 can be given as the $(m - r)$ rows of the following $(m - r) \times m$ matrix B_f:

$$B_f = \left[I_\mu - M_{11}^T \left(M_{12}^T \right)^{-1} \right] \qquad (15.18)$$

where I_μ is the identity matrix of order $\mu = m - r$. Note that $MB_f^T = 0$. That is, the vector spanned by the row vectors of M is orthogonal to the vector spanned by the row vectors of B_f. The matrix B_f corresponds to the fundamental circuit matrix in the case of a marked graph. Now, the condition that Δm is orthogonal to every solution for $My = 0$ is equivalent to the following condition:

$$B_f \Delta m = 0 \qquad (15.19)$$

Thus, if m_d is reachable from m_0, then the corresponding firing count vector x must exist and Equation 15.19 must hold. Therefore, we have the necessary condition for reachability in an unrestricted Petri net.

15.3 Application Example

The Task: A pick-and-place task was carried out with a PUMA560 (Figure 15.2), a number of tools spread on a workbench, and a vision module attached to the gripper of the PUMA. After the tool is decided upon by the vision algorithm, the PUMA proceeds to pick up (or grasp) the object and place it at a specified location. This process repeats until no more tools are available.

Task Sequence: Task planning begins with the decomposition of a task into a cognitively sequential series of elemental subtasks. Depicting a complex task as a series of subtasks allows a task primitive to be assigned to each subtask. For instance, to invert the robot arm, the *InvertRobot* primitive would be used. This task primitive represents an operation and will be used in vector v_s for the job start matrix (S_v). The sequence and respective notations are given in Table 15.2.

Resource Assignment: Before an operation can be executed, each job requires a particular resource or resources. It is appropriate to view each independent entity on the robot arm as a separate resource (for the purpose of this example) to showcase the controller, but it does not necessarily have to be so (a robot arm can be viewed as one resource). Table 15.3 gives the resource assignment. Note that each resource is shared by all five operations. Because the resources are attached to one

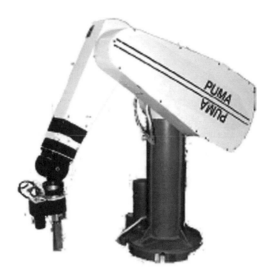

FIGURE 15.2 PUMA with attached gripper and vision module.

another, a resource that is currently carrying out a job must have all the other resources with it. Therefore, the unused resource or resources, albeit idle, cannot be executing any other job and is deemed occupied. For example, if the PUMA is to move, the gripper and camera *must* move together and cannot be carrying out any other job.

State Equation: With reference to Equation 15.1,

$$
\begin{bmatrix} \bar{x}_1 \\ \bar{x}_2 \\ \bar{x}_3 \\ \bar{x}_4 \\ \bar{x}_5 \\ \bar{x}_6 \end{bmatrix} = \begin{bmatrix} 0 & 0 & 0 & 0 & 0 \\ 1 & 0 & 0 & 0 & 0 \\ 0 & 1 & 0 & 0 & 0 \\ 0 & 0 & 1 & 0 & 0 \\ 0 & 0 & 0 & 1 & 0 \\ 0 & 0 & 0 & 0 & 1 \end{bmatrix} \otimes \begin{bmatrix} \overline{IR_c} \\ \overline{Con_c} \\ \overline{G_c} \\ \overline{Move_c} \\ \overline{OG_c} \end{bmatrix} \vee \begin{bmatrix} 1 & 1 & 1 \\ 1 & 1 & 1 \\ 1 & 1 & 1 \\ 1 & 1 & 1 \\ 1 & 1 & 1 \\ 0 & 0 & 0 \end{bmatrix} \otimes \begin{bmatrix} \overline{PR_c} \\ \overline{GRIP_c} \\ \overline{CAM_c} \end{bmatrix} \vee \begin{bmatrix} 1 \\ 0 \\ 0 \\ 0 \\ 0 \\ 0 \end{bmatrix} \otimes \begin{bmatrix} \overline{PI} \end{bmatrix}
$$

$$
\vee \begin{bmatrix} 1 & 0 & 0 & 0 & 0 \\ 0 & 1 & 0 & 0 & 0 \\ 0 & 0 & 1 & 0 & 0 \\ 0 & 0 & 0 & 1 & 0 \\ 0 & 0 & 0 & 0 & 1 \\ 0 & 0 & 0 & 0 & 0 \end{bmatrix} \otimes \begin{bmatrix} \bar{u}_{D1,D6,D11} \\ \bar{u}_{D2,D7,D12} \\ \bar{u}_{D3,D8,D13} \\ \bar{u}_{D4,D9,D14} \\ \bar{u}_{D5,D10,D15} \end{bmatrix}
$$

(15.20)

TABLE 15.2 Task Sequence

Operation Number	Notation	Description
1	IR	Invert the robot arm
2	Con	Connect to the vision module
3	G	Grasp the object
4	Move	Move the end-effector to a specified location
5	OG	Open the gripper

TABLE 15.3   Resource Assignment

	Resource	Operation Notation
rc1	PUMA Robot (PR)	IR, Con, G, Move, OG
rc2	Gripper (GRIP)	IR, Con, G, Move, OG
rc3	Camera (CAM)	IR, Con, G, Move, OG

Each symbolic vector awaits the signals retrieved from the environment by sensor utilization, which will determine the satisfaction of a condition. The number of conditions is always larger than the number of jobs by one, with the last condition signifying task completion.

Petri Net Representation: Figure 15.3 shows the Petri net sequence of the task with its respective resources. A Petri net is nothing but a bipartite digraph described by P, T, I, and O, where P is a set of places, T a set of transitions, I a set of input arcs from places to transitions, and O a set of output arcs from transitions to places. Places (shown as circles or ellipses) represent jobs, and the transitions (shown as rectangles) represent decision or rules for resource assignment/release and starting of jobs.

Place *process start* has one token signifying the start of the process. Places *rc1*, *rc2*, and *rc3* have five tokens each because these resources are used consecutively by each job. The tokens denote the initial number of resources assigned to the resource pools and are user defined. Once the task is complete, the resources are returned and remain idle while place *process out* receives one token signifying the completion and feasibility of the modeled task. For shared resources, the conflict resolution dispatch control (u_D) should be present in the modeling of the task. Places *Vc1* to *Vc5*

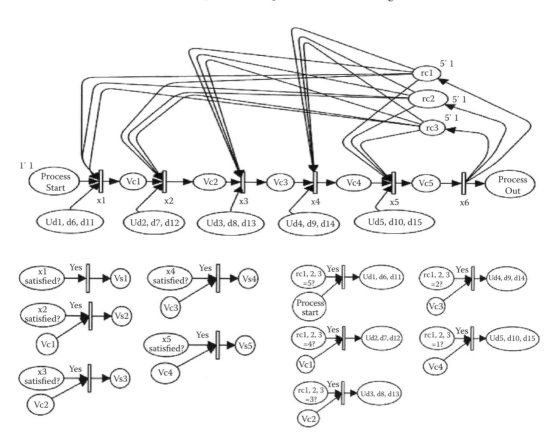

FIGURE 15.3   Petri net representation.

represent the five operations. Recapitulate that when $v_c = 1$ (or when the place has a token), the job is complete.

With reference to equation 15.4, and taking Job 2 as an example, the Petri net can be read as follows: Before Job 2 can start ($Vs2 = 1$), condition x_2 must be satisfied, job one must be complete ($Vc1 = 1$), and the dispatch signal must be present ($u_{d2,d7,d12} = 1$). When job two is complete, $Vc2 = 1$.

From equation 15.12, the marking vector is defined as $m_k = [PI\ IR\ Con\ G\ Move\ OG\ PR\ GRIP\ CAM\ u_{D1,D6,D11}\ u_{D2,D7,D12}\ u_{D3,D8,D13}\ u_{D4,D9,D14}\ u_{D5,D10,D15}\ PO]^T$, and initial marking, which is user specified for the task, is $m_0 = [1\ 0\ 0\ 0\ 0\ 0\ 5\ 5\ 5\ 0\ 0\ 0\ 0\ 0\ 0]^T$. Because the resources are shared and used consecutively over five operations, the initial marking for $m(PR) = m(GRIP) = m(CAM) = 5$. Equation 15.13, used in Equation 15.12, provides a complete dynamical description of any task modeled as a discrete-event system.

Task Execution: With reference to Equation 15.4, one has

$$
\begin{bmatrix} \overline{IR_s} \\ \overline{Con_s} \\ \overline{G_s} \\ \overline{Move_s} \\ \overline{OG_s} \end{bmatrix} = \begin{bmatrix} 1 & 0 & 0 & 0 & 0 & 0 \\ 0 & 1 & 0 & 0 & 0 & 0 \\ 0 & 0 & 1 & 0 & 0 & 0 \\ 0 & 0 & 0 & 1 & 0 & 0 \\ 0 & 0 & 0 & 0 & 1 & 0 \end{bmatrix} \otimes \begin{bmatrix} \overline{x_1} \\ \overline{x_2} \\ \overline{x_3} \\ \overline{x_4} \\ \overline{x_5} \\ \overline{x_6} \end{bmatrix} \vee \begin{bmatrix} 0 & 0 & 0 & 0 & 0 \\ 1 & 0 & 0 & 0 & 0 \\ 0 & 1 & 0 & 0 & 0 \\ 0 & 0 & 1 & 0 & 0 \\ 0 & 0 & 0 & 1 & 0 \end{bmatrix} \otimes \begin{bmatrix} \overline{IR_c} \\ \overline{Con_c} \\ \overline{G_c} \\ \overline{Move_c} \\ \overline{OG_c} \end{bmatrix} \tag{15.21}
$$

The starting of a job is dependent on a set of conditions and on the completion of a previous job. With reference to Equation 15.6, one has

$$
\begin{bmatrix} \overline{PR_s} \\ \overline{GRIP_s} \\ \overline{CAM_s} \end{bmatrix} = \begin{bmatrix} 0 & 0 & 0 & 0 & 0 & 1 \\ 0 & 0 & 0 & 0 & 0 & 1 \\ 0 & 0 & 0 & 0 & 0 & 1 \end{bmatrix} \otimes \begin{bmatrix} \overline{x_1} \\ \overline{x_2} \\ \overline{x_3} \\ \overline{x_4} \\ \overline{x_5} \\ \overline{x_6} \end{bmatrix} \vee [0_{sx3}] \otimes \begin{bmatrix} \overline{PR_c} \\ \overline{GRIP_c} \\ \overline{CAM_c} \end{bmatrix} \tag{15.22}
$$

The release of a resource is dependent on a set of conditions and is not reliant on the availability of any other resource. A resource can only be available a after it is released.

For the case of implementation, equation 15.21 is used to determine the sequence of primitive actions, and Equation 15.22 to release an idle resource for the execution of other jobs. During any iteration for a sequential operation, a single entity of v_s will have an entry of "1" with all other entries being "0." These entries of 1 will appear sequentially, i.e., from v_{s1} to v_{s5} stepwise, provided that there are no external failures. For operations that run in parallel, more than one entry of 1 may be present.

The vectors of v_s are shown in the following for 11 iterations:

$$
\begin{bmatrix} v_{s1} \\ v_{s2} \\ v_{s3} \\ v_{s4} \\ v_{s5} \end{bmatrix} = \begin{bmatrix} 0 \\ 0 \\ 0 \\ 0 \\ 0 \end{bmatrix} \rightarrow \begin{bmatrix} 1 \\ 0 \\ 0 \\ 0 \\ 0 \end{bmatrix} \rightarrow \begin{bmatrix} 0 \\ 0 \\ 0 \\ 0 \\ 0 \end{bmatrix} \rightarrow \begin{bmatrix} 0 \\ 1 \\ 0 \\ 0 \\ 0 \end{bmatrix} \rightarrow \begin{bmatrix} 0 \\ 0 \\ 0 \\ 0 \\ 0 \end{bmatrix} \rightarrow \begin{bmatrix} 0 \\ 0 \\ 1 \\ 0 \\ 0 \end{bmatrix} \rightarrow \begin{bmatrix} 0 \\ 0 \\ 0 \\ 0 \\ 0 \end{bmatrix} \rightarrow \begin{bmatrix} 0 \\ 0 \\ 0 \\ 1 \\ 0 \end{bmatrix} \rightarrow \begin{bmatrix} 0 \\ 0 \\ 0 \\ 0 \\ 0 \end{bmatrix} \rightarrow \begin{bmatrix} 0 \\ 0 \\ 0 \\ 0 \\ 1 \end{bmatrix} \rightarrow \begin{bmatrix} 0 \\ 0 \\ 0 \\ 0 \\ 0 \end{bmatrix} \tag{15.23}
$$

Iteration 1 is a null vector and can be interpreted as "no jobs started." The same interpretation is used for any null vector. Iteration 2, however, has an entry of 1 at v_{s1} and entries of 0 for the rest. This signifies that the system has satisfied all the necessary conditions, and that the job prior to it is complete. Entries of 1 are used in the robot controller as an indication to carry out the corresponding job.

With reference to Equation 15.8, one has

$$\overline{PO} = [0 \quad 0 \quad 0 \quad 0 \quad 0 \quad 1] \otimes \begin{bmatrix} \overline{x}_1 \\ \overline{x}_2 \\ \overline{x}_3 \\ \overline{x}_4 \\ \overline{x}_5 \\ \overline{x}_6 \end{bmatrix}$$

When condition x_6 (task is complete) is satisfied, the task is said to have ended ($PO = 1$).

15.4   Results and Discussions

Dynamic changes of the tokens with time are plotted in Figure 15.4. The first five lines (from the top) represent the five jobs that are defined as "in operation" when a token value of 1 is present over the duration of that operation. The 0 value before and after the 1 depicts "job not started" and "job completed," respectively. It is seen that the five jobs (*IR* to *OG*) are executed sequentially. When a job starts, the token value in the resource lines (lower three lines) decreases by one; this depicts the resource being used. After all the jobs are complete, the resource is returned, i.e., one token value is returned. As the resources are shared, each resource line exhibits the same changes.

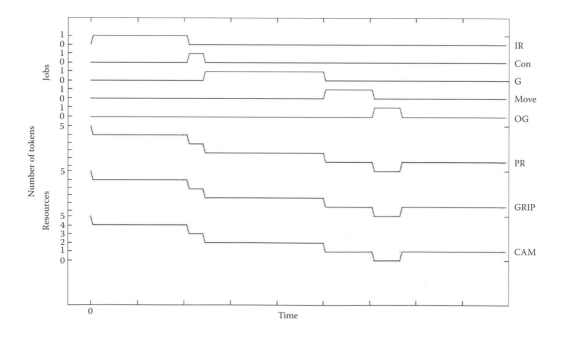

FIGURE 15.4 Token depiction of the operations and resources versus time.

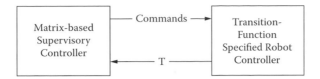

FIGURE 15.5 Matrix-based supervisory controller with the transition function-specified controllers (*T* is the transmission buffer).

The main contribution of the work presented in this chapter lies with the improvement of the job start (Equation 15.4) and resource release (Equation 15.6), which forms a relationship between the starting of a job and the completion of another, and the release of a resource depending on another resource's availability.

$$\overline{v}_s = S_v \otimes \overline{x} \vee U_{v_c} \otimes \overline{v}_c \tag{15.4}$$

$$\overline{r}_s = S_r \otimes \overline{x} \vee U_{v_c} \otimes \overline{r}_c \tag{15.6}$$

In its general form, the preceding equations allow a user to define which job (or set of jobs) should be complete before a new job can start, and which resource (or set of resources) should be idle before releasing a resource. From a broader perspective, it is seen that the equations provide a means to synchronize local and external subsystems in terms of jobs and resources. The applicability of this approach in mechatronic systems ranges from manufacturing systems, mobile manipulation, and multirobot systems, to name a few.

15.5 Formalization of the Framework

Mechatronic control systems (or robot control systems) generally require a layer that manages, in a modular manner, the hardware or low-level control, and a supervisory layer that allows task specification for the execution of tasks. Whether a primitive has completed its execution is dependent on the current and terminal states of the system. Therefore, a framework that caters to the development of these primitives and the transmission of the system's sensor information is mandatory. Supervisory control of a system essentially boils down to the processing of information about the current state of the system and its environment, enabling a defined task to be executed.

This framework adopts a transition-function-based formalism [20,21] that introduces rigor into the design and implementation of a single or multiagent system (Figure 15.5). Once a large mechatronic system is decomposed into modular components, for an operational system, the design and implementation solely by providing the relevant code for the specified functions would suffice. Through their public interfaces, these components provide only the data necessary for the computation of the control of an agent, which is utilized by the transition functions resident in the control subsystem to compute the next state of this subsystem. The formalization process constitutes the current and future work of our research.

15.6 Conclusion

Robotic systems are sophisticated mechatronic systems. Mechatronic (or robotic) tasks in general require a great deal of robustness because the mechatronic system might encounter some unforeseen problems that prevent the task from completion. Information regarding the current and terminal states of a primitive behavior is essential to the supervisory controller of the system for the execution of the task. By employing the transition-function-based approach at the lower level, as presented here, this necessity has been realized.

In conjunction with the robotic pick-and-place task, it was observed that the matrix-based supervisory controller allowed an effective and simple online control by applying a set of discrete event control signals. Together with the Petri net–marking transition equation, the matrix formulation provides a complete dynamical description that can be used for analysis and computer simulation. Task primitives, used to represent each subtask, are assigned by decomposing a complex task into a series of elemental sequences. The novel approach taken for the actual task execution employs the results obtained from the allowable firing vector and is used as an indication of the availability of the necessary resources. Subsequently, a particular operation can be carried out. The results presented in this chapter show the simplicity and power of this approach.

From the viewpoint of task planning, the matrix-based controller proves to be advantageous because the matrix model provides a logical and simplistic approach for job sequencing and resource allocation once the task has been modeled. The controller also provides a check to confirm the feasibility of the user-modeled task.

References

1. Frazzoli, E., Dahleh, M.A., and Feron, E., Maneuver-based motion planning for nonlinear systems with symmetries, *IEEE Trans. on Robotics*, Vol. 21, No. 6, 1077–1091, 2005.
2. Montreuil, V., Duhaut, D., and Drogoul, A., A collective moving algorithm in modular robotics: contribution of communication capacities, *IEEE International Symposium on Computational Intelligence in Robotics and Automation*, Espoo, Finland, 2005, pp. 641–646.
3. Soysal, O. and Sahin, E., Probabilistic aggregation strategies in swarm robotic systems, *IEEE Swarm Intelligence Symposium*, Pasadena, CA, 2005, pp. 325–332.
4. Thomas, U., Wahl, F.M., Maass, J., and Hesselbach, J., Towards a new concept of robot programming in high speed assembly applications, *IEEE/RSJ International Conference on Intelligent Robots and Systems*, Edmonton, AB, Canada, 2005, pp. 3827–3833.
5. Frazzoli, E., Dahleh, M.A., and Feron, E., Maneuver-based motion planning for nonlinear systems with symmetries, *IEEE Transactions on Robotics*, Vol. 21, No. 6, 1077–1091, 2005.
6. Xing, K., Jin, X., and Feng, Y., Deadlock avoidance Petri net controller for manufacturing systems with multiple resource service, *IEEE International Conference on Robotics and Automation*, Barcelona, Spain, 2005, pp. 4757–4761.
7. Tricas, F., Garcia-Valles, F., Colom, J.M., and Ezpeleta, J., A Petri net structure-based deadlock prevention solution for sequential resource allocation systems, *IEEE International Conference on Robotics and Automation*, Barcelona, Spain, 2005, pp. 271–277.
8. Peterson, J.L., *Petri Net Theory and the Modeling of Systems*, Prentice-Hall, Englewood Cliffs, NJ, 1981.
9. Tsinarakis, G.J., Tsourveloudis, N.C., and Valavanis, K.P., Modeling, analysis, synthesis, and performance evaluation of multioperational production systems with hybrid timed Petri nets, *IEEE Transactions on Automation Science and Engineering*, Vol. 3, No. 1, 29–46, 2006.
10. Caccia, M., Coletta, P., Bruzzone, G., and Veruggio, G., Execution control of robotic tasks: a Petri net-based approach, *Control Engineering Practice*, Vol. 13, No. 8, 959–971, 2005.
11. Kim, G., Chung, W., Park, S.-K., and Kim, M., Experimental research of navigation behavior selection using generalized stochastic Petri nets (GSPN) for a tour-guide robot, *IEEE/RSJ International Conference on Intelligent Robots and Systems*, Edmonton, Canada, 2005, pp. 2259–2265.
12. Tacconi, D.A. and Lewis, F.L., A new matrix model for discrete event systems: application to simulation, *IEEE Control Systems*, Vol. 17, No. 5, 62–71, 1997.
13. Bogdan, S., Lewis, F.L., Kovacic, Z., Gurel, A., and Stajdohar, M., An implementation of the matrix-based supervisory controller of flexible manufacturing systems, *IEEE Transactions on Control Systems Technology*, Vol. 10, No. 5, 709–716, 2002.

14. Mireles, J., Jr. and Lewis, F.L., Intelligent material handling: development and implementation of a matrix-based discrete-event controller, *IEEE Transactions on Industrial Electronics*, Vol. 48, No. 6, 1087–1097, 2001.
15. Giordano, V., Lewis, F.L., Ballal, P., and Turchiano, B., Supervisory controller for task assignment and resource dispatching in mobile wireless sensor networks, *International Journal of Advanced Robotic Systems*, Cutting Edge Robotics, Proliteratur Verlag, Germany, Section 5, Part 1, 133–152, 2005 [ISBN: 3-86611-038-3].
16. Koh, N.W., Ang, M.H., Jr., and Lim, S.Y., Implementation of a matrix-based discrete event controller for robotic tasks, *IEEE Asian Conference for Industrial Automation and Robotics*, Bangkok, Thailand, http://www.aciar2005.ait.ac.th/, 2005.
17. Koh, N.W., Ang, M.H., Jr., and Lim, S.Y., Robotic tasks employing an improved matrix-based discrete event controller, *International Symposium on Collaborative Research in Applied Science*, Vancouver, Canada, 2005, pp. 195–202.
18. Koh, N.W., Ang, M.H., Jr., and Lim, S.Y., A theoretical insight to the improved matrix-based supervisory controller, *IEEE 3rd International Conference on Computational Intelligence, Robotics and Autonomous Systems*, Singapore, 2005, pp. 89 (IC5-6), ISSN: 0219–6131. (CD proceedings)
19. Murata, T., Petri nets: properties, analysis and applications, *Proceedings of the IEEE*, Vol. 77, 1989, pp. 542–580.
20. Koh, N.W., Zieliński, C., Ang, M.H., Jr., and Lim, S.Y., Matrix-based supervisory controller of transition-function specified robot controllers, *Proceedings of the 16th CISM-IFToMM Symposium on Robot Design, Dynamics and Control*, Warsaw, Poland, CISM No. 487, 2006, pp. 229–236.
21. Zieliński, C., Formal approach to the design of robot programming frameworks: the behavioral control case, *Bulletin of the Polish Academy of Sciences*, Vol. 53, 1–11, 2005.

16

Fuzzy Modeling and Control

Z. Guo

J. Mao

Y. Yue

Y. Li

Summary

Intelligent mechatronic systems rely on intelligent control methods. Fuzzy logic is widely used for approximating the decision-making process of humans in realizing intelligent mechatronic systems. An approach of adaptive fuzzy modeling based parameter optimization fuzzy tree (POFT) for a class of nonlinear time-invariant systems is presented in this chapter. A fuzzy state equation model based on the adaptive fuzzy tree algorithm is developed, which is used to partition the workspace of a nonlinear mechatronic system adaptively. The model obtained by this method has high accuracy and low computational cost. As an illustrative example, the method is applied to the modeling and control problem of the Furuta pendulum. The numerical simulation presented here shows that the method is effective.

16.1 Introduction

There is a rational need to develop intelligent mechatronic systems. A degree of intelligence may be incorporated into a mechatronic system in several ways; for example, through intelligent sensing, intelligent actuation and, above all, intelligent control. Often the intelligent behavior is realized by representing the decision-making process of a human expert. To this end, fuzzy logic has been proved to be an effective approach. System modeling has many uses, including computer simulation, design, evaluation, and control. In particular, models are indispensable in model-based control. Mechatronic systems are by and large nonlinear, and their modeling is rather challenging, and so is the problem of model-based control.

The adaptive fuzzy tree (FT) learning algorithm [1–4] is useful in solving the problem of fuzzy modeling for a complex nonlinear system, for example, a mechatronic system. The main characteristics of this algorithm are (1) the input data set is partitioned adaptively, and a fuzzy area around every discriminant edge is set up by the membership functions corresponding to every subset of input data to smooth the discontinuities and to reduce the error of approximation; and (2) the parameters of the antecedent (if part) and consequent (then part) in fuzzy rules are learned simultaneously. This algorithm is particularly effective in solving high-dimension problems [1].

In this chapter, a modeling method using POFT is used to adaptively partition and construct the subspace of an FT model. A genetic algorithm with back propagation training (GABPT) is used to optimize the parameters of FT. The model obtained in this manner has a high accuracy. A fuzzy state equation model for a class of nonlinear time-invariant systems based on POFT is developed, which is used to partition the workspace of the nonlinear system adaptively. A controller is designed based on this model. The obtained fuzzy state equation model is known to be more accurate and lower in computational cost than those of conventional fuzzy state equation models based on local linearization.

16.2 POFT Fuzzy Modeling

In this section, the use of POFT in the modeling of complex nonlinear (e.g., mechatronic) systems is presented.

16.2.1 Fuzzy Tree Model

Fuzzy rules of an FT model are defined as follows [1–4]:

$$R^l: \text{if } \boldsymbol{x} \text{ is } N_{t_l}(t_l \in T) \quad \text{then} \quad \hat{y}_{t_l} = (\boldsymbol{c}_{t_l})^T \boldsymbol{x} \tag{16.1}$$

where $\boldsymbol{x} = [1\,x_1\,x_2, \quad , x_n]^T \in R^{n+1}$ is the input data vector, N_{t_l} is a fuzzy set of fuzzy rule R^l, T is the set of leaves, which is a subset of all vertexes T denoted as $T \in T$, $l = 1,2, \quad ,p$, p is the number of if–then rules, and $\boldsymbol{c}_{t_l} = [c_0^{t_l}\,c_1^{t_l}, \quad ,c_n^{t_l}]^T \in R^{n+1}$ is the weighting vector.

The FT model is a special Takagi–Sugeno model. Here, $\mu_{t_l}(x)$ is the membership function of a corresponding fuzzy set N_{t_l}. Then, the final output of FT model is given by

$$\hat{y}(x) = \sum_{l=1}^{M} \mu_{t_l}(x)(c_{t_l})^T x \bigg/ \sum_{l=1}^{M} \mu_{t_l}(x) \tag{16.2}$$

where M is the number of leaves.

For each vertex $t \in T$, there is a corresponding linear approximation $\hat{y}_t = (\boldsymbol{c}_t)^T \boldsymbol{x}$ and fuzzy set N_t with membership function $\mu_t(x): R^{n+1} \to [0,1]$. N_t describes a fuzzy subspace χ_t. If $\chi_{r(T)}$ is the entire input–output data space, then the membership function of the root vertex of fuzzy set $N_{r(T)}$ is given by

$$\forall \boldsymbol{x} \in \chi_{r(T)}, \mu_{r(T)}(\boldsymbol{x}) = 1$$

For each vertex $t \in T$, $t \notin r(T)$, if $\{\boldsymbol{x}^i \,|\, \boldsymbol{x}^i \in R^n, i = 1,2, \quad ,K\}$ is the given input data set, then the membership function $\mu_t(\boldsymbol{x})$ is obtained by

$$\mu_t(\boldsymbol{x}) = \mu_{p(t)}(\boldsymbol{x})\hat{\mu}_t(\boldsymbol{x}) \tag{16.3}$$

where

$$\hat{\mu}_t(\boldsymbol{x}) = \frac{1}{1 + \exp\left[\alpha_t\left(\boldsymbol{c}_{p(t)}^T \boldsymbol{x} - \theta_{p(t)}\right)\right]} \tag{16.4}$$

$$\theta_{p(t)} = \sum_{i=1}^{K} \mu_{p(t)}(\boldsymbol{x}^i)\left(\boldsymbol{c}_{p(t)}^T \boldsymbol{x}^i\right) \bigg/ \sum_{i=1}^{K} \mu_{p(t)}(\boldsymbol{x}^i) \tag{16.5}$$

Here, $\hat{\mu}_t(\boldsymbol{x})$ is an instrument membership function and θ_t is the center of gravity of the output data in the t-th given subset. It is clear that $\mu_t(\boldsymbol{x}) \le \mu_{p(t)}(\boldsymbol{x})$, namely, $N_t \subseteq N_{p(t)}$. Hence, the fuzzy subset of every child vertex is included in those of the parent vertex. The fuzzy subset of input data corresponding to $p(t)$ is partitioned into two child fuzzy subsets by the child vertex. Finally, the workspace of input data is partitioned into the fuzzy subspaces of all leaves.

The total approximate error is given by

$$RMSE = \sqrt{\sum_{i=1}^{K}(\hat{y}^i - y^i)^2 \Big/ K} \tag{16.6}$$

The approximate error of vertex $t \in T$ is

$$e_t = \sum_{i=1}^{K}\left[\mu_t(\boldsymbol{x}^i)\left(y^i - c_t^T x^i\right)\right]^2 \tag{16.7}$$

The growing condition of the binary tree is defined as $e_t > \varepsilon$; ε is given from the expected approximate error *ERMSE* as next. From (16.6) and (16.7), there is

$$\sum_{t \in T} e_t = K \times (RMSE)^2$$

Then, one can choose

$$\varepsilon = \frac{K \times (ERMSE)^2}{M} \tag{16.8}$$

where K is the total number of data samples and M is the total number of leaves. If $e_t < \varepsilon$ for all $t \in T$, then

$$\sum_{t \in T} e_t = K \times (RMSE)^2 < M \times \frac{K \times (ERMSE)^2}{M}$$

This means, we have $0 < RMSE < ERMSE$.

16.2.2 Parameter Optimization

The parameters of FT are $\{c_t \,|\, t \in T\}$ and $\{\alpha_t, \theta \,|\, t \in T\}$. Usually, a constant is chosen for the parameter of fuzzy membership function in the basic FT model [2]. However, for realizing better accuracy, a POFT method-based GABPT algorithm is presented in this chapter, which optimizes the parameters of FT.

1. **Training algorithm of** c_T: Let the input–output data be $\{(\boldsymbol{x}^i, y^i) \,|\, \boldsymbol{x}^i \in \boldsymbol{R}^n, y^i \in R, \ i = 1, 2, \quad K\}$. The *RMSE* is defined by (16.6). The weighting vector $c_T = [c_{t_1}, \quad, c_{t_M}]^T$ can be optimized by minimizing *RMSE* with respect to c_T. Thus, a least-squares solution (LS) is obtained through

$$c_T = (X^T X)^{-1} X^T Y \tag{16.9}$$

where $\quad X = [X^1, \quad, X^K]^T, \quad X^i = \left[\dfrac{\mu_{t_1}(\boldsymbol{x}^i)}{\displaystyle\sum_{l=1}^{M}\mu_{t_l}(\boldsymbol{x}^i)}(\boldsymbol{x}^i)^T \quad \dfrac{\mu_{t_M}(\boldsymbol{x}^i)}{\displaystyle\sum_{l=1}^{M}\mu_{t_l}(\boldsymbol{x}^i)}(\boldsymbol{x}^i)^T \right]^T$

$$Y = [y^1 \quad y^K]^T, \quad t_1, \quad, t_M \in T.$$

2. **Training algorithm of α_t:** Here, α_t denotes the width between the two fuzzy subspaces, which can influence the approximate error. From (16.4) and (16.6) it is seen that the *RMSE* of FT is a complex nonlinear function of α_t. The POFT method-based GABPT algorithm is proposed here for α_t training, which can improve the efficiency and accuracy of global optimization.

Back-propagation training is given by

$$\frac{\partial E}{\partial \alpha_t} = \sum_{i=1}^{K} \frac{\partial e^i}{\partial \alpha_t} = \sum_{i=1}^{K} (\hat{y}^i - y^i) * \frac{\partial \hat{y}(\mathbf{x})}{\partial \mu_t(\mathbf{x})}\bigg|_{\mathbf{x}^i} * \frac{\partial \mu_t(\mathbf{x})}{\partial \alpha_t}\bigg|_{\mathbf{x}^i} \tag{16.10}$$

$$\frac{\partial \hat{y}(\mathbf{x})}{\partial \mu_t(\mathbf{x})} = [(\mathbf{c}_t)^T \mathbf{x} - \hat{y}(\mathbf{x})] \bigg/ \sum_{t=1}^{M} \mu_t(\mathbf{x}) \tag{16.11}$$

$$\frac{\partial \mu_t(\mathbf{x})}{\partial \alpha_t} = \mu_{p(t)}(\mathbf{x}) * [1 - \mu_t(\mathbf{x})] * \mu_t(\mathbf{x}) * [(\mathbf{c}_{p(t)})^T \mathbf{x} - \theta_t] \tag{16.12}$$

$$\alpha(i+1) = \alpha(i) - k \left(\sum_{leftnode} \frac{\partial E}{\partial \alpha_t} - \sum_{rightnode} \frac{\partial E}{\partial \alpha_t} \right) \tag{16.13}$$

where k is the training rate.

Algorithm 1

1. Specify a feasible region $[\alpha_{low}\ \alpha_{up}]$ and a resolving power p_{res}. Every chromosome is coded as a binary string with $(\alpha_{up} - \alpha_{low})/p_{res}$ bits. Here, a different chromosome corresponds to a different parameter α_t to be optimized.
2. Specify the expected approximate error *ERMSE*, probability of crossover p_{res}, number of crossover times p_n, probability of mutation p_m, number of iterative times *ganum* and population N, individual number of back-propagation training m, and training rate k.
3. Initialize N chromosomes randomly. Let the fitness function *Fitness* = $1/RMSE$.
4. Select the m largest *Fitnesses* to be trained by (16.13), whereas the others continue to be trained by GA. Find the optimal α_{opt}. Substitute α_{opt} for α_t in (16.4), and compute (16.3), (16.6), and (16.9) to obtain the new *RMSE*.
5. If $RMSE(\alpha_{opt}) \leq ERMSE$, or the limit of the generation number is met, then stop; the optimal α_{opt} is the obtained solution. Otherwise, go to step 4.

The modeling of POFT is carried out by algorithm 2.

Algorithm 2

1. Provide the set of input–output data pairs $(\mathbf{x}^i, y^i)$, $i = 1, 2,\ ,K$, $\mathbf{x}^i \in R^{n+1}$, $y^i \in R$, and the expected error *ERMSE*.
2. Let the membership function of root vertex $\mu_1 = 1$ and $\mathbf{c}_1 \in R^{n+1}$. Calculate $\mathbf{c}_t$ and e_t by (16.9) and (16.7).
3. Calculate *RMSE* by (16.6). If $RMSE \leq ERMSE$, then stop; the number of leaves is the number of FT fuzzy rules. Otherwise, calculate the parameters e_t and ε of every leaf by (16.7) and (16.8). If $e_t > \varepsilon$, then partition N_t into $N_t(r(t))$ and $N_t(l(t))$, and continue the algorithm.
4. For the vertex t, train the parameter α_t by using algorithm 1 and calculate the parameters $\theta_{p(t)}$, $\mu_t(\mathbf{x})$, $\mu_t(\mathbf{x})$, e_t and c_t by (16.3)–(16.5), (16.7), and (16.9), respectively, and go to step 3.

If the limit of the resolving power is met, then the optimal parameter can be obtained from the feasible region $[\alpha_{low} \ \alpha_{up}]$ by the POFT modeling method. For a given level of precision, the new method can simplify the partitioning of data space, decrease the number of fuzzy rules, and reduce the computational load.

16.3 Fuzzy State Model and Control

In this section, a fuzzy state–space model is developed using the POFT approach. A controller is designed using this model.

16.3.1 The Structure of POFT Fuzzy State Equation

The POFT fuzzy state equation model for a class of nonlinear systems based on the adaptive fuzzy tree algorithm may be expressed as

$$\begin{cases} x_1 = x_2 \\ x_2 = f_2(\pmb{x}) + g_2(\pmb{x}, u) \\ x_3 = x_4 \\ x_4 = f_4(\pmb{x}) + g_4(\pmb{x}, u) \end{cases} \tag{16.14}$$

$$y = \pmb{C}\pmb{x}$$

where f_2, g_2, f_4, and g_4 are continuous functions; $u \in R$ and $y \in R$ are system input and system output, respectively; $\pmb{x} = [x_1, x_2, \ , x_n]^T \in R^n$ is the state vector; and $\pmb{C} \in R^{1 \times n}$.

The i-th rule of POFT fuzzy state equation model for a class of nonlinear time-invariant systems is as follows:

$$R^{(i)}: \quad \text{if } \pmb{x} \text{ is } N_{t_i}(t_i \in T), \quad \text{then} \quad (x_n)_{t_i} = (\pmb{c}_{t_i})^T \pmb{x} + d_i u \quad (i = 1, 2, \ , l) \tag{16.15a}$$

where T is the set of leaves and N_{t_i} is the fuzzy set of leaves t_i. Expression (16.15a) is also given by the set of linear state–space models
$R^{(i)}$: if $\pmb{x}$ is $N_{t_i}(t_i \in T)$ then,

$$\begin{cases} \pmb{x} = \pmb{A}_i \pmb{x} + \pmb{B}_i u \\ y = \pmb{C}\pmb{x} \end{cases} \quad (i = 1, 2, \ , l) \tag{16.15b}$$

where $\pmb{A}_i$ and $\pmb{B}_i$ are the corresponding matrices of system and input gain, respectively. Then, the POFT fuzzy state equation model for a class of nonlinear time-invariant systems is as follows:

$$\begin{cases} \pmb{x}(t) = \sum_{i=1}^{l} \mu_{t_i}(\pmb{x})[\pmb{A}_i \pmb{x}(t) + \pmb{B}_i \pmb{u}(t)] = \pmb{A}\pmb{x}(t) + \pmb{B}\pmb{u}(t) \\ y(t) = \pmb{C}\pmb{x}(t) \end{cases} \tag{16.16}$$

where $\mu_{t_i}(\pmb{x})$ describes the subordinate degree of the state vector $\pmb{x}$ to $N_{t_i}(t_i \in T)$ and $\sum_{t_i \in T} \mu_{t_i}(\pmb{x}) = 1$.

The POFT fuzzy state equation model is of a special type. The fuzzy set of state vector **x** is given in the antecedent of the fuzzy rules.

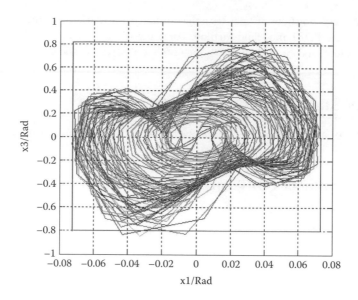

FIGURE 16.1 x_1–x_3 phase plane trajectories.

16.3.2 POFT Fuzzy State Equation Modeling

POFT fuzzy state equation modeling is composed of two parts as given by (16.15b): one is to obtain the data set by partitioning the state space adaptively, and the other is to identify the system matrices A_i and B_i. Then, the POFT approximate models of $f(x_1, \quad , x_n)$ and $g(x_1, \quad , x_n)$ are modeled simultaneously.

Consider Furuta pendulum [5,6] as an example. If the nonlinear dynamic equation (16.14) of the controlled plant is given, a point is chosen as the input sample, which is from the feasible region of the state vector. Let $u(t) = 0$. Then, x_n calculated by (16.14) is chosen as the output sample. The phase plane trajectory is shown in Figures 16.1 and 16.2, as obtained by Furuta pendulum dynamic equation for $u(t) = 0$. The rectangular region is chosen as the feasible region from Figures 16.1 and 16.2. Then the data samples for modeling can be obtained from the region through averaging or somewhat randomly.

Once data samples are chosen, the membership function $\mu_{t_i}(x)$ of fuzzy set $N_{t_i}(t_i \in T)$, A_i in (16.15b) and A in (16.16) can be obtained by the POFT method.

Divide the data samples into L groups. The input matrix B_i is obtained by (16.15a) and (16.17):

$$d_i = \sum_{j=1}^{L} \mu_{t_i}(x(j)) \, g(x(j)) \bigg/ \sum_{j=1}^{L} \mu_{t_i}(x(j)) \tag{16.17}$$

Then B in (16.16) can be determined. The workspace (the feasible region of the state space) is partitioned by the fuzzy membership function $\mu_{t_i}(x)$, $t_i \in T$. The adaptive partition of the workspace has the advantage of automatic adjustment according to the density of output data and the linear approximation error.

Consequently, the POFT fuzzy state equation modeling is given by the following algorithm.

Algorithm 3

1. Select the data samples and an expected error *ERMSE*1 for modeling $f(x_1, \quad , x_n)$ under $u = 0$.
2. Obtain the membership functions $\mu_{t_i}(x)$, $t_i \in T$, and A_i in (16.15) by using algorithm 2 to partition the workspace adaptively.
3. Select a pseudorandom sequence $u(t)$, the data samples, and an expected error *ERMSE*2 for modeling $g(x_1, \quad , x_n)$.
4. Construct an approximate model $\hat{g}(x)$ of the function $g(x_1, \quad , x_n)$ by algorithm 2.

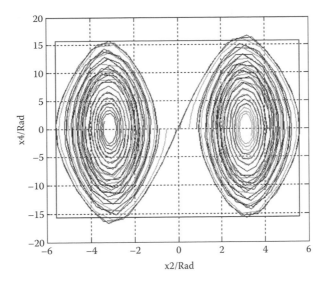

FIGURE 16.2 x_2–x_4 phase plane trajectories.

5. Identify the input matrix B_i by substituting $\hat{g}(x)$ in (16.17).
6. Construct the POFT fuzzy state equation model by (16.16).

16.3.3   Controller Design Based on POFT Fuzzy State Equation

Suppose that every linear subsystem (A_i, B_i) is controllable. Then a state feedback control law L_i can be designed to make $H_i = A_i - B_i L_i$ stable, and the adaptive partition controller can be designed as follows:

$$u(t) = -\sum_{i=1}^{l} \mu_i(x) L_i x(t) = -L x(t) \tag{16.18}$$

Consequently, the POFT fuzzy state equation model of the closed-loop system based on the adaptive partition controller is formulated by

$$\begin{cases} x(t) = \sum_{i=1}^{l} \mu_i(x)(A_i - B_i L_i)x(t) = (A - BL)x(t) \\ y(t) = Cx(t) \end{cases} \tag{16.19}$$

The stability of the system is governed by the first equation of (16.19). It can be represented as follows:

$$x(t) = \left(\sum_{i=1}^{l} \mu_i(x) A_i - \sum_{i=1}^{l} \mu_i(x) B_i \sum_{j=1}^{l} \mu_j(x) L_j \right) x(t)$$

$$= \sum_{i=1}^{l} \sum_{j=1}^{l} \mu_i(x) \mu_j(x)(A_i - B_i L_j)x(t)$$

$$= \sum_{i=1}^{l} \sum_{j=1}^{l} \mu_{ij}(x) H_{ij} x(t)$$

where $\mu_{ij}(x) = \mu_i(x)\mu_j(x)$, $H_{ij} = A_i - B_i L_j$.

Theorem 16.1 gives a sufficient condition for ensuring stability of the POFT fuzzy state equation model of the closed-loop system.

Theorem 16.1 [7]: For fuzzy state equation model of a continuous-time system (16.19), the global closed-loop system is asymptotically stable if there exists a common positive definite matrix P such that the following inequality is satisfied:

$$H_{ij}^T P + P H_{ij} < 0 \quad (i=1,2, \ ,l; \ j=1,2, \ ,l) \tag{16.20}$$

One can obtain the positive-definite matrix P by using MATLAB LMI toolbox because (16.20) is a set of linear matrix inequalities.

16.4 Furuta Pendulum Example

In this section, POFT fuzzy state equation model is applied to the control of Furuta pendulum [5]. The parameters of the Furuta pendulum are given in Table 16.1.

The horizontal arm of the system rotates about a rotating shaft whose rotational angle is θ_0. The pendulum rotates about the pendulum pivot whose rotational angle is θ_1. Also, τ_1 is the input torque applied to the joint of the horizontal arm, and this is the control input. The aim of control is to swing up the pendulum from a pendant position to an upright position. Let $s_i = \sin(\theta_i)$, $c_i = \cos(\theta_i)$, and $i = 0,1$. The kinetic model of Furuta pendulum is given by

$$M q + h(q,q) = \tau \tag{16.21}$$

where

$$M = \begin{bmatrix} I_0 + m_1 \left(L_0^2 + l_1^2 s_1^2 \right) & m_1 l_1 L_0 c_1 \\ m_1 l_1 L_0 c_1 & I_1 + m_1 l_1^2 \end{bmatrix}, \quad q = \begin{bmatrix} \theta_0 \\ \theta_1 \end{bmatrix}, \quad \tau = \begin{bmatrix} \tau_1 \\ 0 \end{bmatrix},$$

$$h(q,q) = \begin{bmatrix} \frac{1}{2} m_1 l_1^2 \sin(2\theta_1)\theta_1 & \frac{1}{2} m_1 l_1^2 \sin(2\theta_1)\theta_0 - m_1 l_1 L_0 \sin(2\theta_1)\theta_1 \\ \frac{1}{2} m_1 l_1^2 \sin(2\theta_1)\theta_0 & 0 \end{bmatrix} q - \begin{bmatrix} -c_0\theta_0 \\ m_1 g l_1 \sin(\theta_1) - c_1\theta_1 \end{bmatrix}$$

TABLE 16.1 List of Measured and Estimated Parameter Values

I_0	I_1	L_0	l_1	m_1	c_0	c_1	g
1.75e–2	1.98e–4			5.38e–2	0.118	8.3e–5	
(kgm^2)	(kgm^2)	0.215 (m)	0.113 (m)	(mg)	(Nms)	(Nms)	9.8 (m/s^2)
Inertia of arm about shaft	Inertia of pendulum	Arm length	Length from pendulum-pivot to gravity center	Weight of pendulum	Friction of arm	Friction of pendulum	Acceleration due to gravity

Source: From Furuta, K. and Yamakita, M., *IECON'91*, Kobe, Japan, 1991, pp. 2193–2198. With permission.

Let $x^T = [x_1 \quad x_2 \quad x_3 \quad x_4] = [\theta_0 \quad \dot{\theta}_0 \quad \theta_1 \quad \dot{\theta}_1]$. Then, the following state–space model of Furuta pendulum is obtained:

$$\begin{cases} \dot{x}_1 = x_2 \\ \dot{x}_2 = f_3(x) + g_3(x,u)\tau_1 \\ \dot{x}_3 = x_4 \\ \dot{x}_4 = f_4(x) + g_4(x,u)\tau_1 \end{cases} \quad (16.22)$$

The POFT fuzzy state equation of Furuta pendulum can be formulated by

$$\begin{cases} \dot{x}(t) = \sum_{i=1}^{4} \mu_{t_i}(x)A_i x(t) + \sum_{i=1}^{4} \mu_{t_i}(x)B_i u(t) \\ y(t) = Cx(t) \end{cases} \quad (16.23)$$

Select 625 groups of state vectors from $x_1 = [-0.07, 0.07]$, $x_2 = [-5.8, 5.8]$, $x_3 = [-0.8, 0.8]$, and $x_4 = [-15, 15]$, which are partitioned into five parts as the input samples by analyzing the feasible regions of the state space. Note that x_4 can be obtained by (16.15b).

Using the POFT fuzzy modeling algorithm to construct a POFT model of four leaves, we can obtain A_i, B_i, $(i = 1, 2, 3, 4)$ in (16.23) as follows:

$$A_1 = \begin{bmatrix} 0 & 1 & 0 & 0 \\ 0.0005 & -0.0412 & -6.4297 & 0.035 \\ 0 & 0 & 0 & 1 \\ -0.002 & -32.998 & -18.775 & 1.848 \end{bmatrix}, \quad A_2 = \begin{bmatrix} 0 & 1 & 0 & 0 \\ -0.0036 & -0.094 & -6.351 & 0.0193 \\ 0 & 0 & 0 & 1 \\ -0.0016 & 40.236 & 5.5089 & -1.246 \end{bmatrix}$$

$$A_3 = \begin{bmatrix} 0 & 1 & 0 & 0 \\ 0.0028 & 0.0680 & -6.530 & -0.086 \\ 0 & 0 & 0 & 1 \\ 0.0024 & 40.236 & 5.5075 & -1.246 \end{bmatrix}, \quad A_4 = \begin{bmatrix} 0 & 1 & 0 & 0 \\ 0.0002 & -0.0438 & -6.427 & 0.036 \\ 0 & 0 & 0 & 1 \\ 0.0001 & -30.998 & -18.78 & 1.848 \end{bmatrix}$$

$$B_1 = \begin{bmatrix} 0 \\ 54.527 \\ 0 \\ -18.214 \end{bmatrix}, \quad B_2 = \begin{bmatrix} 0 \\ 54.527 \\ 0 \\ -7.3925 \end{bmatrix}, \quad B_3 = \begin{bmatrix} 0 \\ 54.527 \\ 0 \\ -7.3925 \end{bmatrix}, \quad B_4 = \begin{bmatrix} 0 \\ 54.527 \\ 0 \\ -18.214 \end{bmatrix}.$$

In the controller design, we choose the pole placement method to control each linear subsystem. Suppose that every linear subsystem (A_i, B_i) is controllable, and the state feedback control law is L_i $(i = 1, 2, 3, 4)$. Then the adaptive partition controller is designed by Theorem 16.1 as

$$u(t) = -\sum_{i=1}^{4} \mu_i(x)L_i x(t) = -Lx(t)$$

Accordingly, the POFT fuzzy state equation model of the Furuta pendulum closed-loop system is formulated by

$$
\begin{cases}
x(t) = \sum_{i=1}^{4} \mu_i(x)(A_i - B_i L_i)x(t) = (A - BL)x(t) \\
y(t) = Cx(t)
\end{cases}
\tag{16.24}
$$

Let the expected poles be $(-3.5 \pm 3i)$, and the rest be $(-1.5 \pm 4i)$. Then the state feedback control laws $L_i (i=1,2,3,4)$ of the subsystems can be founded as

$$L_1 = [-6.383 \quad -27.015 \quad -3.083 \quad -3.246], \quad L_2 = [-6.699 \quad -28.524 \quad -3.234 \quad -3.507],$$

$$L_3 = [-7.548 \quad -32.382 \quad -3.659 \quad -4.188], \quad L_4 = [-10.64 \quad -42.352 \quad -4.689 \quad -6.243].$$

Using Theorem 16.1, we can obtain the following positive-definite matrix P by using MATLAB LMI toolbox:

$$
P = \begin{bmatrix}
109.74 & 102.92 & 25.73 & 38.66 \\
102.92 & 121.26 & 10.59 & 16.19 \\
25.73 & 10.59 & 28.46 & 20.07 \\
38.66 & 16.19 & 20.07 & 39.92
\end{bmatrix}
$$

The control system (16.24) is globally asymptotically stable according to Theorem 16.1. Figures 16.3 and 16.4 are the responses of the Furuta pendulum control system with the same initial condition $x = [0.15 - 0.2 \ 0 \ 0]^T$ as controlled by the present method and that given in Reference [8], respectively. By comparison, it is seen that the adaptive controller based on the POFT fuzzy state equation is better than that in [8].

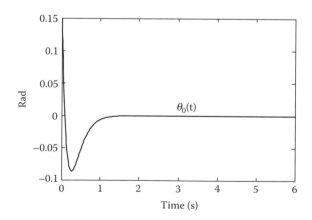

FIGURE 16.3(A) Angle $\theta_0(t)$ of response of adaptive control based on POFT.

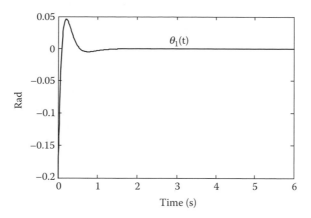

FIGURE 16.3(B) Angle $\theta_1(t)$ of response of adaptive control based on POFT.

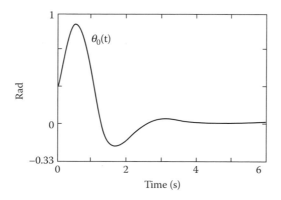

FIGURE 16.4(A) Angle $\theta_0(t)$ of response of flatness-based controller. (From Aguilar-Ibanez, C., *Proceedings of the American Control Conference*, Anchorage, AK, 2002, pp. 1954–1959. With permission.)

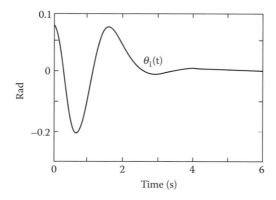

FIGURE 16.4(B) Angle $\theta_1(t)$ of response of flatness-based controller. (From Aguilar-Ibanez, C., *Proceedings of the American Control Conference*, Anchorage, AK, 2002, pp. 1954–1959. With permission.)

16.5 Conclusion

Intelligent mechatronic systems rely on intelligent control methods. Fuzzy logic is widely used for approximating the decision-making process of humans in realizing intelligent mechatronic systems. Because they are nonlinear to varying degrees, model-based control of mechatronic systems will benefit from accurate models. A new fuzzy modeling method based on POFT fuzzy state equation model for a class of nonlinear time-invariant systems has been presented in the chapter. The method partitions the workspace of a nonlinear (mechatronic) system adaptively. As an illustrative example, the Furuta pendulum was simulated for control using the developed approach. It demonstrated the controller design procedure. The controller performance was shown to be better than that from a previous approach for the same problem.

Acknowledgment

This work was supported by the Natural Science Foundation of China under Grants 10276005 and 90205012 and by the National 973 Research Program under Grant 2002cb312205.

References

1. Mao, J., Yue, Y., and Zhang, J., Adaptive-tree-structure-based fuzzy inference system, *IEEE Transactions, Fuzzy System*, Vol. 13, No. 1, 1–12, 2005.
2. Zhang, J., Mao, J., Xia, T., and Wei, K., Fuzzy-tree model: its applications to complex system modeling, *ACTA Automatica Sinica*, Vol. 26, No. 3, 378–381, 2000.
3. Mao, J., Zhang, J., Dai, J., and Wei, K., Approximate limit sampling data using fuzzy-tree model, *Journal of Beijing University of Aeronautics and Astronautics*, Vol. 26, No. 2, 231–234, 2000.
4. Yue, Y., Modeling and control of several classes of complex systems based on computational intelligence, Ph.D. dissertation, Beijing University of Aeronautics and Astronautics, Beijing, China, 2002, pp. 84–106.
5. Furuta, K. and Yamakita, M., Swing up control of inverted pendulum, *IECON'91*, Kobe, Japan, 1991, pp. 2193–2198.
6. Furuta, K., Super mechano-systems: fusion of control and mechanism, *Proceedings of 15th IFAC World Congress*, Barcelona, Spain, 2002, pp. 34–44.
7. Sun, Z., *Intelligent Control Theory and Technology*, Tsinghua University Press, Beijing, China, 1997, pp. 86–87.
8. Aguilar-Ibanez, C., Control of the furuta pendulum based on a linear differential flatness approach, *Proceedings of the American Control Conference*, Anchorage, AK, 2002, pp. 1954–1959.

IV

Mechatronic Design and Optimization

17

Mechatronic Modeling and Design

S. Behbahani

C.W. de Silva

Summary

In view of the presence of components from different engineering domains in a mechatronic system, integrated design of these systems requires a domain-free simulation tool. Bond graphs (BGs) represent a lamped-parameter, domain-free method that provides a core language to represent components in different domains in a unified modeling environment. In addition, it has unlimited growth capability for exploring a wide range of system topologies in an optimization process. This key feature of BGs is used in the development of an evolutionary mechatronic tool, which is illustrated in Chapter 18. The present chapter gives a matrix-based formulation used in the development of a BG simulation tool [1,2]. It is somewhat similar to and inspired by the finite element method. State–space dynamic equations of the system are derived through BG modeling by analyzing the flow of energy between components. The developed tool has the potential to be integrated with SIMULINK® so that the model of the information domain of the system can be integrated into the power domain.

17.1 Introduction

The term *mechatronics* can be traced back to the late 1960s, but it is primarily in the 1990s that a significant growth of the field was seen with regard to mechatronic developments of electromechanical systems. Different definitions have been proposed for mechatronics. The European Union-sponsored Industrial

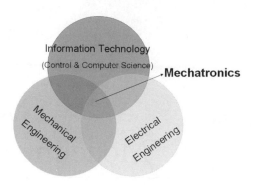

FIGURE 17.1   General representation of mechatronics.

Research and Development Advisory Committee (IRDAC) has suggested the following definition: "Mechatronics is the synergetic combination of precision engineering, electronic control technology, and systems thinking in the design of products and processes" [3]. In some other literature, mechatronics is viewed as the synergistic application of mechanics, electronics, control engineering, and computer science in the development of electromechanical products and systems through integrated design [4]. The general concept is shown in Figure 17.1. Some engineering fields where mechatronic systems and technologies have been incorporated are as follows:

- Machine tools
- Biomechanics
- Robotics
- Automobiles
- Aerospace systems
- Food processing machinery
- Home appliances

Due to widespread application of mechatronic systems and the competition to offer better mechatronic products at lower cost, there has been a renewed attention in the area of optimal design. In conventional design methodologies, first the mechanical part of the machine is designed and built. The next stage is called *instrumentation*, where the electrical components, including sensors and actuators, are designed and added to the system. In the last stage, after the mechanical and electrical hardware are all designed, built, and assembled together, a controller is designed for the machine to perform the expected tasks of the machine. It is easy to see that traditional sequential design methods are not optimal. This is because different subsystems of the machine are designed separately and sequentially, although an integrated and concurrent design approach is ideally necessary due to dynamic interactions and operational dependencies between components [5–8]. Some clear advantages of a concurrent design approach are the following [1,5]:

- Increased efficiency
- Cost effectiveness
- Ease of system integration
- Ease of cooperation with other systems
- Better component matching
- Increased reliability

17.2 Bond Graphs (BGs)

Figure 17.2 shows the schematic diagram of a typical mechatronic system. Basically, there is a dynamic process (or plant) whose dynamic behavior has to be controlled to achieve a desired response. This figure shows the main subsystems/components that are present in a mechatronic system. They are interconnected through the flow of either information or power. The entire system can be divided into two domains:

1. High-power domain (power domain): Some components of the system operate under high power. Their connections to other parts are also associated with high power/energy transfer. The variation of the power variables is an important issue, which must be analyzed in this domain.
2. Low-power domain (information domain): Components in this domain of the system are connected together through low-power energy transfer links. This low-power energy transfer is needed in the transformation of important information about the system. The associated low-power energy transfer does not have a significant effect on the overall energy transfer in the system and can be neglected in comparison with the energy transfer in the power domain; the information transfer associated with it is the main aspect of this domain, which has to be considered in system analysis.

These two domains can clearly be detected in the representation of a mechatronic system, as shown in Figure 17.2. The high-power energy transfers are shown by half arrows, whereas the information transfers are shown by full arrows. The two domains are connected with each other through two basic components: the sensor and the power amplifier. By means of a sensor, an output variable (or a state variable) of the system is measured and fed into the information domain, whereas the power amplifier acts as a power source modulated by its input information signal.

Due to the component interactions, generally, the entire system should be simulated and analyzed in an integrated manner. For this purpose, a common language is required that can effectively function in different engineering fields involved in a mechatronic system. The most commonly used modeling/ simulation tools are appropriate only for a single specific domain. Hence, they are not particularly appropriate for mechatronic systems. BGs have proved to be an effective modeling method for mixed systems [9–11]. They use a unified graphical representation for lumped systems, which provides a common and core language for describing the basic elements and connections in different engineering fields that typically appear in a mechatronic system. Figure 17.3 shows an example of a BG model. The basic elements and connections will be explained in succeeding sections.

BG is based on analyzing the flow of energy, which is the product of a *flow* variable (e.g., velocity, current, and flow rate) and an *effort* variable (e.g., force, voltage, and pressure), through the interconnections of components called *ports*. The final result of a BG model is the state–space dynamic equations of the system. The information domain of the system can be modeled by the block diagram representation and be connected to the power domain by a modulated power source, as shown in Figure 17.3.

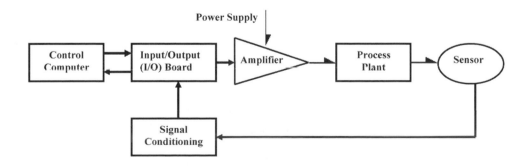

FIGURE 17.2 Schematic diagram of a typical mechatronic system.

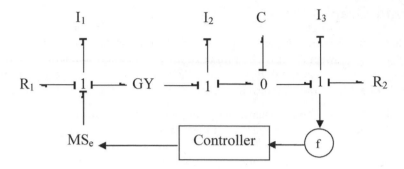

FIGURE 17.3 A simple BG model.

BGs have three embedded strengths in design applications [12]:

- A wide range of systems can be modeled because of the multi- and interdomain nature of BGs. Multidomain systems (combined electrical, mechanical, hydraulic, pneumatic, and thermal systems) can be modeled using a common notation, which is especially important in the design of mechatronic systems.
- It provides high computational efficiency in the evaluation of design alternatives. This feature makes it suitable for use in design optimization.
- BGs possess a graphical methodology to represent mixed systems. Any system model can be generated by a combination of bond and node components, rather than by the specification of equations.

The graphical representation of BGs has an open architecture. This means that a free composition of bonds and nodes can be added to different locations of a model to create a new model with a new topology. In addition, any two system models can exchange a branch of their BG models to create two new models, which have some characters inherited from the initial models. This unlimited growth capability of BGs provides the capability to explore a wide range of topologies in the process of design and optimization of a mechatronic system, especially by integrating with evolutionary algorithms. In other words, the use of BGs for mechatronic design can result in an optimization tool that is not restricted just to the sizing but to optimizing the topology of the model as well. This important feature of BGs has been utilized in the development of the evolutionary mechatronic tool that is further illustrated in Chapter 18.

17.3 BG Terminology

BG modeling is based on the analysis of the flow of energy between components or elements. The connection of an element to another element is called a *port*, where there is power interaction between the particular element with the rest of the system (Figure 17.4). The flow of energy between ports is shown by a line, which is called *bond*. When two elements are interconnected, there are power interactions between them. The power can be in different forms: notably, mechanical, electrical, hydraulic, or thermal. Power is the product of two variables called *power variables* (Table 17.1); mechanical power is the product

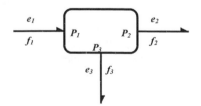

FIGURE 17.4 A multiport element.

TABLE 17.1 Power and Energy Variables in Different Domains

Generalized Variable	Mechanical (Translation)	Mechanical (Rotational)	Electrical	Hydraulic
Effort, e	Force, F	Torque, τ	Voltage, v	Pressure, P
Flow, f	Velocity, v	Angular velocity, ω	Current, i	Flow rate, Q
Momentum, p	Linear momentum, p	Angular momentum, H	Flux linkage, λ	Pressure momentum, P
Displacement, q	Displacement, x	Rotation angle, θ	Charge, q	Volume, V

of force and velocity, whereas electrical power is the product of voltage and current. BGs provide a common language to classify all power variables in different fields. In BG terminology, all power variables are called either *effort variables* or *flow variables*, and are denoted by the symbols e and f, respectively. Force, torque, pressure, and voltage are examples of effort variables, whereas linear velocity, angular velocity, flow rate, and current are examples of flow variables. The flow of power between two connected elements through a port can then be expressed as the product of an effort variable and a flow variable:

$$P(t) = e(t) . f(t) \qquad (17.1)$$

There are two other important variables in describing a dynamic system. These variables are *momentum* p and the *displacement q*, and are called *energy variables* (Table 17.1). Momentum and displacement are the time integrals of effort and flow, respectively.

$$p(t) = \int e(t) . dt \qquad (17.2)$$

$$q(t) = \int f(t) . dt$$

Then, the energy flow can be described as

$$E(t) = \int p(t) . dq = \int q(t) . dp \qquad (17.3)$$

The direction of the power flow is shown by a half arrow on the bond. It is the direction of the power flow at any instant of time when both the effort and flow happen to be positive (or, both negative).

17.4 Basic Elements and Junctions

There are several primary 1-port element types, as follows:

- *Resistance, R*: The 1-port resistance is an element in which there is a static relation between effort and flow variables, like the electrical resistance and the mechanical damper (Figure 17.5). If a linear relation can be expressed between effort and flow variables, the resistance is called linear resistance. Here

$$e = R . f \qquad (17.4)$$

  Otherwise, it is a nonlinear resistance. Instead of defining a resistance, one can define its inverse, called *conductance*.
- *Inductance, I*: The 1-port inductance is an element in which a constitutive law exists between the momentum p and the flow f. In other words, one can express a static relation between the effort

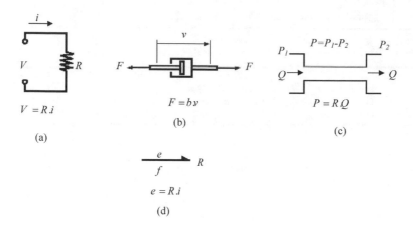

FIGURE 17.5 Resistor element: (a) electrical resistance, (b) mechanical damper, (c) hydraulic connection with pressure drop, (d) general representation of resistor element in BG.

variable and the derivative of the flow variable. Examples are electrical inductance and mechanical mass/inertia (Figure 17.6):

$$p = I.f$$
$$e = I.f$$

(17.5)

- *Capacitance, C:* The 1-port capacitance is an element in which displacement q and effort e are related through a static constitutive law. Examples are electrical capacitor and mechanical spring (figure 17.7). Then,

$$q = C.e$$
$$f = C.e$$

(17.6)

- *Effort source, S_e:* A source of effort is an element that supplies energy to the system with a specified effort value, e.g., force generator, voltage generator, and pressure supply. The effort variable in the

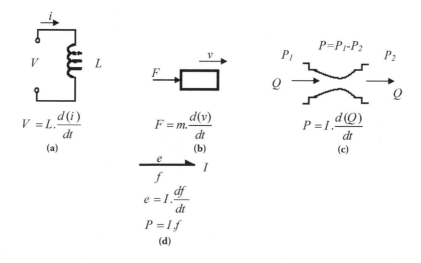

FIGURE 17.6 Inductance element: (a) electrical inductance, (b) mechanical mass/inertia, (c) hydraulic nozzle, (d) general representation of an inductance element in BG.

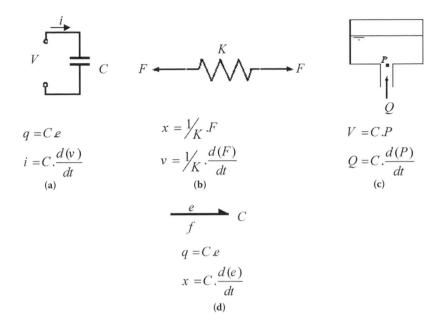

$$q = C\,e$$
$$i = C.\frac{d(v)}{dt}$$

(a)

$$x = \frac{1}{K}\,F$$
$$v = \frac{1}{K}.\frac{d(F)}{dt}$$

(b)

$$V = C.P$$
$$Q = C.\frac{d(P)}{dt}$$

(c)

$$q = C\,e$$
$$x = C.\frac{d(e)}{dt}$$

(d)

FIGURE 17.7 Capacitor element: (a) electrical capacitor, (b) mechanical spring, (c) hydraulic accumulator or gravity tank, (d) general representation of a capacitor element in BG.

output port is impressed by this element. On the other hand, this element accepts a flow variable impressed by the rest of the system to it.

- *Flow source, S_f*: A source of flow is an element that supplies energy to the system with a specified flow value, e.g., mechanical shaker, current generator, and constant-displacement pump. The flow variable in the output port is impressed by this element. On the other hand, this element accepts an effort variable impressed by the rest of the system to it.
- *Modulated effort source, MS_e*: This is a special case of effort source where its value is specified by an input information signal. It is useful in modeling the interconnection between power and information domains of a mechatronic system (see Section 17.2).
- *Modulated flow source, MS_f*: Similar to an effort source, MS_f is a flow source whose value is specified by an input information signal.
- *Effort sensor*: It is an element that measures the effort variable in a port and generates an information signal.
- *Flow sensor*: This is an element that measures the flow variable in a port and generates an information signal.

Also, there are two types of junctions, as follows:

- *1-junction*: In this junction, all connected bonds have a common flow variable, and their effort variables add to zero (compatibility). For the 0-junction shown in Figure 17.8(a), one may write

$$f_1 = f_2 = f_3 = f_4$$
$$e_1 = e_2 + e_3 + e_4 \tag{17.7}$$

- *0-junction*: It represents a junction where all connected bonds have a common effort variable and their flow variables add to zero (continuity). For the 0-junction shown in Figure 17.8(b), one may write

$$e_1 = e_2 = e_3 = e_4$$
$$f_1 = f_2 + f_3 + f_4 \tag{17.8}$$

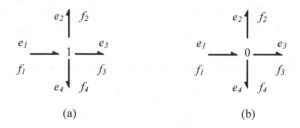

FIGURE 17.8 Multiport junctions: (a) 1-junction, (b) 0-junction.

$$\xrightarrow{\;1\;} TF \xrightarrow{\;2\;} \qquad\qquad \xrightarrow{\;1\;} GY \xrightarrow{\;2\;}$$

$$\text{(a)} \qquad\qquad\qquad\qquad \text{(b)}$$

FIGURE 17.9 Two port elements: (a) the transformer, (b) the gyrator.

Furthermore, there are two types of 2-port elements: the *transformer TF* and *gyrator GY* elements. Figure 17.9(a) shows the representation of a transformer in BGs. If this transformer is an ideal transformer, its constitutive laws are given by

$$e_1 = r_T \cdot e_2$$
$$r_T \cdot f_1 = f_2 \tag{17.9}$$

where r_T is called *transformer modulus*. The transformer can be used in the modeling of rigid levers, gear pairs, electrical transformers, and hydraulic rams.

Figure 17.9(b) shows the representation of a gyrator in BGs. The constitutive laws of the gyrator are

$$e_1 = r_G \cdot f_2$$
$$r_G \cdot f_1 = e_2 \tag{17.10}$$

where r_G is *gyrator modulus*.

17.5   Object-Oriented Modeling

BG modeling of a mechatronic system can be viewed as an object-oriented modeling (OOM) method even though BGs were developed by Professor Henry M. Paynter at Massachusetts Institute of Technology in 1959 [13], long before the concept of object-oriented modeling was introduced. Basically, the concept of OOM means that different subsystems of a mixed machine can be modeled separately and be interconnected to create the overall model. Furthermore, subsystem models can be reused again in modeling other machines [14–16].

A typical machine has a hierarchical structure. The system may be composed of some lower-level subsystems. Each subsystem may also consist of some lower-level subsystems. In object-oriented BG modeling, the model of each subsystem can be considered as an *object*. Consequently, a general model for components that commonly appear in mechatronic systems can be established and reused wherever it is necessary. Each subsystem model (say object) has several inherent parameters, i.e., values of the parameters in the submodel. To be able to reuse the model of a component as an object, its model needs to be generated so that it receives all its inherent parameters from outside through the so-called *interfaces*.

Whenever the model is used as an object, its inherent parameters should be provided by the higher-level submodel (that can be the main model) which has used that particular object.

The following procedure can be used for modeling a mixed system:

- Properly shrink the system into several smaller parts and construct a hierarchical pattern for it.
- If you have access to the model of a component in the lowest level, find the required inherent parameters for it, and simply substitute it into the model. Otherwise, construct a BG model for that particular component. It is recommended that the model of the component as an object be generated; i.e., it receives all its inherent data through interfaces. By this, if a similar component exists in other parts of the machine, you are able to reuse the created object just by defining its actual parameters. In addition, you may add it to a so-called library of objects for modeling other machines.
- Create the model of the higher-level subsystem or the main model by interconnecting the submodels and properly defining their inherent parameters.

17.5.1 Modeling of Mechanical Systems

For systems of primarily mechanical nature, use the following procedure [11]:

- Detect all the mechanical nodes, i.e., nodes with distinct velocity, and find the mass (or moment of inertia) at the node.
- Consider a 1-junction for each mechanical node. Attach the mass/inertia element directly to its corresponding 1-junction.
- Any force-generating element, mechanical capacitor (spring), or mechanical resistor (damper) in the system will appear between two mechanical nodes. If there is such an element between two nodes, insert a 0-junction between the corresponding 1-junctions and join that element to it (Figure 17.10a). If there is more than one element between two nodes, first bundle them together by connecting them to a 1-junction and then connect the unit to the 0-junction that you have inserted between the corresponding nodes (Figure 17.10b).
- Assign power directions to all bonds.
- If one of the 1-junctions represents the zero-velocity reference point, delete that 1-junction and its corresponding bond.

17.5.2 Modeling of Electrical Systems

To model an electrical circuit, use the following procedure [11]:

- Detect all the nodes in the circuit that have a distinct voltage. Consider a 0-junction for each of these nodes.
- If there is a 1-port element (resistor, inductance, capacitor, or source element) between two nodes of the circuit, insert a 1-junction between the corresponding 0-junctions and join that element to

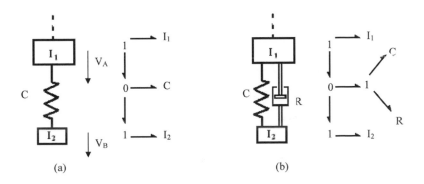

(a)   (b)

FIGURE 17.10 Insertion of mechanical elements between distinct nodes: (a) single element, (b) multiple elements.

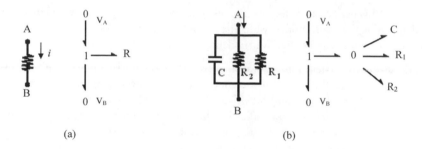

(a)   (b)

FIGURE 17.11 Insertion of electrical elements between distinct nodes: (a) single element, (b) multiple elements.

it (Figure 17.11a). If there is more than one element between two nodes, first bundle them together by connecting them to a 0-junction and then connect the resulting unit to the 1-junction that you have inserted between corresponding nodes (Figure 17.11b).

- Assign power directions to all bonds.
- If there is an explicit ground in the circuit, delete its corresponding 0-junction and its bond; otherwise, choose an arbitrary ground and delete its corresponding 0-junction and bond.

17.5.3 Modeling of Hydraulic Systems

Modeling of hydraulic systems is very similar to the procedure of modeling electrical circuits. The following procedure may be used for hydraulic systems (Karnopp et al., 2000) [11]:

- Consider a 0-junction for any point in the hydraulic circuit that has distinct pressure.
- If there is a 1-port hydraulic component between two distinct nodes, insert a 1-junction between the corresponding 0-junctions and join the element to it (similar to Figure 17.11a). If there is more than one element between two nodes, first bundle them together by connecting them to a 0-junction and then connect the resulting unit to the 1-junction that you have inserted between the corresponding nodes (similar to Figure 17.11b).
- Assign power directions to all bonds.
- Delete the 0-junction corresponding to the reference pressure and its bond.

17.5.4 Simplification of BG Model

Once a BG model has been created, one may be able to simplify it by using the following set of rules:

1. Eliminate redundant junctions: If a junction is the only link between two other junctions or between a junction and a 1-port element, that junction can be eliminated (Figure 17.12).

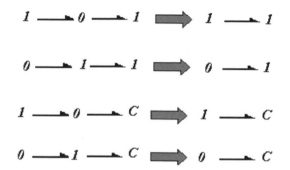

FIGURE 17.12 The conditions under which a junction is redundant and can be eliminated.

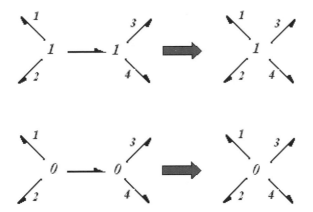

FIGURE 17.13 The conditions under which two junctions can be melted together.

2. Melt equal junctions: If two or more similar junctions are connected together, they can be merged together (Figure 17.13).

   If there is a transformer between two junctions of the same type, then they can be melted, and elements in one of them can be transferred into the other, but a transformation has to be applied on them. If the transformer connects node n to node m, and flow direction is from n to m, then, if elements in node m are transferred to node n, their equalized value will be as follows (Figure 17.14):

$$R_{eq} = R \cdot r_T^2$$
$$I_{eq} = I \cdot r_T^2$$
$$C_{eq} = \frac{C}{r_T^2} \qquad (17.11)$$
$$S_{e_{eq}} = S_e \cdot r_T$$
$$S_{f_{eq}} = \frac{S_f}{r_T}$$

where r_T is the module (ratio) of the transformer. If node n is melted into m, then the inverse of r_T should be substituted into the foregoing equations.

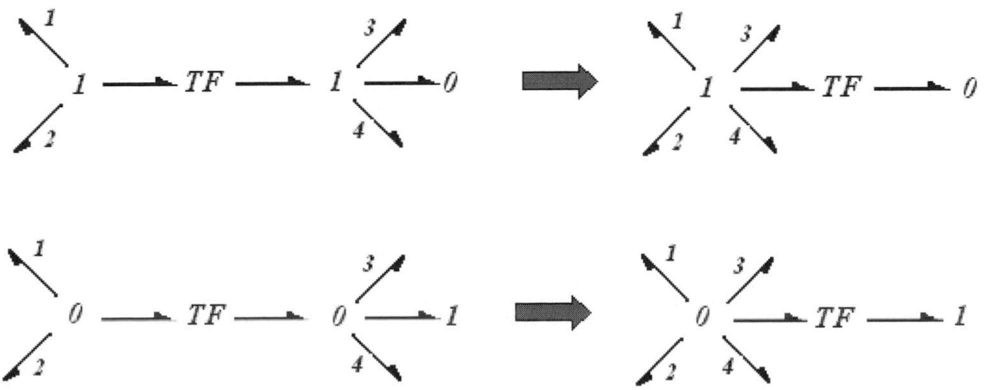

FIGURE 17.14 Melting of equal junctions connected by a transformer.

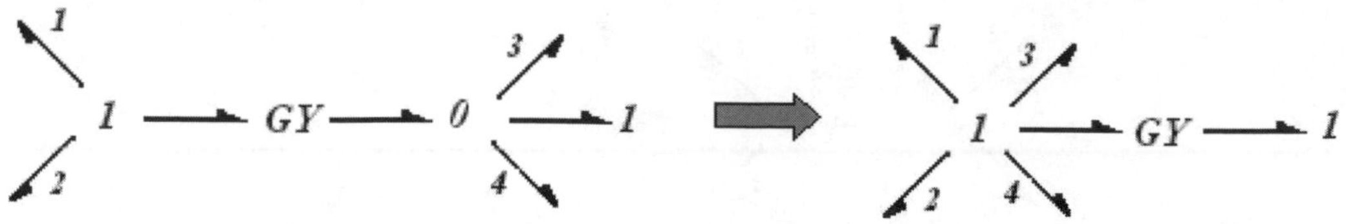

FIGURE 17.15 Melting of equal junctions connected by a gyrator.

If there is a gyrator between two junctions of dissimilar type, then they can be melted together. The transformation of the elements of one of them to the other is more complex than in the case of a transformer. In addition to the change in the magnitude of the elements, their nature also changes. A capacitor appears as an inductance and vice versa. A source of effort becomes a source of flow and vice versa.

If a gyrator connects node n to node m, and the flow direction is from n to m then, if elements in node m are transferred into node n, their equalized value will be as follows (figure 17.15)

$$R_{eq} = \frac{r_G^2}{R}$$

$$I_{eq} = C \cdot r_G^2$$

$$C_{eq} = \frac{I}{r_G^2} \qquad (17.12)$$

$$S_{e_{eq}} = S_f \cdot r_G$$

$$Sf_{eq} = \frac{S_e}{r_G}$$

where r_G is the module of the gyrator. If node n is melted into node m, then the inverse of r_G should be substituted into the foregoing equations.

3. Melt similar elements connected to a junction: A 0-junction is like a parallel connection of electrical circuits. Therefore, if there are similar elements connected to a 0-junction, the following rules are used:

$$\frac{1}{R_{eq}} = \frac{1}{R_1} + \frac{1}{R_2}$$

$$\frac{1}{I_{eq}} = \frac{1}{I_1} + \frac{1}{I_2} + \qquad (17.13)$$

$$C_{eq} = C_1 + C_2 +$$

A 1-junction is like a series connection in electrical circuits. Therefore, if there are similar elements connected to a 1-junction, the following rules are used:

$$R_{eq} = R_1 + R_2 +$$

$$I_{eq} = I_1 + I_2 + \qquad (17.14)$$

$$\frac{1}{C_{eq}} = \frac{1}{C_1} + \frac{1}{C_2} +$$

17.5.5 Causality Analysis

Both flow and effort variables exist at a port, and only one of them can be controlled. One of them will be the input to the port, and the other one will be the output from that port. In other words, when two components are connected together by a bond, the effort variable in the bond causes one of the elements to respond with a flow, and this flow causes the other element to respond with an effort. In BGs, the direction of the effort signal is shown by a short perpendicular line at the end of a bond, which is called a *causality stroke*.

Causality analysis is a very important process in BG modeling because of several reasons. First, it is required for derivation of state–space equations. In each junction, one, and only one, of the connected bonds has dominant causality. For example, all bonds connected to a 0-junction have similar effort variables. Therefore, only one of the bonds connected to a 0-junction can have the effort as its output, and all the other bonds will have opposite causality. It will be explained later that, in writing the equations for each junction, it is necessary to know which connected bond has dominant causality.

Second, causality analysis checks whether the model is feasible or not. Any causality conflict in a model implies that the model is not physically feasible. Chapter 18 explains an evolutionary mechatronic design tool that integrates a BG simulation tool with genetic programming [1,2,12]. Genetic programming is a random search tool. As a result, there exists the possibility of generating a large number of unfeasible models. It follows that causality analysis is useful in preventing time wastage by avoiding the analysis of models that are not feasible.

Table 17.2 gives the possible causal forms and causal relations for basic 1-port elements [11]. Source elements have a restricted causal form because they impress either effort or flow output, regardless of the system connected to them. In storage elements (e.g., capacitor and inertia), the integral causality is preferred.

Tables 17.3 and 17.4 give the possible causal forms for 2-port elements and multiport junctions [11]. In a 0-junction, it is important that only one connected port has effort output. Others accept the associated effort as input and impress flow outputs. 1-Junctions have opposite causality. The port that has the dominant causality specifies the flow value of all other connected ports. Other ports accept this flow variable as input and impress effort outputs.

TABLE 17.2 Possible Causality Forms for 1-Port Elements and Associated Equations

Elements	Causal Forms	Causal Relations
Effort source	S_e	$e(t) = E(t)$
Flow source	S_f	$f(t) = f(t)$
Resistor	R R	$e = R.f$ $f = e/R$
Capacitor	C C	$e = \dfrac{1}{C}\int_0^t f \,.\, dt = \dfrac{q}{C}$ $f = \dfrac{d}{dt}(C.e)$
Inertia	I I	$f = \dfrac{1}{I}\int_0^t e.dt = \dfrac{p}{1}$ $e = \dfrac{d}{dt}(I.f)$

TABLE 17.3 Possible Causality Forms for 2-Port Elements and Associated Equations

Elements	Causal Forms	Causal Relations
Transformer	$\longrightarrow\!\!\mid TF \longrightarrow\!\mid$	$f_1 = \dfrac{f_2}{r_T},\ e_2 = {e_1}\Big/{r_T}$
	$\mid\!\!\longrightarrow TF \mid\!\!\longrightarrow$	$e_1 = r_T.e_2,\ f_2 = r_T.f_1$
Gyrator	$\mid\!\!\longrightarrow GY \longrightarrow\!\mid$	$e_1 = r_G.f_2,\ e_2 = r_G.f_1$
	$\longrightarrow\!\!\mid GY \mid\!\!\longrightarrow$	$f_1 = {e_2}\Big/{r_G},\ f_2 = {e_1}\Big/{r_G}$

The underlying process for causality analysis of a BG model is as follows [11]:

1. Source elements have a restricted causality form. Consider a source element. Assign its required causality. Extend the causal implications as far as possible using the causality requirement of 2-port and multiport elements.
2. Repeat step 1 for all sources. If any causal conflict happens in this stage, it can be implied that the model is not physically feasible.
3. Consider a storage element whose causality is not specified in the previous stages. Assign its preferred causality. Extend the causal implications as far as possible using the causality requirement of 2-port and multiport elements.
4. Repeat step 3 for all unassigned sources.
5. If the causality of the model is not complete yet, choose any unassigned R-element. Assign an arbitrary causality to it. Extend the causal implications as far as possible using the causality requirement of 2-port and multiport elements.
6. Repeat step 5 for all unassigned R-elements.
7. If the causality of the model is still not complete, choose an unassigned bond and assign arbitrary causality to it. Extend the causal implications as far as possible using the causality requirement of 2-port and multiport elements.
8. Repeat step 7 for all unassigned bonds.

If a conflict appears during the causality analysis process, the model is unacceptable and therefore makes dynamic simulation unnecessary.

TABLE 17.4 Possible Causality Forms for Multiport Junctions and Associated Equations

Elements	Causal Forms	Causal Relations
0-Junction	$\xrightarrow{1}\!\!\mid 0 \xrightarrow{3}$ with 2 branch	$e_2 = e_3 = e_1,$ Bond 1 is dominant $f_1 = -\ (f_2 + f_3)$
1-Junction	$\xrightarrow{2}\!\!\mid 1 \mid\!\!\xrightarrow{3}$ with 1 branch	$f_2 = (f_3 + f_1),$ Bond 1 is dominant $e_1 = -\ (e_2 + e_3)$

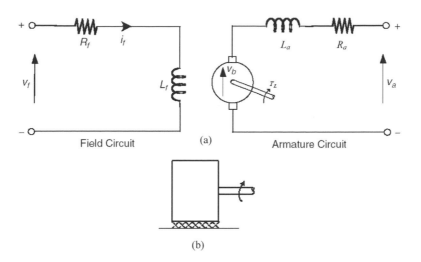

FIGURE 17.16 Equivalent circuit of a DC motor.

17.5.6 Incorporation of the Information Domain

Components of a mechatronic system are linked with each other through the flow of energy and/or flow of information. The controller part of a mechatronic system usually has very low levels of energy transfer that are negligible compared to the energy flow between other parts. However, the information flow associated with this low-level energy flow is not negligible. As a result, a mechatronic system may be divided into two domains: power domain and information domain. The BG modeling considers the flow of energy between components and derives state–space dynamic equations of the system. In other words, it accounts for the high energy part of the system. The resultant dynamic equations from a BG tool can be linked to SIMULINK® to incorporate the control and information domains of the system into the power domain and simulate the equations. Another benefit of this is that numerical subroutines of Matlab® and SIMULINK® can be used to simulate the state–space equations.

Example 1: A Servo System

The process of modeling a DC servo system is presented as an example. Consider a conventional DC motor with separate windings in the stator and the rotor. The equivalent circuit for this example is shown in Figure 17.16 [4].

The diagram of the mechanical part is shown in Figure 17.16(b). The mechanical loading of the armature can be considered as inertia J_m, and the mechanical resistance can be accounted for by an equivalent damping b_m. Flow of the current in the stator circuit generates a magnetic field, which in turn generates a torque to rotate the rotor. The rotor speed causes the flux linkage of the rotor coil with the stator filled, thereby generating a voltage (back e.m.f) in the rotor coil. The generated torque in the rotor is proportional to the current in the rotor coil. On the other hand, the back e.m.f. voltage in the rotor armature is proportional to the rotor speed [4]. In the case of ideal electrical-to-mechanical energy conversion, by considering consistent units, it can be shown that

$$\frac{T_m}{i_a} = \frac{V_b}{\omega_m} = k_t$$

Because the effort variable in the mechanical part is proportional to the flow variable in the electrical part, and also the effort variable in the electrical part is proportional to the flow variable in the mechanical part, the interconnection between the mechanical part and the electrical part can be viewed as a gyrator. Figure 17.17(a) shows the BG model for this DC servo motor.

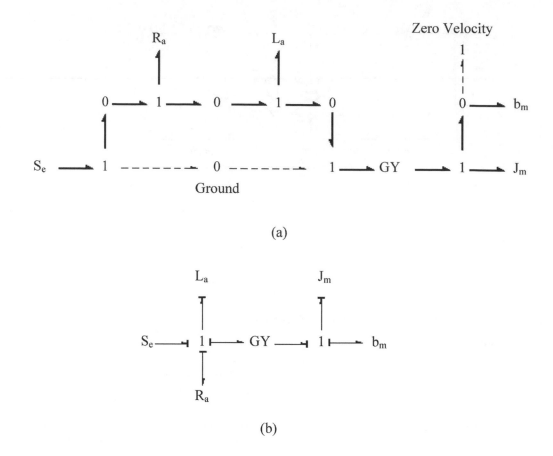

FIGURE 17.17 BG model of DC motor: (a) before simplification, (b) after simplification and causality analysis.

Now we can simplify the obtained model using the rules outlined in Section 17.5.4. The model, after simplification and causality analysis, is shown in Figure 17.17(b).

Example 2: Piezoelectric Accelerometer

A piezoelectric crystal has the unique property that it generates a voltage when a load is applied on it. This phenomenon has been employed in piezoelectric accelerometers to measure an acceleration by measuring the inertia force caused by the acceleration. Piezoelectric accelerometers are preferred over other types of accelerometers in many engineering applications mainly because of their light weight and high frequency response (up to 1 MHz) [4]. The light weight of these sensors reduces the undesirable loading effects from the sensor to the main machine on which the accelerometer is mounted. However, loading effects from the outside world to the sensor can be significant because the generated voltage is very small (on the order of 1 mV). This is because these sensors have high output impedance. To reduce loading effects, special impedance-transferring amplifiers have to be employed [4].

Figure 17.18 shows the schematic diagram of a piezoelectric accelerometer [9]. The crystal and the inertial mass are restrained by a high-stiffness spring. This high-stiffness spring, along with the low inertia of the system, causes a high fundamental natural frequency for the sensor.

Now we illustrate the process of BG modeling of this system [9,17]. In the mechanical part, two nodes with distinct velocities can be detected: one is for the housing motion and the other for the oscillator. As the housing motion is the input to this system, we model it as a source of flow. The mass of the oscillator is modeled as an inductor element connected to the 1-junction with a velocity of y. The piezoelectric effect is produced by the relative displacements between the oscillator and the housing; hence, the connection to the piezoelectric accelerometer has been considered from the 1-junction, which

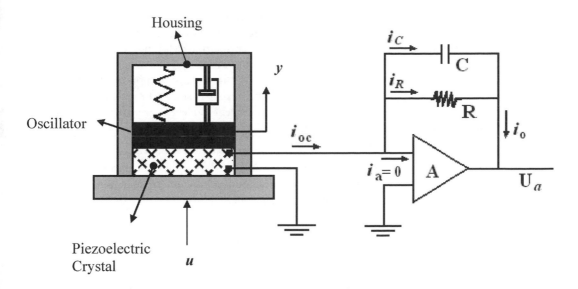

FIGURE 17.18 Schematic diagram of piezoelectric accelerometer.

represents the velocity $u - y$. The next part of the system is the piezoelectric material that converts mechanical power to electrical power. It can be simply modeled as a transformer.

The BG modeling of the operational amplifier is more difficult than the other parts. Figure 17.19 shows the equivalent electrical circuit of an operational amplifier. For a practical op amp, the input impedance Z_i is very high (typically 2 $M\Omega$ [4]); hence, the current between inverting and noninverting input ports is almost zero. The amplifier gain is also very high (typically $10^5 - 10^9$ [4]), and the output impedance is very low. These characteristics indicate that to have a feasibly finite output voltage, the voltage difference between inverting and noninverting input ports should also be zero. In the context of electrical circuit analysis, these twin conditions are modeled using the concept of *virtual earth*. The following two main assumptions are considered in op amp analysis [4]:

- The current between input ports of an op amp is zero.
- The voltage difference between input ports is zero.

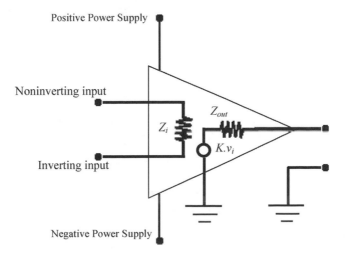

FIGURE 17.19 Equivalent circuit of an op amp.

Now, let us look at these conditions in the BG modeling context. These twin conditions of voltage difference and current between input ports of op amp being zero lead to a causality problem. It requires that both the effort and the flow variables be controllable in this port, which is not feasible according to causality rules. There is a causality conflict in the output port of the op amp as well. The voltage dictated by the op amp circuit is generated by a voltage source. It means that two distinct bonds define the voltage in the output port. However, this is physically possible because it is similar to making two identical voltage sources parallel to each other. Because the ideal equivalent model of the op amp does not have a causality conflict, it may seem that we can include the ideal model in the BG model to solve this causality conflict problem. However, this raises two problems [17]:

- When a very high gain is used, the dynamic behavior of the system may be ill conditioned.
- The second-mentioned property of the op amps (i.e., the voltage difference between input ports is zero) is derived by the assumption that the output voltage is finite. However, the BG itself cannot provide this condition.

The causality conflict raised from the use of the simplifying concept of virtual earth leads to the expression of *bicausality* concept in BGs [17]. In particular, we define a new nonstandard and nonrealistic BG element that is both a source of flow and a source of effort, and is shown by *SS* in the BG representation. Note that such a nonstandard element does not exist in reality. As a consequence, one has to consider a new nonstandard causality condition where there are causality strokes on both sides of a bond. To make a distinction between this nonstandard condition and the normal conditions, we indicate a bicausal bond by half causality strokes on both sides [17].

Figure 17.20 shows the BG model of the piezoelectric accelerometer and its associated causality analysis. The state–space equation for the op amp may be written as

$$q_C = i_{oc} - i_R = i_{oc} - \frac{v_C}{R} = i_{oc} - \frac{q_C}{R \cdot C}$$

$$v_{out} = v_C - 0 = \frac{q_C}{C}$$

Consequently, one obtains

$$C\frac{dv_{out}}{dt} + \frac{v_{out}}{R} = i_{oc}$$

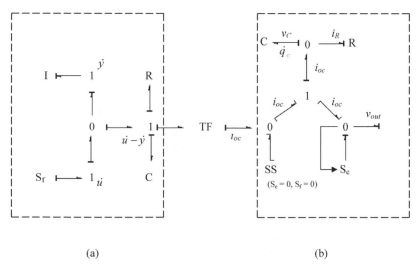

(a)   (b)

FIGURE 17.20 BG model of the piezoelectric accelerometer.

17.6   Derivation of State–Space Equations

BG modeling leads to straightforward derivation of the state–space dynamic equations of a system. These equations can be solved/simulated in the time domain by using numerical methods, or they can be used to find the transfer function of the system. In this section, a new BG formulation is described, which facilitates the creation of a BG program [1,2]. It is based on the matrix representation of the governing equations in local nodes and then creation of global state–space equations by assembling them. This is somewhat similar to and inspired by the finite element representation of equations.

The process of obtaining the state–space equations using BGs is accomplished in several stages, including the following:

1. Local matrix formation
2. Assembly
3. Final reduction of the matrices

17.6.1   Local Matrix Formation

The first step is to analyze the power flow and derive the local equations for each junction. Energy exchange from a junction can be considered in two parts:

1. Energy exchange to the basic elements connected to the junction
2. Energy exchange to other junctions connected to it

In each junction, one connected bond has dominant causality, either a bond connected to a 1-port element or a bond connected to another junction. As mentioned before, all bonds connected to a 1-junction have a common flow variable, and all bonds connected to a 0-junction have a common effort variable. The bond that has the dominant causality is the bond that specifies this common magnitude in the junction, and it has to flow through all the other bonds. Effort and flow variables in the dominant bond are power variables, which are used to express the governing equations.

Three local equations are written for each junction [1,2]:

1. Compatibility/continuity equations: For 0-junction flow variables, add to zero (continuity equation), and for 1-junction effort variables, add to zero (compatibility equation).
2. The equation that expresses the common variable in the junction (i.e., flow variable in 1-junction, or effort variable in 0-junction): It is actually the constitutive law of the element that specifies the common variable in the junction. If an interconnecting bond has dominant causality, no equation is written in this stage.
3. The equation describing the variation of the coenergy variable in the junction.

Consider a 1-junction. The most common case is that a bond connected to an inductor has dominant causality in the junction (Figure 17.21). In this case, the momentum in this bond is one of the state variables of the system. If a capacitor is also connected to the junction, the displacement in the bond connected to this capacitor is also a state variable of the system. If no capacitor is connected, the third equation is left empty.

The equations in the junction are as follows:

$$\dot{p}_1 = e_1 = -e_2 - e_3 + S_e + \sum e_{bonds} = -R \cdot f - \frac{q_2}{C} + S_e + \sum e_{bonds}$$

$$f = \frac{p_1}{I}$$

(17.15)

$$\dot{q}_2 = f$$

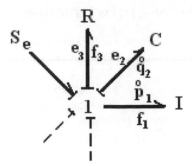

FIGURE 17.21 A typical 1-junction.

or

$$
\begin{Bmatrix} p_1 \\ f \\ q_2 \end{Bmatrix} = \begin{bmatrix} 0 & -R & -1/c \\ 1/I & 0 & 0 \\ 0 & 1 & 0 \end{bmatrix} \cdot \begin{Bmatrix} p_1 \\ f \\ q_2 \end{Bmatrix} + \begin{Bmatrix} S_e \\ 0 \\ 0 \end{Bmatrix} + \begin{Bmatrix} \sum e_{bonds} \\ 0 \\ 0 \end{Bmatrix}
\tag{17.16}
$$

The energy exchange with other junctions will be taken into account in the assembly step, which comes next. If it is ignored temporarily, the equations for this typical 1-junction can be written in the following form:

$$
\begin{Bmatrix} p_1 \\ f \\ q_2 \end{Bmatrix} = K_{mat} \begin{Bmatrix} p_1 \\ f \\ q_2 \end{Bmatrix} + F_{mat}
\tag{17.17}
$$

The matrix K_{mat} may be termed a *local stiffness matrix*, and vector F_{mat} may be viewed as a *local force vector*. It should be mentioned that equation 17.17 is still not valid. It will become valid only after bringing the effect of the other bonds into account.

This is the case when an I-element has dominant causality and has assigned the flow variable. If no I-element exists, then an R-element may assign the flow. In this case, one has

$$
\begin{Bmatrix} e_1 \\ f \\ q_2 \end{Bmatrix} = \begin{bmatrix} 0 & 0 & -1/C \\ 1/R & 0 & 0 \\ 0 & 1 & 0 \end{bmatrix} \cdot \begin{Bmatrix} e_1 \\ f \\ q_2 \end{Bmatrix} + \begin{Bmatrix} S_e \\ 0 \\ 0 \end{Bmatrix} + \begin{Bmatrix} \sum e_{bonds} \\ 0 \\ 0 \end{Bmatrix}
\tag{17.18}
$$

If the flow is assigned by a source of flow, then one has

$$
\begin{Bmatrix} e_1 \\ f \\ q_2 \end{Bmatrix} = \begin{bmatrix} 0 & 0 & -1/C \\ 0 & 0 & 0 \\ 0 & 1 & 0 \end{bmatrix} \cdot \begin{Bmatrix} e_1 \\ f \\ q_2 \end{Bmatrix} + \begin{Bmatrix} S_e \\ S_f \\ 0 \end{Bmatrix} + \begin{Bmatrix} \sum e_{bonds} \\ 0 \\ 0 \end{Bmatrix}
\tag{17.19}
$$

Finally, if it is assigned by an interconnecting bond, the equation of the second row is left empty at this stage. Consequently, one has

$$
\begin{Bmatrix} e_1 \\ f \\ q_2 \end{Bmatrix} = \begin{bmatrix} 0 & 0 & -1/C \\ 0 & 0 & 0 \\ 0 & 1 & 0 \end{bmatrix} \cdot \begin{Bmatrix} e_1 \\ f \\ q_2 \end{Bmatrix} + \begin{Bmatrix} S_e \\ 0 \\ 0 \end{Bmatrix} + \begin{Bmatrix} \sum e_{bonds} \\ 0 \\ 0 \end{Bmatrix}
\tag{17.20}
$$

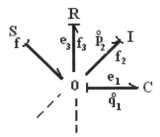

FIGURE 17.22 A typical 0-junction.

Now consider a typical 0-junction. Suppose that a bond connected to a capacitor has dominant causality in the junction (Figure 17.22).

The displacement in this bond is one of the state variables of the system. The momentum in the bond connected to the inductor is another state variable of the system (if it exists). Writing equations for this junction yields

$$\dot{q}_1 = f_1 = -f_2 - f_3 + S_f + \sum f_{bonds} = -\frac{e}{R} - \frac{p_2}{I} + S_f + \sum f_{bonds}$$

$$e = \frac{q_1}{C} \tag{17.21}$$

$$\dot{p}_2 = e$$

or

$$\left\{ \begin{array}{c} \dot{q}_1 \\ e \\ \dot{p}_2 \end{array} \right\} = \left[\begin{array}{ccc} 0 & -1/R & -1/I \\ 1/C & 0 & 0 \\ 0 & 1 & 0 \end{array} \right] \cdot \left\{ \begin{array}{c} q_1 \\ e \\ p_2 \end{array} \right\} + \left\{ \begin{array}{c} S_f \\ 0 \\ 0 \end{array} \right\} + \left\{ \begin{array}{c} \sum f_{bonds} \\ 0 \\ 0 \end{array} \right\} \tag{17.22}$$

The energy exchange is ignored temporarily and will be taken into account in the assembly step, which comes next. The equations can then be written in the form

$$\left\{ \begin{array}{c} \dot{q}_1 \\ e \\ \dot{p}_2 \end{array} \right\} = K_{mat} \cdot \left\{ \begin{array}{c} q_1 \\ e \\ p_2 \end{array} \right\} + F_{mat} \tag{17.23}$$

The matrix K is analogous to a local stiffness matrix, and vector F is analogous to a local force vector. As the energy exchange has been ignored temporarily, Equation (17.23) is still not valid. It will become valid only after completing the assembly step that accounts for interconnecting bonds.

This is the case when a C-element is assigned the effort variable. In the absence of a capacitor, an R-element may assign the flow. In this case,

$$\left\{ \begin{array}{c} \dot{f}_1 \\ e \\ \dot{p}_2 \end{array} \right\} = \left[\begin{array}{ccc} 0 & 0 & -1/I \\ R & 0 & 0 \\ 0 & 1 & 0 \end{array} \right] \cdot \left\{ \begin{array}{c} f_1 \\ e \\ p_2 \end{array} \right\} + \left\{ \begin{array}{c} S_f \\ 0 \\ 0 \end{array} \right\} + \left\{ \begin{array}{c} \sum f_{bonds} \\ 0 \\ 0 \end{array} \right\} \tag{17.24}$$

If the effort is assigned by a source of effort (not usual), one has

$$
\begin{Bmatrix} f_1 \\ e \\ p_2 \end{Bmatrix} = \begin{bmatrix} 0 & 0 & -1/I \\ 0 & 0 & 0 \\ 0 & 1 & 0 \end{bmatrix} \cdot \begin{Bmatrix} f_1 \\ e \\ p_2 \end{Bmatrix} + \begin{Bmatrix} S_f \\ S_e \\ 0 \end{Bmatrix} + \begin{Bmatrix} \sum f_{bonds} \\ 0 \\ 0 \end{Bmatrix}
\tag{17.25}
$$

Finally, if it is assigned by a bond, one has

$$
\begin{Bmatrix} f_1 \\ e \\ p_2 \end{Bmatrix} = \begin{bmatrix} 0 & 0 & -1/I \\ 0 & 0 & 0 \\ 0 & 1 & 0 \end{bmatrix} \cdot \begin{Bmatrix} f_1 \\ e \\ p_2 \end{Bmatrix} + \begin{Bmatrix} S_f \\ 0 \\ 0 \end{Bmatrix} + \begin{Bmatrix} \sum f_{bonds} \\ 0 \\ 0 \end{Bmatrix}
\tag{17.26}
$$

17.6.2 Assembly

Now, each junction has created a 3×3 local stiffness matrix and a 3×1 local force vector. These local matrices are assembled in a global stiffness matrix in their proper locations. For example, the first to third rows and columns are allocated for the first joint. The stiffness matrix of the second joint is placed between the fourth and sixth rows and columns. Its force vector is also placed between the fourth and sixth rows.

The next step is to take into account the energy flow through the connected joints and assemble all local equations into a global set of equations. It should be mentioned that, after the simplification step (explained in Section 17.5.4), all connected junctions are in different forms. Two conditions may apply for this process:

1. If the associated bond is not dominant for causality of the connected joints (Figure 17.23a), the following items should be added to the global stiffness matrix. (If, for example, the associated bond connects node n to node m and the flow of the energy is from node n to node m):
   - The second energy variable of node m should be subtracted from the first equation of node n.
   - The second energy variable of node n should be added to the first equation of node m. It means

$$
K_{3n-2,3m-1} = 1
$$
$$
K_{3m-2,3n-1} = -1
\tag{17.27}
$$

2. If the associated bond is dominant for the causality of the connected joints (Figure 17.23b), then the following items are added to the stiffness matrix:
   - The second energy variable of node n is equal to the first energy variable of node m.
   - The second energy variable of node m is equal to the negative of the first energy variable of node n.

$$
K_{3n-1,3m-2} = 1
$$
$$
K_{3m-1,3n-2} = -1
\tag{17.28}
$$

(a)   (b)

FIGURE 17.23 Connection of junctions: (a) connecting bond is not causality dominant, (b) connecting bond is causality dominant.

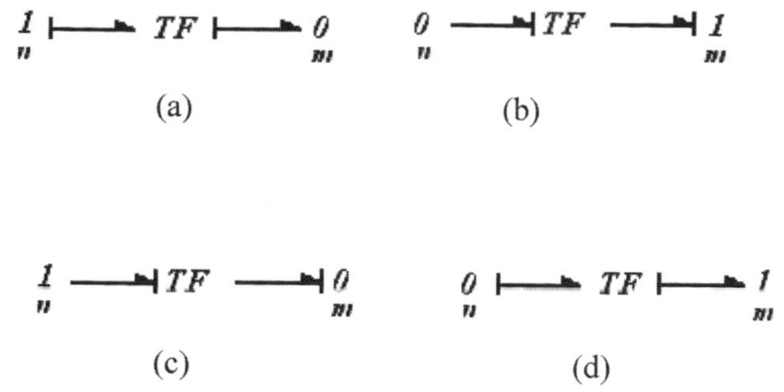

FIGURE 17.24 Assembly conditions for a transformer.

Four conditions may arise if two junctions are connected together through a transformer (Figure 17.24). After the simplification step, transformers can only be placed between dissimilar junctions.

If the associated bonds are not dominant for causality of the connected joints, then,

(a) If the flow of energy is from 1-junction to 0-junction (Figure 17.24a),

$$K_{3n-2,3m-1} = r_T$$
$$K_{3m-2,3n-1} = -r_T$$

(17.29)

(b) If the flow of energy is from 0-junction to 1-junction (Figure 17.24b),

$$K_{3n-2,3m-1} = \frac{1}{r_T}$$
$$K_{3m-2,3n-1} = \frac{-1}{r_T}$$

(17.30)

If the associated bonds are dominant for the causality of the connected joints, then,

(c) If the flow of energy is from 1-junction to 0-junction (Figure 17.24c),

$$K_{3n-1,3m-2} = \frac{1}{r_T}$$
$$K_{3m-1,3n-2} = \frac{-1}{r_T}$$

(17.31)

(d) If the flow of energy is from 0-junction to 1-junction (Figure 17.24d),

$$K_{3n-1,3m-2} = r_T$$
$$K_{3m-1,3n-2} = -r_T$$

(17.32)

Four similar conditions may occur when two junctions are connected together through a gyrator (Figure 17.25). After the simplification step, a gyrator can only be placed between similar junctions.
If the associated bonds are not dominant for causality of the connected joints, then,

(a) If two 1-junctions are connected (Figure 17.25a),

$$K_{3n-2,3m-1} = r_G$$
$$K_{3m-2,3n-1} = -r_G$$

(17.33)

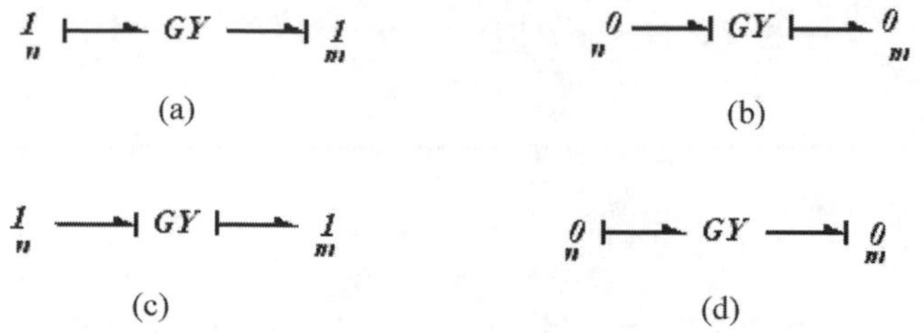

$$\underset{n}{\underline{1}} \; \longmapsto \; GY \; \longmapsto \; \underset{m}{\underline{1}} \qquad\qquad\qquad \underset{n}{\underline{0}} \; \longmapsto \; GY \; \longmapsto \; \underset{m}{\underline{0}}$$

(a)   (b)

$$\underset{n}{\underline{1}} \; \longmapsto \; GY \; \longmapsto \; \underset{m}{\underline{1}} \qquad\qquad\qquad \underset{n}{\underline{0}} \; \longmapsto \; GY \; \longmapsto \; \underset{m}{\underline{0}}$$

(c)   (d)

FIGURE 17.25 Assembly conditions for a gyrator.

(b) If two 0-junctions are connected (Figure 17.25b),

$$K_{3n-2,3m-1} = 1/r_G$$
$$K_{3m-2,3n-1} = -1/r_G \tag{17.34}$$

If the associated bonds are dominant for the causality of the connected joints, then,

(c) If two 1-junctions are connected (Figure 17.25c),

$$K_{3n-1,3m-2} = 1/r_G$$
$$K_{3m-1,3n-2} = -1/r_G \tag{17.35}$$

(d) If two 0-junctions are connected (Figure 17.25d),

$$K_{3n-1,3m-2} = r_G$$
$$K_{3m-1,3n-2} = -r_G \tag{17.36}$$

17.6.3 Final Reductions

After performing the assembly step, redundant variables can be eliminated from the equations. Only storage elements in each junction will provide a state variable for the system. Other variables appear in an algebraic form and can be eliminated simply by algebraic manipulations. After this step, only the state–space variables will remain.

Example: A Servo System

To illustrate the outlined procedure, the state–space equations of the example of the servo system are derived manually. The BG modeling of this example was presented in Section 17.5, and the resultant BG model is shown in Figure 17.17. Here we explain the steps of derivation of the state–space equations:

Step 1—Local Matrix Formation
For the left 1-junction, the local stiffness and force matrices are as follows:

$$K_1 = \begin{bmatrix} 0 & -R_a & 0 \\ 1/I_a & 0 & 0 \\ 0 & 1 & 0 \end{bmatrix} \quad F_1 = \begin{Bmatrix} V_a \\ 0 \\ 0 \end{Bmatrix}$$

For the right node, we have

$$K_2 = \begin{bmatrix} 0 & -R_m & 0 \\ 1/I_m & 0 & 0 \\ 0 & 1 & 0 \end{bmatrix} \quad F_2 = \begin{Bmatrix} 0 \\ 0 \\ 0 \end{Bmatrix}$$

Step 2—Assembly

Place the local matrices within the global matrices at proper locations. Because there is a gyrator connecting two nodes and the gyrator module is k_t, we have

$$K_{1,5} = k_t$$

$$K_{4,2} = -k_t$$

Then, the global equations will be in the following form:

$$\begin{Bmatrix} \dot{p}_1 \\ \dot{f}_a \\ \dot{f}_a \\ \dot{p}_2 \\ \dot{f}_m \\ \dot{f}_m \end{Bmatrix} = \begin{bmatrix} 0 & -R_a & 0 & 0 & k_t & 0 \\ 1/I_a & 0 & 0 & 0 & 0 & 0 \\ 0 & 1 & 0 & 0 & 0 & 0 \\ 0 & -k_t & 0 & 0 & -R_m & 0 \\ 0 & 0 & 0 & 1/I_m & 0 & 0 \\ 0 & 0 & 0 & 0 & 1 & 0 \end{bmatrix} \begin{Bmatrix} p_1 \\ f_a \\ f_a \\ p_2 \\ f_m \\ f_m \end{Bmatrix} + \begin{Bmatrix} V_a \\ 0 \\ 0 \\ 0 \\ 0 \\ 0 \end{Bmatrix}$$

Step 3—Reduction

There are only two state variables in this system, which are presented in the first and fourth equations. Other equations express algebraic parameters and can be illuminated. After illumination of the algebraic equations, we obtain

$$\begin{Bmatrix} \dot{p}_1 \\ \dot{p}_2 \end{Bmatrix} = \begin{bmatrix} -R_a/I_a & k_t/I_m \\ -K_t/I_a & -R_m/I_m \end{bmatrix} \begin{Bmatrix} p_1 \\ p_2 \end{Bmatrix} + \begin{Bmatrix} V_a \\ 0 \end{Bmatrix}$$

17.7 Conclusion

Methodologies of design and modeling for mechatronic systems must employ integrated and concurrent approaches in view of the presence of different types of interacting components from different engineering fields in these systems. A typical mechatronic system entails two main domains, namely, the power domain and the information domain. BG modeling is a graphical, domain-free representation for lumped systems, which can lead to the derivation of the state–space dynamic equations of a mechatronic system through the analysis of the flow of energy. It is particularly appropriate for the simulation of the power domain of a mechatronic system. An important strength of BGs is that system modeling can be simplified using the concept of object-oriented modeling. A new matrix-based formulation for the derivation of the state–space equations of a system was presented in this chapter, which provides straightforward steps for BG modeling of a mechatronic system, and may be applied for the generation of a BG tool and also for training purposes. A synergic integration of BG modeling and block diagram representation, which is commonly used for the controller modeling in tools such as SIMULINK®, can provide an integrated tool for the modeling and design of mechatronic systems.

References

1. Behbahani, S., Practical and Analytical Studies on Development of Formal Evaluation and Design Methodologies for Mechatronic Systems, Ph.D. thesis, Department of Mechanical Engineering, The University of British Columbia, Vancouver, Canada, 2006.
2. Behbahani, S. and de Silva, C.W., Identification of a mechatronic model using an integrated bond-graph and genetic-programming approach, *Proceedings of International Symposium on Collaborative Research in Applied Science (ISOCRIAS)*, Vancouver, Canada, 2005, pp. 158–165.
3. Van Brussel, H.M.J., Mechatronics—a powerful concurrent engineering framework, *IEEE/ASME Transaction on Mechatronics*, Vol. 1, No. 2, 127–136, 1996.
4. de Silva, C.W., *Mechatronics—An Integrated Approach*, Taylor and Francis, CRC Press, Boca Raton, FL, 2005.
5. Behbahani, S. and de Silva, C.W., Use of mechatronic design quotient in multi-criteria design, *Proceedings of International Symposium on Collaborative Research in Applied Science (ISOCRIAS)*, Vancouver, Canada, 2005, pp. 214–221.
6. de Silva, C.W., Sensing and information acquisition for intelligent mechatronic systems, *Proceedings of the Symposium on Information Transition*, Chinese Academy of Science, Hefei, China, November 2003, pp. 9–18.
7. Zhang, W.J., Li, Q., and Guo, L.S., Integrated design of mechanical structure and control algorithm for a programmable four-bar linkage, *IEEE/ASME Transactions on Mechatronics*, Vol. 4, No. 4, 354–362, 1999.
8. Li, Q., Zhang, W.J., and Chen, L., Design for control—a concurrent engineering approach for mechatronic system design, *IEEE Transactions on Mechatronics*, Vol. 6, No. 2, 161–169, 2001.
9. Granda, J.J., The role of bond graph modeling and simulation in mechatronics systems, *Mechatronics*, Vol. 12, No. 9–10, 1271–1295, 2002.
10. Amerongen, J.V., Mechatronic design, *Mechatronics*, Vol. 13, No. 10, 1045–1066, 2003.
11. Karnopp, D., Margolis, D.L., and Rosenberg, R.C., *System Dynamics: Modeling and Simulation of Mechatronic Systems*, Wiley, New York, 2000.
12. Seo, K., Fan, Z.N., Hu, J., Goodman, E.D., and Rosenberg, R.C., Toward a unified and automated design methodology for multi domain dynamic systems using bond graphs and genetic programming, *Mechatronics*, Vol. 13, No. 8–9, 851–885, 2003.
13. Paynter, H.M., *Analysis and Design of Engineering Systems*, MIT Press, Cambridge, MA, 1961.
14. Borutzky, W., Bond graph and object-oriented modeling—a comparison, *Proceedings of the Institution Mechanical Engineers, Part I: Journal of Systems and Control Engineering*, Vol. 216, 2002, pp. 21–33.
15. Broenink, J.F., Introduction to physical systems modeling with bond graphs, available at: http://www.ce.utwente.nl/bnk/papers/BondGraphsV2.pdf.
16. Cellier, F.E. and McBride, R.T., Object-oriented modeling of complex physical systems using the Dymola bond graph library, available at http://www.scs.org/scsarchive/getDoc.cfm?id=2002.
17. Gawthrop, P.J. and Palmer, D., A bicausal bond graph representation of operational amplifiers, *Proceedings of the Institution Mechanical Engineers, Part I: Journal of Systems and Control Engineering*, Vol. 217, No. 1, 49–58, February 2003.

18

Evolutionary Mechatronic Tool

S. Behbahani

C.W. de Silva

Summary

This chapter presents the development of a bond-graph-based evolutionary system tool to model and identify mechatronic systems in a rather optimal manner. The integration of bond graphs and block diagrams in modeling provides a domain-independent simulation environment suitable for multi-domain mechatronic systems. Genetic programming is linked to this environment to create a multi-purpose evolutionary tool. This tool can be used for model identification in a variety of problems, provided that a suitable evaluation scheme can be established for that problem.

18.1 Introduction

Design, modeling, control, and analysis of machines are typical activities of an engineer. For a mechatronic system, these activities tend to be complex because of the presence of components from different domains and dynamic interactions between them. Approaches for these activities for a mechatronic system should somewhat deviate from those for conventional, nonmechatronic systems, in part due to the need for the integrated design and analysis in the former. In a typical mechatronic problem, one needs to find the optimum structure of a machine that best satisfies the user-defined requirements of the solution. This entails finding the best topology and size of the machine. The topology is defined as the gross number of components, their type, and the way they are interconnected, whereas sizing refers to the finding the numerical value of the elements [1]. If the topology of the best solution were known, one could use a variety of optimization tools to find the corresponding size. The mechatronic problem is usually complex because the topology of the best solution is not necessarily obvious. As the topology of a system has naturally an unlimited growth capability and consists of unlimited conditions, finding the topology of a mechatronic system opens a huge search space for the designer. Due to this complexity, topology design is usually viewed as a problem that needs human intelligence and should be performed by experts.

The main challenge in creating a tool for autonomous design and modeling of mechatronic systems is to develop a method that effectively explores all possible topologies for a mechatronic system. This should have two main characteristics:

1. As the topology can grow unlimitedly, the exploration method needs to have a topologically open-ended structure. Genetic programming (GP) is a branch of evolutionary algorithms that has an open-ended growth possibility for solutions [1–4].

2. Generally, a mechatronic system contains components from different engineering fields; mechanical, electrical, hydraulic, and so on. Although these fields are physically different, their basic components and governing equations are analogical. To facilitate topology exploration, one needs a core language to express components from different domains in a unified language. Bond graph modeling (BG) is a domain-free graphical representation of multidomain systems [5–8]. All components from different domains can be modeled in a unified environment. This also satisfies the requirement of integrated design for mechatronic systems.

This chapter explains the underlying process of integration of BG with GP in the creation of an autonomous evolutionary mechatronic tool [3–5,9,10]. BG has been employed as a powerful and domain-free modeling tool with an embedded open-ended representation, and GP has been used as a powerful and open-ended tool for search and optimization. The software tool developed here is applicable to any mechatronic problem, provided that a fitness evaluation scheme is established for the application of GP in the particular problem.

18.2   Genetic Programming (GP)

Evolutionary algorithms—genetic algorithms, in particular—are optimization methodologies that are inspired by the evolution of biological species in nature. They are stochastic search methods based on Darwin's principle of survival of the fittest. GP is a branch of evolutionary computing as is genetic algorithms (GAs). The main difference between GP and conventional GAs comes from the particular representation of the solution [11].

There are many different branches of GA, but the essence of all of them is almost the same. Here we present a general procedure for GAs. It has five main steps:

1. Before starting to apply GAs to the problem at hand, one needs to develop a methodology for representing the possible trial solutions for the particular problem in a chromosome-like structure. This is the main contribution in extending the application of evolutionary algorithms to a field where evolutionary algorithms have not been applied. Although the rest of the process can still be challenging, it is almost similar for all applications.

2. Generate an initial population of random solutions. Being random, these initial solutions are probably not acceptable.

3. For each trial solution, evaluate how well it satisfies the predefined desired specifications. Assign a fitness score for each trial solution. This step may require linking the GA program to some simulation tools for analyzing the behavior of trial solutions.

4. Create a new population of solutions by imitating the natural evolution of biological species. A new generation is formed based on Darwin's principle of survival of the fittest through the following genetic operations:
   - Copy the best existing solutions (selection).
   - Create new solutions by "mutation."
   - Create new solutions by "crossover".

5. Return to step 3 until a satisfaction condition is met or a specified number of iterations is completed. The best program that results from this approach in any generation (the best-so-far solution) is designated as the *outcome* of GP.

To solve a problem by GAs, the following four elements have to be incorporated:

1. A representation scheme is needed for possible solutions. It should have a chromosome-like structure, so that each part of it can be viewed as a gene. A particular gene in a chromosome stores some characteristics of that chromosome. If this gene is transferred to the next generation, the particular characteristic will be inherited in that next generation.
2. Reproduction operations have to be developed so that each reproduced individual inherits some specifications of its parents.
3. A fitness evaluation methodology is necessary to reflect how well a trial solution satisfies the predefined performance requirements.
4. An evolution strategy is necessary to insure that good solutions with higher fitness have a greater chance to participate in the reproduction operations in creating the next generation.

These four elements are further explained now, particularly in the context of GP.

18.2.1 Representation Scheme

In conventional GA, each individual (trial) solution is coded by a string of zeros and ones (Figure 18.1a), whereas in GP a tree formed by *construction functions* represents a solution (Figure 18.1b). Each tree-like combination of construction functions creates an individual solution when the functions are applied to a predefined *embryo* model. In fact, a mapping between each optional solution and a combination of defined construction functions should be established.

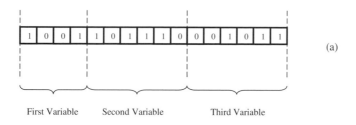

(a)

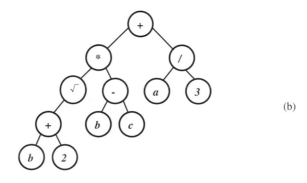

(b)

FIGURE 18.1 Chromosome-like representation of individual solutions: (a) conventional GA, (b) tree-like representation in GP.

This method of representation was initially introduced in order to represent mathematical equations in a tree-like format [2]. An example is shown in figure 18.1b. This tree-like representation makes GP suitable for concurrent sizing and topology optimization. Each tree generates a solution alternative and is considered a chromosome, whereas its branches are considered genes. As clear from the figure 18.1, a change in the genes of a chromosome can result in a new topology, in addition to a new size of the result.

Two main concepts in this methodology are the embryo model and the construction functions. Embryo includes very basic specifications of solution and remains unchanged in all alternative solutions. Embryo needs to have several *modifiable sites*. These are places where the embryo can be extended through them.

Construction functions are the alphabet of creation of new models. They are the functions that are applied to the embryo in order to create new and extended models. They have open architecture. This means that they can create new modifiable sites in the model. This feature makes them grow like branches of a tree. Proper termination functions are also necessary to prevent unlimited growth of branches.

18.2.2 Reproduction Operations

A new generation of trial solutions is created from the previous generation by reproduction operations, including:

- Crossover
- Mutation
- Survival (or selection)

The most important of these, which has the highest percentage of reproduction probability, is the crossover operation. In the crossover operation, two solutions are mated to form two new solutions or offspring. The parents are randomly selected from the population but in such a way that the members with higher fitness have a greater chance to be selected. The creation of the offspring from the crossover operation is accomplished by exchanging a gene branch between parents (Figure 18.2). An important improvement of GP over conventional GAs is the ability of the former to accomplish crossover from the same solution [11]. In Figure 18.3 the same parent is used twice to create two new children.

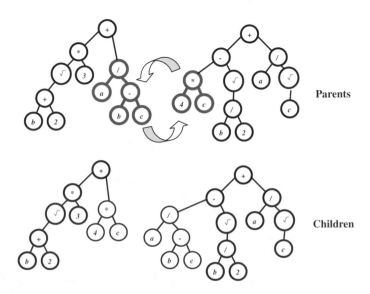

FIGURE 18.2 Crossover genetic operation in a tree-like GP representation.

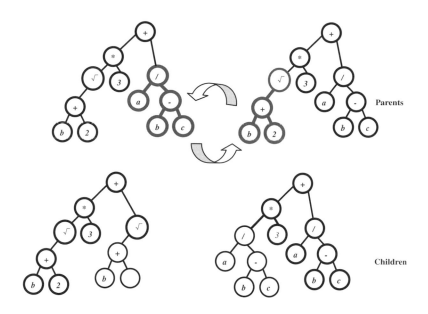

FIGURE 18.3 Crossover reproduction operation between identical parents.

Mutation is another operation of genetic programming, in which an individual is selected randomly but in such a way that the solutions with higher fitness have a greater chance to be selected. In this operation, a gene branch is selected randomly and replaced by a new randomly created gene branch, as shown in Figure 18.4. Mutation gives an important strength to GP. The process of hill climbing is the basis of most numerical optimization techniques. In this process, the optimization path may get locked into a local optimum and be unable to escape. The mutation operation of GP avoids, or at least reduces, the possibility of locking into a local minimum. It works like a gunshot on the optimization process [12].

In addition to the above-mentioned operations, there is a critical operation where the best solutions are always copied to the next generation in order to guarantee that the best solutions are retained. This operation also guarantees that no degradation will happen in the best solutions of the population and, in turn, guarantees the convergence of the optimization process.

18.2.3 Fitness Function

The most difficult and also most important concept of GP is the fitness function. The destiny of each trial solution depends on its fitness. The best and the fittest solutions have a greater chance for survival in the new generation and also for participation in a larger number of reproduction operations. On the other hand, solutions with a lower level of fitness are more likely to die without producing any offspring in the new generation. The fitness function is a function reflecting the degree of satisfaction of each trial solution. It should vary greatly from one type of solution to the next and should be such that the farther a model is from the ideal solution, the more it will be penalized.

18.2.4 Selection Method

The essence of GP is the selection of good solutions from a trial set of solutions and making them participate in genetic operations. Three methods have been used in literature for the selection of solutions that have good fitness values [13]:

1. Probability method
2. Tournament method
3. Ranking method

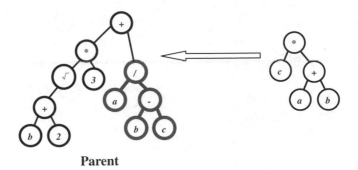

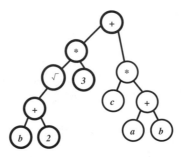

FIGURE 18.4 Mutation reproduction operation in a tree-like representation of GP.

The first method specifies a selection probability for each individual solution based on its fitness. If $f(S_i)$ is the fitness of the solution S_i, and $\sum_{j=1}^{N} f(s_j)$ is the total sum of fitness of all the members in the population, then the probability that the solution S_i will be selected is

$$P(s_i) = \frac{f(s_i)}{\sum_{j=1}^{N} f(s_j)} \tag{18.1}$$

The listed second method for selecting a solution to be copied is known as the tournament selection. Typically, GP chooses two solutions randomly, and the solution with the higher fitness will win. This method simulates a biological mating patterns in which two members of the same sex compete to mate with a third one of a different sex.

In the third method, the selection is done by the rank (not the numerical value) of fitness of the solutions in the population. This method may be used with various modifications [4,5]. Once a generated set of solutions is evaluated, they are ranked in the ascending order of their fitness. Two parent solutions are necessary to perform the crossover operation. Two random numbers are created by this operation, and then two parents are selected, based on these random numbers. To give greater opportunity for the solutions with high fitness, a mapping function is employed such that the random numbers ranging between 0 and 1 are mapped mostly to high ranked members. In [4], the following mapping function has been employed:

$$\frac{n}{N} = r^{\xi} \tag{18.2}$$

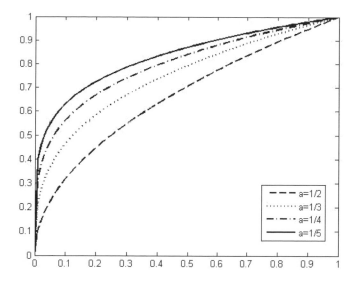

FIGURE 18.5 The mapping that enforces a higher chance of selection into better solutions.

where, r is a random number, N is the total population, n is the rank of the selected solution, and ξ is a discrimination factor that is less than one. This mapping is shown in Figure 18.5. The lower the value of ξ, the higher the probability of death of a weak member (i.e., higher discrimination). In the GP process it is useful to tune this number.

Each solution in the form of a GP tree is presented as a "chromosome." Very long chromosomes may be generated in the evolution process [4]. A change in the genes of a long chromosome may not push it toward the optimal solution in an effective manner. A long chromosome may eventually attain a rather high fitness and create many more long chromosomes in the next generation. These chromosomes may lie at the top of the generation and prevent further improvement of the solution, or at least reduce the efficiency of the GP. In view of this, it is useful to introduce a penalty for having a long gene. However, the members that are ranked very high are made immune to this penalty [4].

The age of a chromosome is another factor that needs to be considered. Specifically, a penalty is introduced for the age of a chromosome. If a chromosome is quite old and yet it does not have a high rank, it is desirable to delete it and let the younger solutions continue in the competition. However, members having very high ranks are made immune to this penalty [4].

18.2.5 Genetic Programming Procedure

Figure 18.6 shows the flowchart of GP. To solve a problem by GP, first an *embryo* solution should be provided. The embryo includes the very basic specifications of the solution, such as inputs and outputs. It has some *modifiable sites* and some *open sites* for modification and extension of the embryo. *Construction functions* provide possible modifications and extensions available for different types of modifiable sites. The first population of possible solutions is created by applying different random combinations of construction functions to the embryo. This first population will hardly include good solutions, yet different solutions can be ranked in terms of their fitness. The next generation of solutions is formed by GP operations that are based on the Darwin's principle of survival of the fittest. Selection

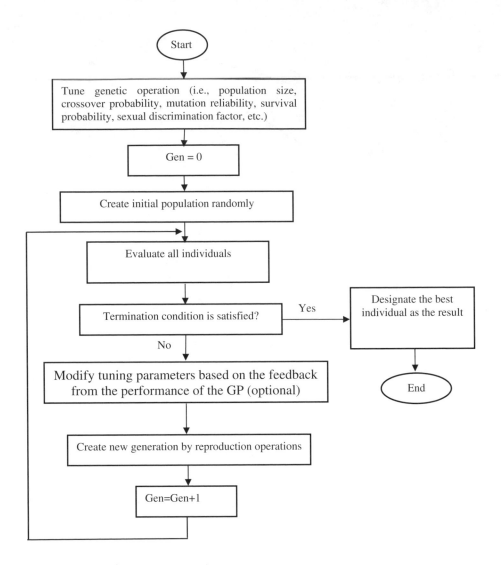

FIGURE 18.6   Flowchart of genetic programming.

of the parents for a GP operation and the place in the chromosome for applying the particular operation are all performed with some level of randomness to mimic the natural evolution. Because of Darwin's principle, the next generation is likely to be better than the current generation, and eventually an optimum solution will be achieved by continuing this process.

18.3   Integration of Bond Graphs (BG) and Genetic Programming

Modeling of mixed systems requires a common language that can effectively function in the different engineering fields that are involved in a mechatronic system. BG are proved to be an effective and accepted modeling method for such systems. They provide a common and core language for describing basic elements and connections across different fields.

Representation of the mixed systems by BG provides an embedded strength for BG when used in design applications [3,4]. The graphical representation of BG allows their generation by combining bond and node components that can represent any system model. This graphical representation of BG also has an open architecture. This means that a free composition of bonds and nodes can be added to different locations of a model, thereby creating a new model with a new topology. This unlimited growth capability of BG provides the capability to explore a wide range of topologies in the process of design and optimization. In other words, the use of BG for mechatronic design can result in an optimization tool that will not be restricted to the system sizing aspect but will optimize the topology of the model as well. This graphical flexibility of BG and the extensive search ability of GP have provided the rationale for integration of these two powerful tools. The main requirement in this context is to develop a methodology that represents BGs with tree-like structures used in GP. In other words, a mapping should be established between tree-like representation methods of GP with BGs. Two main elements of this mapping are the embryo model and the construction functions that are explained next.

18.3.1 Bond Graph Embryo Model

To solve any problem by integrating BG and GP, an embryo solution should be provided by the user. This BG embryo model should contain the fundamental information about the solution, such as inputs and outputs. The embryo model is the common part of all the trial solutions that would be created in the process of optimization by GP. It is problem dependent, and it is the user who should decide the level of detail in the embryo. As a general rule, the inputs and outputs of the problem that are involved in the fitness evaluation should be available in the embryo model.

In the embryo model, several places should be specified as modifiable sites. Considering the typical BG embryo model shown in Figure 18.7, three types of modifiable sites can be observed [3]:

1. Modifiable joints, shown by a dashed circle
2. Modifiable bonds, shown by a dashed square
3. Arithmetic sites, shown by a dashed elliptic

Modifiable joints allow for extension of the embryo by adding new elements to nodes. A basic one-port element or a useful compound object can be added to a node. To retain the generality of the method, an element may be either linear or nonlinear.

Modifiable bonds allow extension of the connections between nodes. Arithmetic sites correspond to modifiable numerical values of the elements in the embryo and also to elements that are added to it. Arithmetic sites are necessary for the sizing optimization of a created model. If an element is linear, a

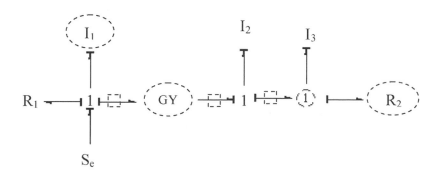

FIGURE 18.7 A BG embryo model example.

constant number is created by the arithmetic functions to represent the physical magnitude of the element. For a nonlinear element, a scheme for evolutionary function generation is used to find the best function that represents the element.

18.3.2 Construction Functions

The next necessary item is the establishment of construction functions and terminals to provide the mapping between a GP tree-like representation method and all possible BG models.

Each of the already explained modifiable sites has its own possible construction functions. In this section, the construction functions developed in [3] for linear systems and functions developed in [4,5] for nonlinear systems are explained. All construction functions have an open architecture, which means that they introduce new modifiable sites and also keep old modifiable sites for further improvement. This feature makes the individual chromosomes to grow like branches of a tree. It also makes it necessary to consider an *end function* for each type of modifiable site, thereby preventing unlimited growth.

Six types of functions are developed for the integration of GP and BG:

1. Add-Element: Adds a basic one-port element.
2. Insert-junction: Inserts a junction (either 1 or 0 junction).
3. Arithmetic: Performs arithmetic operations for. finding the value of an element.
4. Compound: A useful combination of nodes and bonds is added.
5. Add-Friction function.
6. Add-Backlash function.

Next, these functions are illustrated.

18.3.2.1 Add-Element Functions

Consider a modifiable node. A basic one-port element (i.e., resistor, capacitor, or inductor) may be added to this node [3,4]. Figure 18.8 shows a modifiable node before and after adding a one-port element—a resistor in this case. When the new element is added, first of all the old modifiable site should be kept modifiable for further change. Second, the added element needs a numerical value, either by specifying its value if it is linear or by providing an arithmetic function. This opens a new modifiable arithmetic site in the model. In addition, the bond which connects the element to the initial node is also a site where the model can be further extended. It means that when an add-element function is applied to a modifiable mode, two new modifiable sites are created, and the old

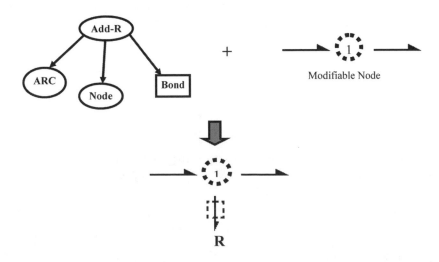

FIGURE 18.8   Add-element construction function.

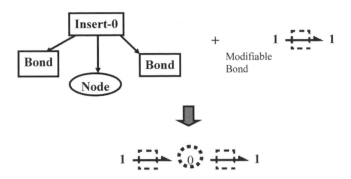

FIGURE 18.9 Insert-Junction construction function.

modifiable site is also kept for further modification (Figure 18.8). Consequently, this function has three arguments. The first one indicates the value of the added element with arithmetic functions, the second one is the old modifiable site for further modification, and the last one is a bond-type site for inserting a junction at the connection point of the added bond. There are three types of Add-Element functions: *Add-R*, *Add-C*, and *Add-I*. As these functions may give rise to an unending sequence of add operations, it is necessary to have the End-Node function to terminate the operation.

18.3.2.2 Insert-Junction Functions

Consider again a modifiable bond. A junction may be inserted between bonds [3,4]. This function has three arguments and can create three new modifiable sites. Two of them are bond type, and the third one is node type. They are shown in Figure 18.9. As these functions may result in an open sequence of operations, it is necessary to provide a termination operation for them. Accordingly, for each modifiable bond, three options are available: Insert-0, Insert-1, and End-Bond.

18.3.2.3 Arithmetic Functions for Linear Elements

Each added new element or a modifiable element in the embryo model needs a value. If the element is linear, its magnitude can be represented by a constant number. Otherwise, it has to be represented as a function of other variables in the system. In the linear case, each arithmetic modifiable site may be a random number or a combination of arithmetic functions and random numbers. Two kinds of arithmetic functions are used: addition and subtraction [3]. Each function has two arguments, which may also be either random numbers or a set of arithmetic functions. Regardless, the result of an arithmetic subtree will be a number, which represents the numerical value of the element. The resulting number may be interpreted in various ways, depending on the designer. For example, it may be interpreted as the logarithmic value of an element [3] or as the change in an initial default value [4]. The value of parameter ζ is computed by

$$\zeta = \zeta_d(1+\rho.\varepsilon) \tag{18.3}$$

Here, ε is the result of the arithmetic sub-tree, ρ is a modification factor, and ζ_d is the default value for parameter ζ. It is found desirable to apply a somewhat high ρ in the beginning of a GP operation for coarse modification and then reduce it later for fine modification.

18.3.2.4 Arithmetic Functions for Nonlinear Elements

The value of a nonlinear element is a function of other parameters in the system. In particular, it can depend on the state–space variables of the node to which the particular element is connected. Here, several construction functions are developed to create different functions representing the value of a nonlinear element [4]. These construction functions have been used in the genetic programming optimization

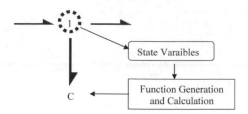

FIGURE 18.10   Function generation for nonlinear elements.

technique to explore different functions and find the optimum one for representing the value of a particular nonlinear element [4]. If the nonlinear element is in the embryo model, the user can determine how many variables may appear in the variation of the value of the element. If the nonlinear element is created in the process of optimization, then it is considered to depend only on the state variables of the node connected to the particular nonlinear element (Figure 18.10).

Table 18.1 gives a list of arithmetic operators and the number of arguments needed for each of them. Each argument can be a function itself, a random number, or one of the variables of the function. Figure 18.11 shows an example of creation of a function by using these construction functions.

TABLE 18.1   Arithmetic Operators and Associated Number of Arguments

Operation	Number of Arguments
Addition	2
Subtraction	2
Product	2
Division	2
Power	2
Sign	1
Exp	1
Log	1

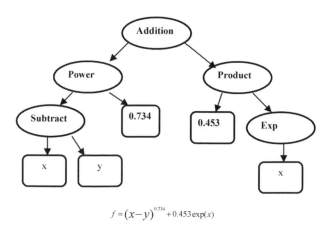

$$f = (x-y)^{0.734} + 0.453\exp(x)$$

FIGURE 18.11   An example of function generation.

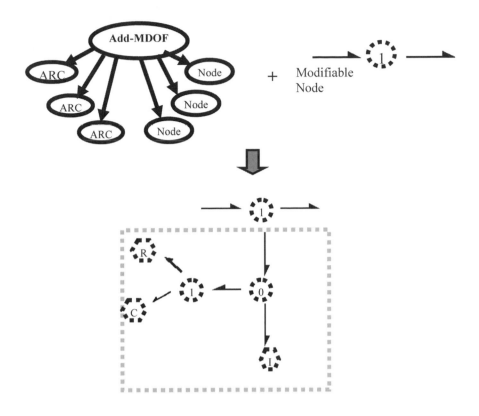

FIGURE 18.12 Add–mechanical degree of freedom (MDOF) construction function.

18.3.2.5 Functions for Compound Object Creation

It is useful to have functions for creating compound objects, which may be commonly needed in the modification operations [4]. Although it is possible that these compound elements are created through the normal operation of a GP by means of the basic functions which have been developed, it is more effective to establish compound functions to facilitate creation of these objects. For example, one useful compound object that may be commonly needed in building the mechanical model of a machine is a mechanical degree of freedom (MDOF). The BG model of an MDOF is shown in Figure 18.12. It adds one mechanical degree of freedom to a mechanical system, which corresponds to adding two state–space variables. It may be added to a modifiable node; consequently, a new function that is available for a modifiable node is Add-MDOF. It has six arguments, which are shown in Figure 18.12 [4].

18.3.2.6 Add-Friction Function

Friction introduces complexities into a mechatronic system. In constructing the model of an existing machine, friction may exist in some location of the system that has not yet been detected by the user, or the user has not been able to identify it. The Add-friction construction function may be used in the course of genetic programming to account for the existence of friction in a modifiable joint [4]. The friction is modeled as a source of effort; however, it is a negative source and should be determined based on the direction of motion. The Stribeck model shown in Figure 18.13 [14] is used for friction. If the speed is zero (or very small in numerical computations), the friction is equal to the force applied to the slider but in the opposite direction. However, the friction model needs to be identified, because the corresponding parameter values are unknown. This function has five arguments (Figure 18.14) [4]. The first four are arithmetic sites to determine the required numerical values for the Stribeck model (see Figure 18.13). The last one is the old modifiable node site for further modification.

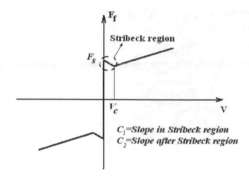

FIGURE 18.13   Stribeck friction model.

18.3.2.7   Add-Backlash Function

Backlash is another source of nonlinearity in a mechatronic system that is commonly neglected during analytical modeling. When neglected, it can cause undesirable responses, chattering, and even unstable behavior.

A construction function is developed in [4] in order to examine the possible existence of backlash at a junction of a model. The Add-backlash function introduces backlash into a modifiable joint. The backlash between two objects can be modeled as a nonlinear spring (Figure 18.15). The Add-backlash function is treated similar to the Add-C function which adds a capacitor element (like spring), but the added element is nonlinear in this case. If the relative displacement between the two nodes in backlash is less than the backlash limit, the spring stiffness is zero (no capacitor), and if the relative displacement exceeds the backlash value, a high stiffness spring is considered between the corresponding nodes. Therefore, it has four arguments similar to the Add-C function. The first argument is an arithmetic site indicating the amount of backlash and the other arguments are similar to those in the Add-C function.

18.3.3   First Generation

The first generation of the population is created by applying random combinations of construction functions on the embryo. Tables 18.2 through 18.4 show the available functions for each type of

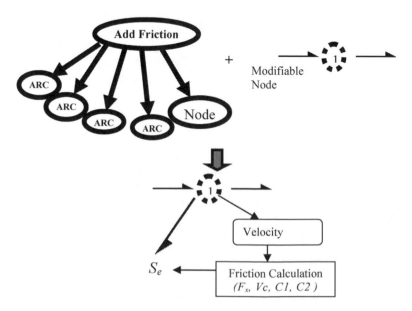

FIGURE 18.14   Add-friction construction function.

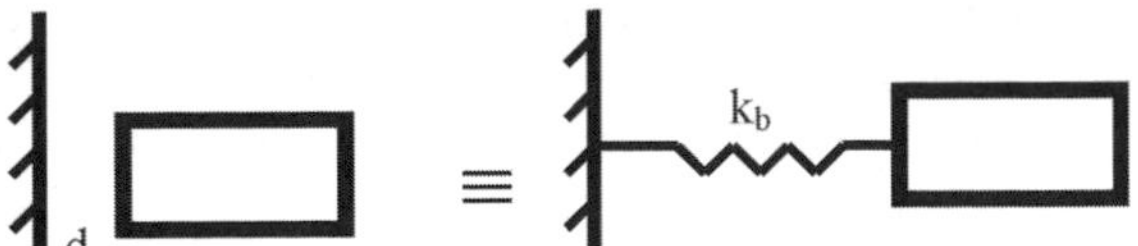

$$k_b = \begin{cases} 0 & x < d \\ \infty & x \geq d \end{cases}$$

FIGURE 18.15 Equivalent system of backlash.

TABLE 18.2 Construction Functions for a Modifiable Node

Function	Code	Arguments
Add-R	41	4
Add-I	42	4
Add-C	43	4
Add-Backlash	44	4
MDOF	61	6
End-node	2	0

TABLE 18.3 Construction Functions for a Modifiable Bond

Function	Code	Arguments
Insert 0	30	3
Insert 1	31	3
End-bond	–2	0

TABLE 18.4 Construction Functions for an Arithmetic Modifiable Site

Function	Code	Arguments
Addition	21	2
Subtraction	22	2
Product	23	2
Division	24	2
Power	25	2
Sign	11	1
Exp	12	1
Log	13	1
A random number	—	0

modifiable site. The required arguments and the code for each function are also shown in these tables. A code is available for each function [4]. The left digit in the code corresponds to the number of arguments of the function. This is useful for vector presentation of tree-like construction functions and for finding the start and the end of each branch.

For each modifiable site in the embryo, a random number is created by the program. Depending on the value of the random number and predefined ranges for each available function, one of the available functions is selected and is applied on the modifiable site. For example, if there is a modifiable node in the embryo model, one of the Add-R, Add-C, Add-I, Add-MDOF, Add-Friction, Add-Backlash, or End-Node functions is selected, based on the value of the randomly generated number and is applied to the modifiable node. If End-node is selected, then the process for this modifiable site is complete and the program can go to the next modifiable site. If one of the other options is selected, then the applied function creates some inner modifiable sites. This reproduction can continue similar to the multiplying of branches on a tree. However, the probability for End-node should be high enough to prevent an infinite loop [4].

This tree-like expansion is the reason that each created GP model is usually called a *genetic programming tree* and, combined with mutation and crossover genetic operations, is the reason for high searchability in genetic programming.

18.4   Autonomous System Identification

Integration of BG and GP provides a powerful approach that can be used for design, optimization, and identification of a mechatronic system. The motivation behind the use of this integration for system identification stems from the utilization of the power of GP in order to automatically create the model of a system. Most of the time, it is very difficult to find the magnitude of some physical parameters in a mechatronic system. It may even be difficult to exactly specify the topology of the system. The search power of GP, flexibility of BG, and their integration in the tool developed in [4], can be used to find the model of a system automatically. The main steps of the process are as follows:

1. An embryo model for the system is created.
2. Some experiments are performed on the system, and the behavior of the system is saved.
3. A fitness evaluation method is selected. Here, the fitness should indicate the degree of closeness of the response of each model to the actual behavior of the system.
4. GP is tuned for this particular task.
5. GP is used to evolve the embryo to an accurate model.

GP is based on Darwin's principle of survival of the fittest. Therefore, fitness evaluation plays a fundamental role in genetic programming. It should adequately represent the degree of satisfaction of each optional solution. If the fitness evaluation is not satisfactory, then the genetic program may delete good solutions during the GP operation, resulting in inaccurate results.

The approach of fitness evaluation is specific to the problem at hand. For system identification in particular, the fitness of each candidate model may be taken as the degree of closeness of the response of a candidate model to the response of the real system. This comparison can be made either in the time domain or the frequency domain, or a combination of the two.

The idea of least-squares difference has been used in [4,5], for model comparison. Specifically, to evaluate a parameter denoted by ω, the following expression is used as the fitness function:

$$f = \frac{M}{M + \sqrt{M} \cdot \sqrt{\sum_{k=1}^{M} \left(\frac{\omega_k - \omega'k}{\bar{\omega}} \right)^2}} \tag{18.4}$$

where M is the number of comparison points, ω is the actual response of the system, ω' is the response of the model, $\bar{\omega}$ is the average of the absolute values of ω, and the subscript i is an index denoting a particular comparison point. This expression is found to be somewhat independent of the number of comparison points used. If, for example, all data have a constant deviation from the actual measurement, then this equation gives a unified f, regardless of the number of comparison points.

18.4.1 Evolutionary System Identification of a Vibratory System

The objective of the example given here is to clarify the power of the method presented in this chapter for both sizing and topology optimization. In this example, the method is employed to find the BG model of a nonlinear vibratory system. The system has been tested with a step excitation force and the vibration has been measured at the same point where the force was applied [4].

In this illustrative example, time domain comparison is performed for both displacement and velocity profiles of a vibratory system, using following equation:

$$f = a.\frac{M}{M+\sqrt{M}.\sqrt{\sum_{k=1}^{M}\left(\frac{xk-x'k}{\bar{x}}\right)^2}} + b.\frac{M}{M+\sqrt{M}.\sqrt{\sum_{k=1}^{M}\left(\frac{vk-v'k}{\bar{v}}\right)^2}} \tag{18.5}$$

where, a and b are two weighting factors, and $\bar{x}$ and $\bar{v}$ are the averages of the absolute values of the displacements and the velocities of the real system, respectively.

The embryo model used here is a simple one-degree-of-freedom (1-dof) mass-spring-damper system, with initially linear elements. The spring is considered as the element with possible nonlinearity. The method is used to identify the actual model of the system and also establish a function that can describe the nonlinear element in the system [4]. Initially, the response of the embryo model is quite far from that of the real system. The developed tool is used to evolve this embryo model into a suitable model for describing the real system.

There is a significant difference between using GP for system identification and using it for design purposes. In a typical design problem, GP has to search for a solution that has a response close to the actual response expected from the particular system. There can be several trial designs that have responses very close to each other, whereas their topologies may be quite different. In a design problem, these solutions will have fitness values close to each other. In contrast, GP has to search for a unique solution in a system identification problem. It has been found in [4] that GP may converge to a solution that has a response close to the real system, although its topology is quite different. This situation is more obvious in the identification of a nonlinear system. In fact, GP may converge to a rather complicated solution to create a response close to the response of the real nonlinear system. One reason for this situation is that all the construction functions have an additive effect.

In addressing this problem, it has been found helpful to incorporate knowledge in the GP optimization process [4]. Any knowledge from the user can assist GP to converge easier and faster to a better solution. In the present example, it is assumed that the user knows that the number of mechanical degrees of freedom of the system is unlikely to exceed three. This knowledge is incorporated into the GP process by incorporating a penalty factor for high degrees of freedom in a trial solution.

A result obtained by applying the developed tool to the present example is shown in Figure 18.16, which is quite satisfactory [4]. Figure 18.17 shows the fitness history and its improvement. The interesting point to note is that the tool has effectively determined that a two-degrees-of-freedom (2-dof) system is necessary to model the system, and has optimized both size and topology of the

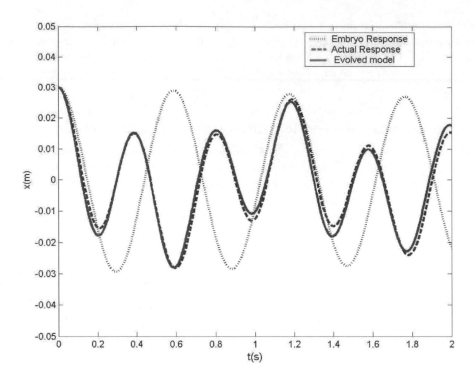

FIGURE 18.16   Performance of the developed tool in evolutionary modeling of a system.

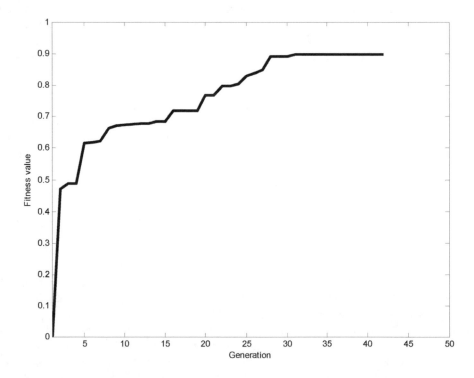

FIGURE 18.17   Fitness improvement using genetic programming.

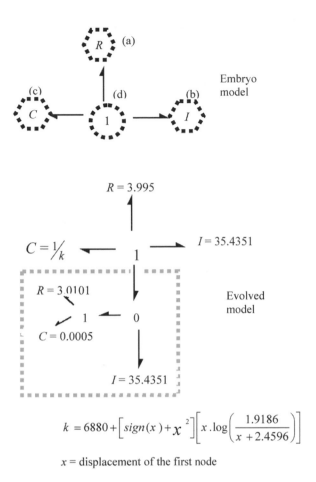

$$k = 6880 + \left[sign(x) + x^{2} \right] \left[x \cdot \log\left(\frac{1.9186}{x + 2.4596} \right) \right]$$

x = displacement of the first node

FIGURE 18.18 Embryo and evolved model in evolutionary system identification of a vibratory system.

embryo model to produce good results. Figure 18.18 shows the embryo model and the evolved model. This example illustrates the power of the method for both sizing and topology optimization for a nonlinear problem.

18.4.2 System Identification of an Electrohydraulic Manipulator

A novel machine has been designed and built in the Industrial Automation Laboratory of the University of British Columbia (UBC) to automatically cut the heads of fish with maximum meat recovery [15]. The location of the gill of a fish is found by a vision system and is sent to a positioning table to optimally locate the cutter. The positioning table is actuated by two electrohydraulic manipulators with two servo valves. A schematic diagram of the electrohydraulic system is shown in Figure 18.19. This electrohydraulic system is also used as the test bed for several research projects that are aimed at developing advanced control technologies to cope with such factors as nonlinear friction, system parameter variation, and communication time delays. An accurate and comprehensive model is quite useful in controlling this system, which is highly nonlinear. Important parts of the system are:

1. Variable displacement pump
2. Servo valve
3. Hydraulic cylinder
4. Mechanical subsystem including the cutter carriage table and the guideways

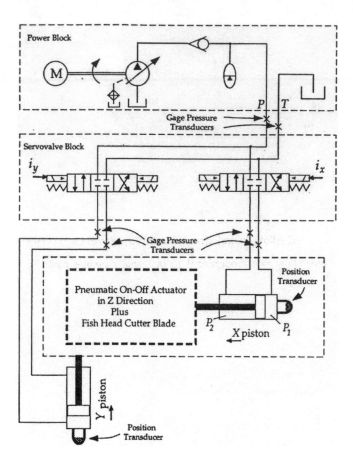

FIGURE 18.19 Schematic diagram of the electrohydraulic manipulator.

Let us now consider the development of a bond graph model for this system. The procedure for BG modeling of mechanical, electrical, and hydraulic components is explained in Chapter 17. Figure 18.20a shows a schematic diagram of the hydraulic cylinder. Following the standard procedures, the BG model of this component is developed, which is shown in Figure 18.21a. The leakage between the two sides of the piston, through the cylinder wall, is accounted for by placing a resistance in the leakage path between these two sides [16,18]. The resistance should be fairly large to restrict the leakage. The capacitors in the two sides of the cylinder come from the compressibility of the fluid and the flexibility of the cylinder, including possible spring loading on the piston [15,18]. In BG terminology, there exists a constitutive law between the effort variable *e* and the flow variable *f*; for example, voltage and current of an electrical capacitor or force and velocity of a mechanical spring (flexibility corresponds to capacitance):

$$C\frac{de}{dt} = f \Rightarrow e = \frac{q}{C} \qquad (18.6)$$

In BG terminology, *q* is the time integral of a flow variable and is called the *displacement variable* (for example, electrical charge, mechanical displacement, and fluid volume in electrical, mechanical, and hydraulic systems, respectively). In a hydraulic system, pressure (*P*) represents the effort variable, the volume flow rate is the flow variable, and fluid volume is the displacement variable. Considering the

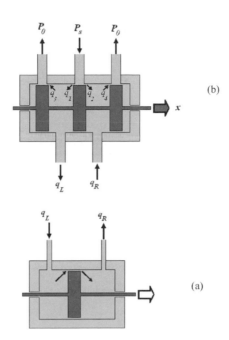

FIGURE 18.20   Schematic diagram of different parts of the electrohydraulic manipulator.

parameters shown in Figure 18.22, the capacitor coefficients can be estimated from the compressibility equation [17] as

$$C = \frac{A.S + V_L}{4A.E} \tag{18.7}$$

Figure 18.20b shows the schematic diagram of the servo valve. Depending on the position of the tongue, four orifices are formed in the servo valve [16,18]. These four orifices can be modeled as four nonlinear resistances. The BG model of the servo valve is shown in Figure 18.21b. Depending on the position of the tongue of the servo valve, the magnitude of the four generated orifices should be estimated and be substituted into the model. In other words, these four orifices should be represented by functions of the position of the tongue of the servo valve.

The mechanical subsystem is simply a sliding mass subjected to friction. The BG model of the mechanical subsystem is shown in Figure 18.21c. The friction has been modeled by the Stribeck model as a source of effort; however, it is a negative source and should be determined based on the direction of motion.

The tongue of the servo valve has a controller, which is a low-power electrical circuit. It does not have significant interaction with the other parts of the system, and can be modeled separately. Figure 18.23 shows the schematic diagram and the BG model of this subsystem [16].

The embryo model of the system is obtained (Figure 18.21) by assembling different subsystem models. Nonlinearity of this system mainly comes from the friction under the slider and from the hydraulic resistance of the four orifices in the servo valve. In this example, the evolutionary tool was used to find the magnitudes of all the elements of the model shown in Figure 18.21, identify friction, and find functions representing the variation of the nonlinear resistances of the orifices in the servo valve. An experiment was carried out on the machine by applying a sine wave excitation. The fitness evaluation was performed by comparing a combination of pressure and displacement profiles, both in the time domain and the frequency domain.

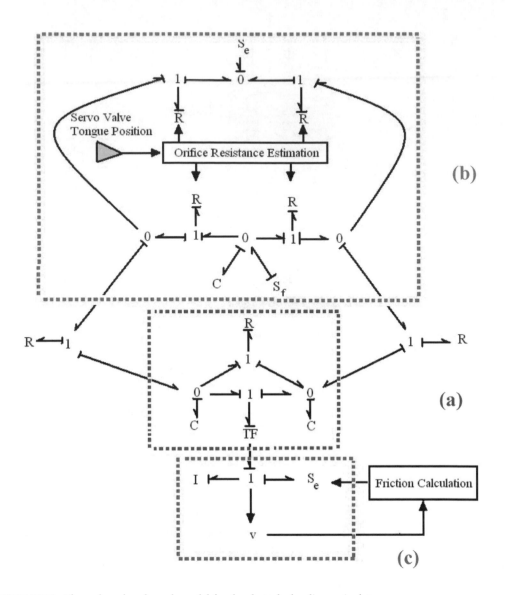

FIGURE 18.21 The embryo bond graph model for the electrohydraulic manipulator.

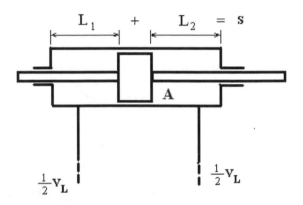

FIGURE 18.22 Schematic diagram of the hydraulic cylinder.

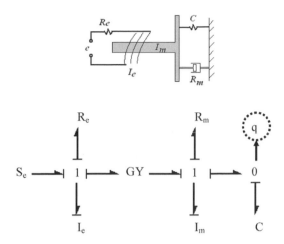

FIGURE 18.23 Schematic diagram and the bond graph model of the servo-valve tongue controller.

Figures 18.24 and 18.25 compare the displacement and the pressure responses of the evolved model with the responses of the actual physical system. It is noticed that the agreement is quite satisfactory, thus validating the model. The main reason for the slight deviation between them is the random nature of the friction. In this system it is found that the friction represents a higher order than the inertia forces (nearly 20 times more). It means that a small deviation in the prediction of the friction is expected to cause a significant deviation in the displacement and velocity profiles. As the real friction has a random characteristic, a deviation between the real friction and the predicted friction is inevitable. Therefore, the agreement between the responses of the created model and the real system is quite satisfactory, despite the slight deviation that is observed between them.

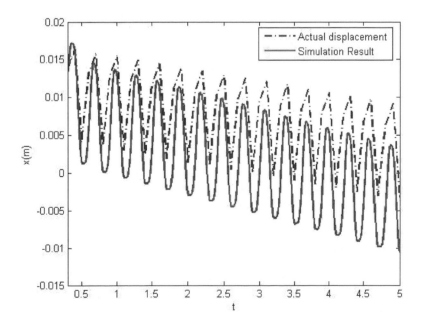

FIGURE 18.24 Displacement comparison for system identification of the electrohydraulic manipulator.

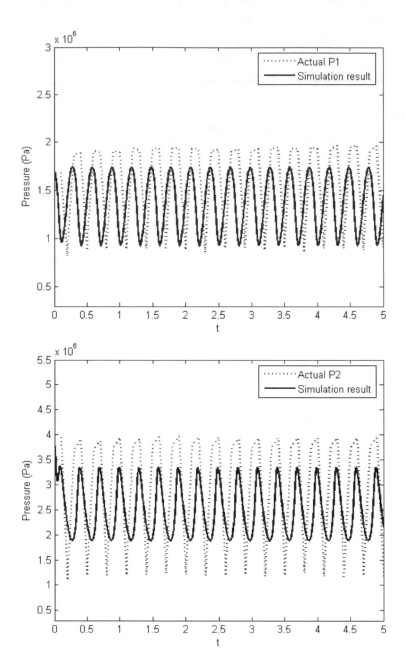

FIGURE 18.25 Pressure comparison for system identification of the electrohydraulic manipulator. (P1 and P2 are the pressures at two output points of the servo valve.)

18.5   Conclusion

Modeling of a mechatronic system is a challenging task due to the presence of complex subsystems in different domains and the need to integrate different engineering fields in representing the overall system. A BG-based evolutionary system tool was developed to model and identify multidomain (i.e., mixed) systems in a rather optimal manner. Integration of BG and block-diagram modeling provided a domain-independent simulation environment suitable for a multidomain system. Genetic programming was linked to this environment to create a multipurpose evolutionary tool. This tool can be used for model identification in a variety of problems, provided that a suitable evaluation scheme can be established for that problem. In this chapter, the successful application of this tool for system identification was illustrated. An electrohydraulic manipulator of a fish processing machine, which is a highly nonlinear mechatronic system, was identified by this tool to show its feasibility and performance. The obtained results were quite encouraging, leading to a rationale for extending the tool for the development of a more general mechatronic design tool.

References

1. Koza, J.R., Bennett, F.H., Andre, D., and Keane, M.A., Synthesis of topology and sizing of analog electrical circuits by means of genetic programming, *Computer Methods in Applied Mechanics and Engineering*, Vol. 186, No. 2, 459–482, 2000.
2. Koza, J.R., Bennett, F.H., Andre, D., and Keane, M.A., *Genetic Programming III, Darwinian Invention and Problem Solving*, Morgan Kaufman Publication, San Francisco, California, 1999.
3. Seo, K., Fan, Z., Hu, J., Goodman, E.D., and Rosenberg, R.C., Toward a unified and automated design methodology for multi domain dynamic systems using bond graphs and genetic programming, *Mechatronics*, Vol. 13, No. 8–9, 851–885, 2003.
4. Behbahani, S., Practical and Analytical Studies on Development of Formal Evaluation and Design Methodologies for Mechatronic Systems, Ph.D. thesis, Department of Mechanical Engineering, The University of British Columbia, Vancouver, Canada, 2007.
5. Behbahani, S. and de Silva, C.W., Identification of a mechatronic model using an integrated bond-graph and genetic-programming approach, *Proceedings of International Symposium on Collaborative Research in Applied Science (ISOCRIAS)*, Vancouver, Canada, 2005, pp. 158–165.
6. Granda, J.J., The role of bond graph modeling and simulation in mechatronics systems, *Mechatronics*, Vol. 12, No. 9–10, 1271–1295, 2002.
7. Amerongen, J.V., Mechatronic design, *Mechatronics*, Vol. 13, No. 10, 1045–1066, 2003.
8. Karnopp, D., Margolis, D.L., and Rosenberg, R.C., *System Dynamics: Modeling and Simulation of Mechatronic Systems*, Wiley, New York, 2000.
9. Fan, Z., Seo, K., Hu, J., Goodman, E., and Rosenberg, R., A novel evolutionary engineering design approach for mixed-domain systems, *Engineering Optimization*, Vol. 36, No. 2, 2004.
10. Wang, J., Fan, Z., Terpenny, J.P., and Goodman, E.D., Knowledge interaction with genetic programming in mechatronic systems design using bond graphs, *IEEE Transactions on Systems, Man, and Cybernetics, Part C: Applications and Reviews*, Vol. 35, No. 2, 172–182, 2005.
11. GP tutorial, available at: http://www.geneticprogramming.com/Tutorial/.
12. Introduction to genetic algorithms, available at: http://cgm.cs.mcgill.ca/~soss/cs644/projects/marko/introduction.html.
13. Karray, F. and de Silva, C.W., *Soft Computing and Intelligent Systems Design*, Addison-Wesley, New York, 2004.
14. Tafazoli, S., de Silva, C.W., and Lawrence, P.D., Tracking control of an electrohydraulic manipulator in the presence of friction, *IEEE Transactions on Control Systems Technology*, Vol. 6, No. 3, 401–411, 1998.
15. De Silva, C.W, *Mechatronics—An Integrated Approach*, Taylor and Francis, CRC Press, Boca Raton, FL, 2005.
16. Cellier, F.E. and McBride, R.T., Object-oriented modeling of complex physical systems using the Dymola bond graph library, available at: http://www.scs.org/scsarchive/getDoc.cfm?id=2002.
17. Viersma, T.J., *Analysis, Synthesis and Design of Hydraulic Servo Systems and Pipelines*, Elsevier, New York, 1980.
18. De Silva, C.W., *Sensors and Actuators—Control System Instrumentation*, Taylor and Francis, CRC Press, Boca Raton, FL, 2007.

19

Mechatronic Design Quotient (MDQ)

S. Behbahani

C.W. de Silva

Summary

This chapter presents a new concept on mechatronic design called *mechatronic design quotient* (MDQ) [6,8]. MDQ is an evaluation index for multicriteria design, reflecting the degree of overall satisfaction of the design criteria for a mechatronic system. It is computed by the aggregation of different criteria using Choquet fuzzy integral. The application of this mechatronic design methodology is illustrated using examples. The MDQ-based design methodology is concurrent, integrated, and system-oriented, resulting in improved design performance [1–3,6,8].

19.1 Introduction

Suppose that you wish to evaluate the design quality of a mechatronic product with reference to some ideal expectations of a user. You will need to consider multiple criteria in the evaluation. Accuracy, durability, intelligence, robustness, stability, reliability, cost, and efficiency are some typical examples of criteria that you may consider. In evaluating the system, these criteria will have interactions with each other. For example two criteria may have some level of redundancy with each other or two criteria may have negative correlation with each other, so that a high score in one implies a low score in the other one, and so on. Then, the desirability of having both criteria together will not be equivalent to the summation of having each of them separately. A criterion may have a veto effect, meaning that its dissatisfaction results in a large negative effect on the final evaluation. Its mere satisfaction is considered as a basic requirement and does not have much positive impact on the final score. To address these interactions, you have to think about the best compromise between these criteria in achieving the overall objectives of the system. You need to aggregate the criteria properly to assess a general sense of user satisfaction of the system. In essence system-based thinking is needed for a realistic evaluation of a mechatronic system.

This discussion can be extended to the mechatronic design process as well. Due to the involvement of multiple interactive criteria in realistic evaluation of a mechatronic system, a system-based and multiobjective design method is ideally needed. On the other hand, the complexity of dynamic interactions between components of a mechatronic system calls for a concurrent and integrated design methodology. In other words, mechatronic design has to be concurrent, integrated, and system-based to achieve the best overall system performance [1–8].

The development of a concurrent and integrated design approach is not easy, however, because of the diversity of design objectives, complexity of possible interactions which may happen between components, need of knowledge in a variety of fields, and the lack of a formal and systematic design approach. Due to these difficulties, most design engineers still follow traditional sequential design methods, which obviously can not lead to optimal designs in most cases.

As most of the existing multidomain systems have been designed through a sequential approach, they may not be optimal. Thus, there is potential for improvement through concurrent design. MDQ can be used for existing mixed systems to help the designer evaluate how well the product has been designed and also find out components or subsystems that can be further improved through design modification. It can also be used for commercial purposes to rank different products from a point of view of technical design.

This chapter presents the concept on mechatronic design called MDQ, which was first introduced in [8]. The application of this mechatronic design methodology is illustrated using examples.

19.2   Design Model

An engineering design problem may be modeled as a mapping from a required behavior space to a parameter space without violating a specified set of constraints. Alternatively, this may be considered as progressing from a specification phase to a realization phase. For a mechatronic design, one can divide the requirement space into two subspaces (Figure 19.1) [4]:

- Real-time behaviors (RTB)
- Non-real-time behaviors (Non-RTB)

Following this division, the system parameters in the structural space can also be divided into two subspaces [4]:

- Real-time and controllable parameters (RTP)
- Non-real-time and uncontrollable parameters (Non-RTP)

Figure 19.2 shows some examples of these concepts as applicable to a robot. Here, "real time" denotes parameters and specifications that may change with time after the machine is built; for example, controller gains, time constants, accuracy, and speed. In contrast, "non-real-time" refers to parameters and specifications that may not change with time after the machine is built; for example, structural material, weight, and workspace.

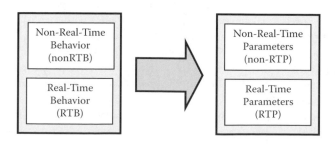

FIGURE 19.1   Mechatronic design mapping.

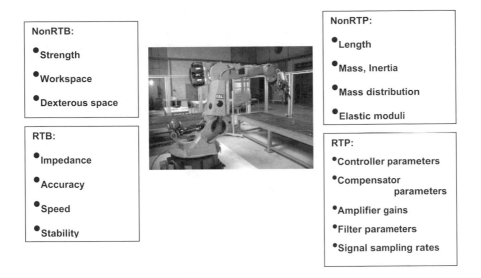

FIGURE 19.2 Mechatronic design concepts for an example of a robot.

19.2.1 Traditional Sequential Design

In a traditional sequential design process (termed *electromechanical design*), Non-RTP design is first based on the Non-RTB specifications (figure 19.3). This process itself is performed in two stages:

1. The mechanical structure of the machine is designed (e.g., configuration of the machine, sizing of the structure, structure material) according to some Non-RTB specifications (e.g., allowable stress, allowable workspace).
2. Electrical components are added to the machine (e.g., actuators, sensors) based on the Non-RTB requirements (e.g., maximum required acceleration, minimum acceptable accuracy).

A weakness of the sequential design process is that different subsystems are designed and optimized with regard to the particular criterion that is important for a subsystem. If a component is purchased from the market, it is selected merely according to its capacity limitations, whereas other performance specifications are not considered in the selection. For example, when an electrical motor is purchased,

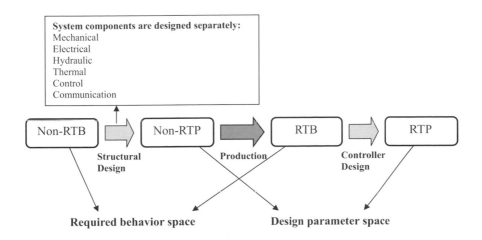

FIGURE 19.3 Model of traditional sequential design (electromechanical design).

it is usually selected merely based on the required power and price limitations. This is because in the design process of the mechanical subsystem, an electrical motor is usually viewed as a constant source of power. However, there are many other characteristics in an electrical motor that affect the performance and limitations of the controller such as torque constant, voltage constant, mechanical inertia, resistance, and inductance. These specifications, which are usually disregarded in the motor selection, can make two motors quite different in terms of the controller design for the expected task, although they may have identical power. After the machine is built, it would be costly to vary a parameter in the mechanical structure or change an electrical component. Therefore, all parameters related to mechanical structure and electrical hardware are hardly variable and can be treated as time invariant (Non-RTP).

After the mechanical and electrical subsystems are designed and built, RTP (e.g., controller algorithm, controller gain, and signal conditioning algorithm) are determined based on RTB specifications (e.g., desired path, speed, accuracy, stability) to control the already-established structure. The parameters in the driving and control structure are by and large changeable (RTP), leading to the programmability of the electromechanical system. The evolution of control engineering and computer science has resulted in the creation of a school of thought that considered the design of the mechanical structure and electrical hardware to no longer be the main design focus in some electromechanical systems, and that the inadequacies of the system mechanics could be compensated for by sophisticated control schemes and software. This thinking can be self defeatist because a perfect control action may be hardly achieved due to hardware limitations and dynamic interactions, regardless of the effort devoted to the design of the control structure. This does not mean, however, that the performance of a machine cannot be improved by better control. It is the "adequacy" and the "optimality" that are in question.

19.2.2　Ideal Mechatronic Design

Mechatronic systems ideally need integrated, concurrent, and system-based design methodology [6]. Figure 19.4 depicts the model of the design methodology presented in this chapter [1–3]. It is believed that controllability and programmability of RTP should be viewed as an opportunity to further improve the design after the machine is built, unlike in the traditional electromechanical design where programmability of RTP is considered an excuse to postpone the design until after the machine is built. In particular, RTP and Non-RTP are designed concurrently, considering both RTB and Non-RTB. After production of the machine, Non-RTP may not come out to be exactly as expected during the design process. This results in deviations of RTB and Non-RTB from their desired and expected conditions. To compensate for these deviations, system identification is then performed to find out the actual Non-RTP. The programmability of RTP is then exploited to compensate for the deviation of the behavior of the system from desired behavior, which intuitively should not be very deep.

An ideal mechatronic design should also be system based. Each component of the system has a particular objective and has a role within the objectives of the overall system. It can have interactions as

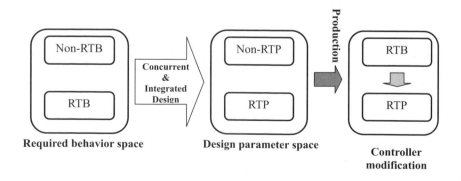

FIGURE 19.4　Model of the mechatronic design methodology.

well with other components, thereby affecting their objectives. A system-based design tries to improve the product performance from different views, considered simultaneously. In other words, it tries to improve the overall degree of satisfaction of the system through an intuitive aggregation of different and sometimes opposite objectives. MDQ is the basis of the design methodology presented in this chapter [1–3]. A system-based design is performed by using an intuitive aggregation of different criteria. MDQ is used as a goal function, and the design problem is treated as an optimization problem [8]. The result of this design methodology provides an optimum combination of the criteria involved in the calculation of MDQ. In other words, it provides the design that has highest global sense of satisfaction.

19.2.3 MDQ Formulation

Desired behavior of a system is represented by some design criteria. In general, the behavior of the system may be treated as a function of the system parameters (design parameters/variables):

$$Y = f(X) \tag{19.1}$$

where $X = \{x_1, x_2, \dots, x_m\}$ is the set of design variables/parameters or decision variables/parameters and $Y = \{y_1, y_2, \dots, y_n\}$ is the set of behavior specifications that represent the design criteria. The design process can then be considered as a procedure of finding X_d so as to satisfy the required behavior specifications Y_d, without violating a set of constraints:

$$Y_d = f(X_d) \tag{19.2}$$

Desired behavior can take three forms:

- Maximizing a function, $y_d \equiv Maximize[c_{max}(X)]$
- Minimizing a function, $y_d \equiv Minimize[c_{min}(X)]$
- Satisfying an equality, $y_d \equiv [c_{eq}(X) = \Omega]$

Two kinds of constraints may exist:

- Inequality constraints, $g_i(X) > 0$
- Equality constraints, $g_e(X) = 0$

Suppose that we want to design a beam from a specified material to carry a specified amount of load. If the required safety factor and the maximum acceptable deflection of the beam are given, its cross section can be computed simply by using standard elasticity equations. In other words, if there is a direct and explicit relation between the parameter space and the behavior space, the parameters can be realized simply by plugging the required behavior into the equations. But, in most engineering problems and particularly in mechatronic design, such a straightforward relation between design variables and behavior specifications cannot be found. Then, the designer encounters a relatively extensive search space resulting from the multiplicity of feasible conceptual choices and the search space for the realization of inherent parameters associated with each choice. The task of the designer is to evaluate the feasible choices, make proper conceptual decisions, and, consequently, find the design parameters not only to satisfy the desired behavior without violating the constraints but also to achieve the best satisfaction. Effective evaluation of possible solutions is the key requirement in an optimum mechatronic design. MDQ represents a multicriteria design evaluation index that can be effectively used in this context [8].

Suppose that n design criteria and r constraints exist. Then, MDQ can be expressed as

$$MDQ(X) = H[s_1(X), s_2(X), \quad , s_n(X)] . \prod_{k=1}^{r} G[g_k(X)] \tag{19.3}$$

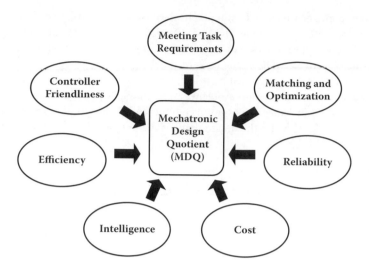

FIGURE 19.5 MDQ attributes.

where H is an aggregation operator, $s_i(X)$ is the partial score between zero and one from the i-th criterion showing its degree of satisfaction, and $G[g(X)]$ is a function indicating whether a constraint has been met. It is equal to one if the constraint has been satisfied and zero, otherwise. A mechatronic design problem can be treated as optimization of MDQ.

A useful definition for MDQ should span a wide range of features and needs of mechatronic systems. Following criteria are some general objectives of a mechatronic design (Figure 19.5):

1. Satisfaction of the task requirements
2. Component matching
3. Efficiency
4. Intelligence
5. Reliability
6. Controller friendliness
7. Cost

19.2.4 Fuzzy Aggregation of Criteria

An appropriate approach for criteria aggregation is presented now. For each design alternative, a partial score between zero and one is assigned to each criterion. Then the MDQ is computed by aggregating these partial scores. The common aggregation technique—weighted average method—is not intuitive because it is essentially a linear integral. It is suitable only if the involved criteria are independent, and hence their weighted effects can be added together. There are some interactions between criteria that affect the human expert's inference, and they cannot be represented by traditional aggregation tools.

There are two nonlinear fuzzy integrals, Choquet and Sugeno integrals, which have been successfully used in literature for aggregation of criteria [9,10]. In particular, Choquet integral fits intuitive requirements for decision making in the case of interacting criteria [10–12].

Considering a finite set of criteria $Y = \{y_1, y_2, \ldots, y_n\}$ in a multicriteria evaluation problem, Choquet integral provides a weighting factor not only for each criterion but also for each subset of criteria. The weighting factor of a subset of criteria is represented by a fuzzy measure in the universe Y satisfying

$$\mu(\phi) = 0, \quad \mu(Y) = 1$$

$$A \subset B \subset Y \Rightarrow \mu(A) \le \mu(B) \tag{19.4}$$

The useful purpose of defining a weighting factor for each subset of criteria is that interaction between criteria and an expert designer's attitudes can be represented mathematically and can be taken into account in the aggregation. As an example, some common semantic interactions and the way they can be represented by fuzzy measures are explained now.

19.2.4.1 Positive Correlation

Two criteria, $y_i, y_j \in Y$, have positive correlation if they have some degree of redundancy; hence, a good score in one of them is usually simultaneous with a good score in the other. For example, for the proposed criteria, reliability and intelligence have positive correlation to some extent. Once a machine is provided with intelligence—for example, in the form of self tuning and self diagnosis—these factors result in increased reliability as well. Then these two criteria possess some degree of redundancy. Positive correlation can be modeled by the following inequality [10]:

$$\mu(i,j) < \mu(i) + \mu(j) \quad y_i, y_j \in Y \tag{19.5}$$

Here $\mu(i,j)$ is the fuzzy measure of the set of criteria which consists of only y_i and y_j.

19.2.4.2 Negative Correlation

Two criteria $y_i, y_j \in Y$ have negative correlation if it is not common to have high score in both of them simultaneously. In this case, a good score in one of them usually implies a bad score in the other one. For example, cost has a negative correlation with other criteria in the MDQ. High scores in reliability and intelligence, for example, result in a low score in cost (high cost). Negative correlation can be modeled by [10]

$$\mu(i,j) > \mu(i) + \mu(j) \quad y_i, y_j \in Y \tag{19.6}$$

19.2.4.3 Substitution

When two criteria are parallel to each other and are interchangeable, the following substitution property is satisfied [10]:

$$\mu(T) < \begin{Bmatrix} \mu(T \cup i) \\ \mu(T \cup j) \end{Bmatrix} \approx \mu(T \cup i \cup j) \quad T \subseteq Y \setminus y_i, y_j \tag{19.7}$$

19.2.4.4 Complementarity

This occurs when two criteria are prerequisites (i.e., complementary) of each other to achieve their desired effect [10]. For example, "meeting task requirements" and "cost" have complementarity to some extent. One has

$$\mu(T) \approx \begin{Bmatrix} \mu(T \cup i) \\ \mu(T \cup j) \end{Bmatrix} < \mu(T \cup i \cup j) \quad T \subseteq Y \setminus y_i, y_j \tag{19.8}$$

19.2.4.5 Preferential Dependence

This is the opposite of the preferential independence, which implies that the competition between $H(s_i, s_j)$ and $H(s_k, s_j)$ is not affected by the common part y. Here, $H(s_i, s_j)$ denotes the aggregated score of partial scores s_i and s_j [10]. Mathematically, one has

$$H(s_i, s_j) \geq H(s_k, s_j) \Leftrightarrow H(s_i, s_l) \geq H(s_k, s_l) \quad y_i, y_j, y_k, y_l \in Y \tag{19.9}$$

Here s_i indicates the partial score of a design choice against criterion y_i.

19.2.4.6 Veto Effect

A criterion y_i has a veto effect if a bad score in this criterion results in a bad global score, regardless of the degree of satisfaction of the other criteria [11]. Specifically,

$$\mu(T) \approx 0 \quad \text{if} \quad T \subset Y, y_i \notin T \tag{19.10}$$

For example, meeting task requirements has a veto effect on the MDQ.

19.2.4.7 Pass Effect

A criterion y_i has a pass effect if a good score in this criterion results in a good global score, regardless of the degree of satisfaction of other criteria [11]. Specifically,

$$\mu(T) \approx 1 \quad \text{if} \quad T \subset Y, y_i \in T \tag{19.11}$$

After specifying the weighting factors for all subsets of criteria, Choquet integral can be used to compute the global score [12]:

$$S = H(s_1, s_2, \ldots, s_n) = \sum_{i=1}^{n} s_{(i)} \cdot [\mu(A_{(i)}) - \mu(A_{(i+1)})] \tag{19.12}$$

where (.) indicates the criteria, which should be sorted in the ascending order based on their partial scores such that $s_{(1)} \leq s_{(2)} \leq \ldots \leq s_{(n)}$ and $A_{(i)} = \{(i), \ldots, (n)\}$ and $A_{(n+1)} = \phi$ [12].

The main difficulty of the Choquet method lies in the identification of the 2^n coefficients of fuzzy measures. The overall importance of a criterion i is not solely determined by the value of $\mu(i)$. Indeed, $\sum_{i=1}^{n} \mu(i)$ is not necessarily equal to one [12]. Intuitive notions expressed in equations 19.5 through 19.11 can serve as a guide here. Another useful concept is the index of overall importance of a criterion, computed by the Shapley value, which is defined as [12]

$$\Psi(\mu, i) = \sum_{T \subseteq Y \backslash i} \frac{(n-t-1)! t!}{n!} [\mu(T \cup i) - \mu(T)] \tag{19.13}$$

where $t = |T|$. Then,

$$\sum_{i=1}^{n} \Psi(\mu, i) = 1 \tag{19.14}$$

Another useful parameter is the interaction index, which is computed by [12]

$$\Theta(\mu, ij) = \sum_{T \subseteq Y \backslash i,j} \frac{(n-t-2)! t!}{(n-1)!} [\mu(T \cup ij) - \mu(T \cup i) - \mu(T \cup j) + \mu(T)] \tag{19.15}$$

A positive correlation leads to a negative interaction index, and vice versa [11]. The fuzzy measures should be specified so as to satisfy the desired overall importance and the interaction indices.

Difficulties can arise in the specification of fuzzy measures if the number of criteria is high. For example, the seven criteria mentioned before will form $2^7 = 128$ possible subsets of criteria, which will need the specification of 128 fuzzy measures. As two of them are predefined (Equation 19.4), 126 values should be specified. Specification of these values can be quite challenging because all issues addressed in equations 19.5 through 19.15 should be reflected in these values. To facilitate the specification of fuzzy measures, a hierarchical procedure has been given in [1,2]. It is based on the fact that for three criteria, six weighting factors should be specified, and it is not difficult to come up with a weighting factor for each subset that will reflect the existing interactions. In the developed method, a hierarchical scheme of criteria is created so that each branch has no more than three sub-branches. The division is such that the items in each category are somewhat similar to each other. By considering this hierarchical pattern, there are only three high-level criteria and at most three subcriteria for each of them.

The process of specification of fuzzy measures will become simple by this hierarchical procedure. The designer looks at the high-level criteria and decides on the overall importance of each criterion and the degree of interaction between them. High level fuzzy measures are then specified to represent the viewpoint of the designer. As there are no more than three high-level criteria, a maximum of six fuzzy measures have to be specified. For these six unknowns, six equations can be written to help the computation of the fuzzy measures so as to satisfy the designer attitudes (three for overall importance and three at the interaction level). However, they can be used only for guidance, because the set of equations is singular and at least one value of overall importance should be established beforehand in order to be able to solve the equations. The same process is repeated for the lower-level criteria. Here, the fuzzy measure of each set of criteria expresses the contribution of the set in the high-level criterion. If the seven criteria mentioned before are divided into three branches with 3, 3 and 1 subcriteria, then only 18 fuzzy measures need to be specified instead of 126, and the specification process becomes more intuitive.

After evaluation of criteria, the computation of the global score also becomes more intuitive and faster. The partial score of each high level criterion is first computed by aggregating the low-level partial scores of its subcriteria. Then an aggregation between the computed scores of high-level criteria gives the global evaluation score of each particular design.

19.3 Mechatronic Design Methodology

A systematic design methodology is presented in this section based on the concept of mechatronic design quotient [1–3,6,8]. A product realization can be treated in four stages:

- Conceptual design
- Detailed design
- Production
- Final improvements

The concept of MDQ is used in each of these stages to help the designer in making decisions. Figure 19.6 gives a flowchart of the design approach.

Conceptual design is often important in the mechatronic design process. In fact, it is not practical to search a complex design space in just one stage in finding the best solution among a variety of possible configurations. In conceptual design, a complex design space is divided into several subspaces, which correspond to conceptual options. Through proper evaluation of all these subspaces, the designer then narrows down the search space to one or two of the subspaces. This approach has two main advantages. First, the search space becomes smaller and less complex. Second, after this stage it becomes easier to process the remaining search space using search tools such as genetic algorithms.

Effective evaluation of possible conceptual choices is the key requirement in the conceptual design of mechatronic systems. MDQ is used as an index to help the designer in proper decision making. The main activities of this stage are generation and evaluation of a conceptual design, and making decisions about the fundamental structure of the product. Decisions in this stage are basic but quite fundamental. Each decision can have a deep effect on other parts of the system, can significantly change the rest of the design process, and also can change the performance of the product with respect to a wide range of criteria. Because of these considerations, a multicriteria decision making is desired. The concept of MDQ

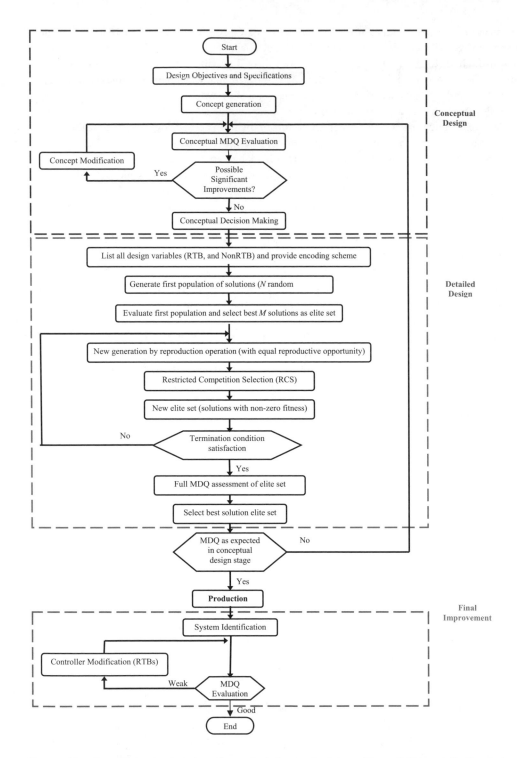

FIGURE 19.6 Flowchart of the developed mechatronic design methodology. (From Behbahani, S., Practical and Analytical Studies on Development of Formal Evaluation and Design Methodologies for Mechatronic Systems, Ph.D. thesis, Department of Mechanical Engineering, The University of British Columbia, Vancouver, Canada, 2007; Behbahani, S. and de Silva, C.W., *Proceedings of the International Symposium on Collaborative Research in Applied Science (ISOCRIAS)*, Vancouver, Canada, 2005, pp. 214–221; Behbahani, S. and de Silva, C.W., *Proceedings of IEEE World Congress on Evolutionary Computation*, Vancouver, Canada, 2006, pp. 1031–1036. With permission.)

is used in this stage to guarantee that the overall satisfaction of the product is considered in the decisions. It can also highlight issues that can make significant improvement in the performance of the product from an early stage of the design.

In the detailed design stage, the designer lists all the design parameters/variables including both RTP and Non-RTP. Design variables in this stage are either continuous (e.g., controller gains, length of a structural member) or discrete (e.g., motor selection, sensor selection). A search universe is then created by listing available options for each discrete variable, and assigning estimated range and possible constraints for continuous variables. An optimization tool is then applied in order to find the solution. As this problem generally does not have a precise and analytical solution approaches, evolutionary techniques such as genetic algorithms are appropriate for the optimization. As MDQ computation can be complex and time consuming, simple genetic algorithms may not be sufficiently strong for the problem. In the developed technique, the optimization process is performed in two stages. A niching genetic algorithm is used in the first stage to narrow down the search space to a limited number of elite solutions. A full and comprehensive MDQ competition is performed in the second stage to find the optimum design.

After the machine is built, the changeability of RTP provides the opportunity for further design improvement. The RTP only affect the RTB of the system, whereas the Non-RTP can affect both RTB and Non-RTB [4]. As the Non-RTP may not be exactly what they should be, the RTB may not be satisfied with the computed RTP in the detailed design stage. For example, structural properties of a system (e.g., mass, damping, and stiffness matrices) may not be identical to what are calculated and expected in the detailed design stage. As a consequence, desired speed, accuracy, and stability (RTB) would not be achieved by the computed controller gains (RTP). In the last design stage, the designer carries out a system identification to practically specify the Non-RTP and then modify the RTP to compensate for any unexpected variations that may arise from the detailed design stage.

There also exist techniques to improve the functionality of a machine, such as the consideration of component redundancy, fault detection and diagnosis, and supervisory control. In the final modification stage, the designer considers all controller parameters and possible improvement techniques. MDQ is used to find the optimum controller parameters and to trade off possible modifications, and decide upon them [1].

19.4 Conceptual Design

The main activities in the conceptual stage of a system design are generation and evaluation of conceptual choices and making fundamental decisions about the structure of the system [13]. This stage has some important features, specifically [1],

- Design decisions are limited mostly to a discrete space, with several limited options.
- Usually there is no explicit analytical relationship or rule to express the preference of one design option over another option in the decision making.
- Decisions in the conceptual stage are related to the fundamental structure of the final product. Their consequences should be reflected in all the lower-level decisions. In practice, if one design parameter is changed from one conceptual option to another, the entire set of design parameters and specifications may change as a result.
- Decisions in this stage affect a wide range of features and specifications. Therefore, a multicriteria decision making is necessary to select the best option.

In the conceptual stage, the designer should try to generate whatever feasible design choices that may exist for the particular design problem. The designer can search for feasible subsystem choices and conceptual options in information sources such as the Internet, similar available machines, catalogs, handbooks, technical reports, and by consulting experts to generate conceptual design alternatives. The objective of the conceptual design stage is not to complete a final design but rather to identify performance-limiting factors of the design proposals as early as possible with the aim to choose satisfactory specifications for these factors [14]. The generated choices should then be evaluated and compared against the required specifications and desired criteria to select the best design. In a mechatronic design, the evaluation of the generated concepts should cover all the subsystems in an integrated manner, considering both real-time and non-real-time issues concurrently, and evaluating the overall sense of satisfaction with regard to a set of multiple criteria (i.e., system-based). Because of these features, the designer can use MDQ to facilitate multicriteria decision making in this stage. In addition to its use in the evaluation and decision making, MDQ also has the ability to indicate weaknesses of design alternatives, which may be improved by some conceptual modifications. The generation, modification, and evaluation of a conceptual design are repeated until the designer is able to justify that the final decision is feasible and is the best available one. Practically, the following steps are taken in this phase [1]:

1. Review the design objectives, basic requirements, possible constraints, axiomatic design specifications, and the general plan of the system.
2. List the conceptual subsystems or components of the system to be designed using multicriteria decision making. Many types of subsystems are involved in the design of a mechatronic system, but not all of them will require multicriteria decision making. Some components can be decided easily with axiomatic requirements. For example, from a practical viewpoint, mechatronics is directly applicable to a motion control system. In a typical motion control system, the basic subsystems (components) are the actuators, sensors, mechanical structure, and the control techniques [19], which may be expressed as:

$$X = \{x_1, x_2, \ldots, x_n\} \tag{19.16}$$

3. Estimate the technical requirements for each conceptual subsystem such as capacity, power, bandwidth, accuracy, weight, size, and cost.
4. In this step, a top–bottom approach is carried out to generate and appropriately present the conceptual choices. For each conceptual subsystem, list the available and feasible design alternatives, which can be called subsystem choices. For example, electrical and hydraulic actuators can represent two choices for the actuator part. Different control techniques can represent the choices for the controller part. Each of these choices may have corresponding subchoices. All these choices can be presented in a tree-like structure. This step basically includes the concept generation by the help of such means as looking at similar machines available in industry, searching through the Internet and catalogs, consulting with experts, reviewing handbooks, and other sources of information.
5. List all possible and feasible combinations that lead to a design alternative. Not all possible combinations of component choices will require multicriteria decision making. Some combinations may be clearly weaker than others and can be deleted without evaluation. Retain only those design alternatives that justifiably need multicriteria decision making.
6. Decide on the MDQ attributes that are considered important for the particular design. Generally, the seven criteria indicated in Figure 19.5 are important in a mechatronic system. Assign a fuzzy measure for each subset of criteria, indicating the degree of importance of that subset. Interactions between criteria should be taken into account and can be used as a guide for choosing these fuzzy measures.
7. Evaluate each design alternative, according to all MDQ attributes and assign a partial score for each attribute. However, accurate assessment may not be possible in this stage due to lack of information. For the offered criteria, the following guidelines can be used for design evaluation in the conceptual phase:
   - *Meeting the task requirements:* Required bandwidth, estimated required force, available space, and required accuracy are basic attributes that may be used for evaluating this criterion.
   - *Reliability:* Basically the number of components in a system has an inverse effect on the system reliability. As a simple estimate, the failure possibility can be considered as the sum of the failure

rates of the components in the system. Reliability is then equal to one minus the failure probability. Note that dynamic effects are not taken into account in this simple assessment.

- *Intelligence:* Intelligent features of a machine are incorporated in the subsequent design stages; for example, in the detailed design stage or through programmability of the machine after the machine is built. In the conceptual stage, features such as self calibration, self tuning, self diagnosis, fault tolerance, biologically inspired behavior, human–machine interaction, and having non-model-based conventional control techniques can provide an assessment of the level of intelligence of a machine [15].

- *Matching:* Dynamic interactions between components, bandwidth issues, capacity of components, and environmental issues can lead to a conceptual assessment of the criterion of matching. In a mechatronic design, a bandwidth matching is required between the frequency content of the desired motion, digital controller frequency (both hardware and software), sampling period, and bandwidth of sensors and actuators [6,19]. On the other hand, it is desired that all components of a mechatronic system operate at their optimal capacities. Overdesigned and underdesigned components degrade the design quality of the system. Some components may not be suitable for a particular environment; for example, mechanical parts of a machine may create an environment that may not be suitable for the proper function of some electrical elements, although in simulation and analysis they may appear to match for that task.

- *Control friendliness:* An important attribute in this context is the system nonlinearity, and it can be assessed by estimating the critical nonlinearities of the system; for example, friction. Possible disturbances and system uncertainties, parameter variation, and estimated order of the closed-loop system are other issues that may be important in the assessment of controller friendliness in the conceptual stage.

- *Efficiency:* This criterion can be evaluated by estimating the probable energy dissipation or wastage in the system, particularly due to friction.

It should be noted that the assessment in the conceptual phase will be tested again in the subsequent stages. However, if a large gap between the estimated MDQ values is found in different stages, the designer should return and review the previous design stages and design assessments again to make sure that the final design is optimal.

8. Aggregate the partial scores by using Choquet integral to determine the global score of each design alternative.
9. Keeping in mind the MDQ attributes of each design alternative, investigate whether there is possibility to improve the system MDQ by some conceptual modifications. Motivated by evolutionary techniques, crossover between good design solutions is a suitable way to search for better conceptual designs. Some architectural modifications such as component sharing, fault detection and diagnosis, and component redundancy can also lead to better design alternatives. If so, consider modifications and go back to Step 6.
10. Select the best design alternative and proceed to the next design stage, which is the detailed design.

19.5 Detailed Design

Preliminary yet fundamental decisions about the structure and the architecture of a product are made in the conceptual design stage. In the conceptual design, the components and subsystems of the product are specified. For example, it is decided whether the actuators are electrical or hydraulic. The control technique is also selected, and only its parameters remain to be determined. Once the overall structure of the product is designed, the designer should compute the corresponding design parameters, or choose the corresponding components from available options in the market. This stage of design involves computation and specification of design parameters; hence, term *detailed design*.

Some design parameters are changeable and controllable even after the machine is built, and some others are not. Some parameters are continuous and should be computed, and others are discrete and are limited to a finite number of available options. Regardless of these classifications, all design variables should be computed and optimized in a concurrent and integrated manner with respect to multiple criteria that are important in the performance of the product. In other words, a mechatronic system ideally needs an integrated, concurrent, and system-based design approach.

Complicated design of a multidomain system can be treated as an optimization problem by using a proper design evaluation index. In the design approach explained in this section, MDQ serves to evaluate the fitness of design trials in an optimization process [1,3]. The optimization process is performed in two stages because of the complexity of the problem. In the first stage, a niching genetic algorithm is used to find local and global optimal design alternatives with respect to some essential MDQ attributes. In the second stage, these local optima will compete with each other with respect to all criteria involved in MDQ.

19.5.1 Niching Genetic Algorithm

There is a problem associated with implementation of genetic algorithms for optimization of MDQ. Fitness evaluation of trial solutions is the key procedure in the genetic algorithms, which is repeated many times. It is very important that the fitness evaluation be fast and does not require complicated analysis. A general and appropriate representation of MDQ should include a wide range of criteria. The accurate assessment of these criteria usually needs rather complicated analysis, which can be time consuming and costly. In particular, the main objective of the developed design approach is to provide a comprehensive and multi-criteria view to the evaluation of design trials in the design process. This, however, conflicts with the efficient utilization of optimization approaches. A comprehensive evaluation would be too time consuming and complicated for performing in the course of a reasonable optimization process.

To counter this problem, it has been proposed to perform the optimization process in two stages [1,3]. In the first stage, only a combination of essential and more important criteria—those which have a veto effect on the MDQ assessment—is applied for fitness evaluation. In addition, a simplified and approximate evaluation is performed in this stage in order to save time and cost of computation. In contrast, the optimization process is implemented so as to retain not only the global optimum, but also local optima, which could potentially represent better designs if a full and accurate assessment were applied. These optimal design trials can be considered as candidate solutions to be optimum. The variations of evaluation indices due to the change of design parameters are expected to be rather smooth in a real physical system. Hence, different optima are expected to have significant differences from each other, representing different possible configurations for the system. By retaining all optimal solutions, each possible configuration is allowed to represent its elite solution for a final competition, even if it has rather low fitness in comparison to other configurations. In the second optimization phase, a full and accurate MDQ assessment is performed as a competition between candidates to find the global optimum design.

A desirable feature of a simple genetic algorithm is its ability to escape local peaks due to its random variations. It means that even a solution corresponding to a local maximum will not likely survive if its fitness is significantly less than the solutions near the global peak. Even in the case of approximately equal peaks, a simple genetic algorithm will randomly converge to one of the peaks, because it does not have any control on the competition between different peaks. This desirable feature of simple genetic algorithms makes them unsuitable for the particular problem considered here, where the designer is interested to find local peaks as well as the global peak. In other words, simple genetic algorithms are not suitable for the optimization of a multipeak function when the designer is interested to know all the peaks.

To address this problem, it has been proposed to use niching genetic algorithms [1,3]. The niching genetic algorithms represent a branch of evolutionary algorithms dedicated for multiobjective and multimodal optimization problems. They are used when the function to be optimized has several peaks and the designer is interested in all of them, regardless of whether it is a local peak or the global peak of the function [16–18].

Analogous to other terms derived from genetic science of biological species, in biology *niche* means a unique ecological role, location, or job of an organism in a community for which a species is well suited within its community, including its habitat, what it eats, its activities, and its interaction with other living things. The niche of an organism permits it to survive in its environment. In other words, the particular task, formation and resources of an ecosystem reduce the interspecies competition with other ecosystems which, in turn, helps a stable survival of this ecosystem. Different members of an ecosystem, however, undergo competition with each other to survive, resulting in a gradual evolution of the ecosystem.

In a niching genetic algorithm, a collection of solutions with similar configuration is considered as an ecosystem. Several techniques are available to implement a niching genetic algorithm, including sharing, deterministic crowding, and restricted competition selection (RCS) [18]. The RCS technique is the best technique for the specific problem of the detailed mechatronic design, because other techniques maintain many solutions around each niche. In this particular problem, only the best solution is needed per niche, because individuals in a niche have similar characteristics. The RCS approach has also better matching with the mystery of the survival of the ecosystems in nature. In this approach, the particular configuration of a collection of solutions with similar configuration (ecosystem) reduces the chance of competition with the solutions with different configurations. Two solutions will compete with each other only if their difference is less than a threshold, called *niche radius*. By restricting competition among dissimilar individuals and performing competition only between similar individuals, a stable subpopulation will find an opportunity to form around a local optimum. Due to competition between similar individuals, each subpopulation evolves and converges to its best. The best solution in a niche represents the elite of that subpopulation. The set of elite solutions is finally created that represents the best solutions with different possible configurations. The corresponding procedure of this technique is as follows [16]:

Step 1: Generate N random trial solutions and evaluate their fitness. Select M fittest solutions as an elite set.

Step 2: Create N new random solutions by applying reproductive operations on the previously created generation. Note that, unlike in a simple genetic algorithm, in the RCS niching algorithm one does not have to consider any discrimination in reproductive opportunity between individuals.

Step 3: Add the elite set to this population to generate a competing set with $N + M$ solutions.

Step 4: This is the fundamental step of RCS. Confront each trial solution to all other trial solutions. If the difference between them is less than the nominal difference (niche radius), perform a competition between them and set the loser's fitness to zero.

Step 5: Generate the new elite set as the set of solutions with non-zero fitness.

Step 6: Randomly select N solutions from the competing set to be the parents for the next generation. It is desirable to include the elite set in this population.

Step 7: If the terminating condition is not satisfied, go to step 2; otherwise, stop the process.

Based on the described RCS procedure, an optimization tool has been developed for multipeak functions [1,3]. To verify the developed tool, it has been tested on several multipeak functions; for example, the function shown in Figure 19.7. The obtained results are quite satisfactory with respect to the number of the peaks found by the tool and their accuracy.

In the detailed mechatronic design, a niching genetic algorithm is used in the first optimization stage, not to find the optimal solution but to limit the search space to some optimal candidates. In the second optimization stage, the designer is able to add other criteria into MDQ assessment and carry out an accurate analysis to make a practical competition between these optimum candidates and find the true optimum.

19.5.2 MDQ Optimization

Practically, the following steps are taken in the detailed design stage [1]:

1. List all essential subsystems and components which should be designed. Each of these components may have subcomponents. If all these subcomponents are arranged together, a tree-like structure is formed. For each subcomponent or design parameter in the lowest level of this tree, decide whether it needs a concurrent, integrated, and multicriteria design methodology. Not all the components and variables in a mechatronic system need to employ a multicriteria and integrated design approach. For instance, some components do not impose any limitation on the performance of the entire system or do not have considerable interaction with other components in the system;

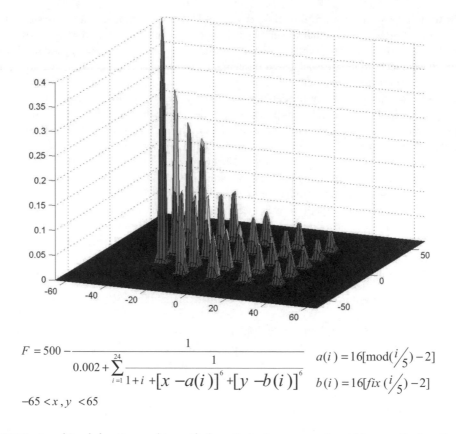

$$F = 500 - \cfrac{1}{0.002 + \displaystyle\sum_{i=1}^{24} \cfrac{1}{1 + i + [x - a(i)]^6 + [y - b(i)]^6}} \qquad \begin{aligned} a(i) &= 16[\mathrm{mod}(i/5) - 2] \\ b(i) &= 16[\mathit{fix}\,(i/5) - 2] \end{aligned}$$

$$-65 < x, y < 65$$

FIGURE 19.7   A multipeak function used to verify the optimization program by niching genetic algorithm.

hence, they can be designed or selected separately. List all the design parameters, determination of which justifiably will need a concurrent design approach, due to their interaction with other parts of the system.

2. For each component in the list, decide whether it should be built or selected from available options in the market. If it has to be built, list design variables that specify the structure of that component and provide a rough estimation of the range which is reasonable for each design parameter. If it has to be purchased from available options in the market, provide a rough estimation or a reasonable range of technical requirements for that component, and prepare a list of available options in the market by searching information sources such as the Internet, and manufacturers' catalogs.

3. List all design parameters that can describe a trial design solution for the system, and develop an encoding scheme to represent each trial design by a string of binary values. This means that each design parameter is discretized to a finite number of segments, expressed as a power of 2. The discretization process is not particularly a problem for a continuous parameter as far as it provides enough bits to achieve the desired level of precision in the feasible range of that parameter. For a discrete parameter, for example, in the case where a part is selected from available options in the market, the discretized value refers to the order of each option in the set of options. If the number of available options is not expressible as a power of 2, then some unnecessary bit patterns will exist. In the genetic algorithm process, a trial solution which addresses to one of these extra values can be discarded, or some parameters may be represented twice so that all binary strings result in a legal set of design variables.

4. Decide on the MDQ attributes that are considered important for the particular design. Generally, the seven criteria indicated in Figure 19.5 are important in a mechatronic system. For the first stage of optimization, consider only those criteria that are more important. Specifically, the criteria that have a veto effect or a pass effect on the evaluation of the system should be considered for this optimization phase.

5. Provide simple and fast routines to analyze each trial solution with respect to the above criteria. In this stage, designers may need to use computer software or develop their own programs to model the system.

6. Provide a relationship or develop a small program which can evaluate how two different trial solutions are related, and decide if they are considered as two different configurations or not.

7. The stage is now set to use the niching genetic algorithm to find local optima, which are called elite solutions. The flowchart of the niching genetic algorithm with restricted competition selection (RCS) is explained in Section 19.5.1. Use this strategy and determine elite solutions that have fairly high fitness levels. The first stage of optimization is completed here and now the designer should perform a more detailed and comprehensive evaluation between these elite solutions to find the best one, which can be considered the optimum design.

8. Consider all the MDQ attributes and assign a fuzzy measure for each subset of criteria indicating the degree of importance of that subset. Interactions between criteria should be taken into account and can be used as a guide for choosing these fuzzy measures.

9. Evaluate each elite design according to all MDQ attributes and assign a partial score for each attribute. In this stage, the designer may need to develop computer programs and use available simulation tools for the design tasks. The analysis in this stage should be as accurate as possible. It is not a critical problem if the run-time of the simulations is high, as it is not repeated many times.

10. Aggregate the partial scores by using Choquet integral to determine the global score of each elite design. The design with the highest global score is considered the optimum design.

19.6 Case Study

The operation of an industrial fish cutting machine called Iron Butcher is explained in Section 18.4.2 of Chapter 18. Now the manipulator of this machine is redesigned through the mechatronic design process presented in the chapter.

19.6.1 Conceptual Design

The main steps of the conceptual design of the manipulator of the Iron Butcher are given below:

1. The objective is to design a 2-D positioning table, which is able to move through a maximum stroke of 50 mm in less than 0.4 sec with an error not exceeding 3 mm. The basic plan is to design a Cartesian table (as for a milling machine) with two motion sensors and two actuators for each direction.

2. The following conceptual components should be designed or selection decisions have to be made about them:

$$X = \{Mechanical\ structure,\ actuator,\ sensor,\ controller\}$$

3. The required power, force, and bandwidth can be estimated by estimating the mass, motion trajectory and friction [6,19]. Table 19.1 gives the estimated values for different conditions.

4. For presenting the feasible design options for each conceptual subsystem and subsystem choices, a tree-like structure (Figure 19.8) is formed.

5. The main objective of the conceptual decision making is to decide on the actuator type (electrical or hydraulic), surface contact condition (lubrication, bearing, or direct contact), and material (steel or aluminum). Other issues do not need multicriteria decision making and can simply be

TABLE 19.1   Estimated Technical Specifications of the Machine for Different Conditions

Material	Contact Condition	Power (W)	Force (N)	Bandwidth (Hz)
Steel	Direct contact	1000	1100	20
Steel	Roller bearing	250	320	20
Aluminum	Direct contact	800	900	20
Aluminum	Roller bearing	200	280	20

selected after these issues are decided upon. For example, usually there are embedded position sensors in modern motors and hydraulic cylinders [6,19]. As the present machine does not have any severe restriction on sensor selection (nearly all sensors meet the required accuracy), the sensor is automatically selected when the actuator is selected.

6. Meeting task requirements, matching, reliability, controller friendliness, efficiency, and cost are the important MDQ attributes for this decision making. These six criteria form $2^6 = 64$ possible subsets of criteria, which need the specification of 64 fuzzy measures. Two of them are predefined

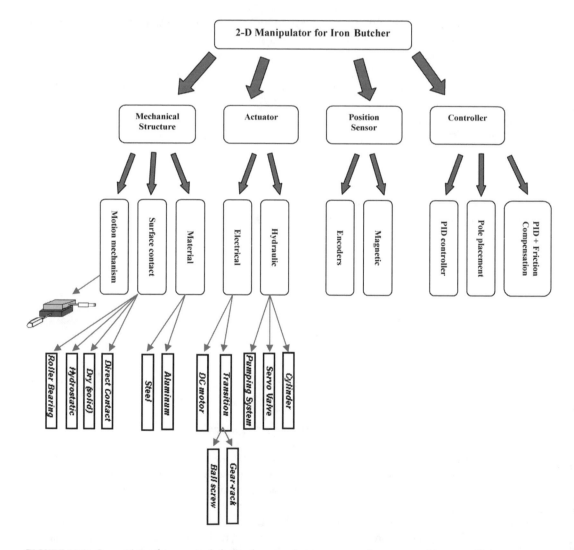

FIGURE 19.8   Generation of conceptual choices by a top–bottom approach represented in a tree-like structure.

TABLE 19.2 High Level Fuzzy Measures Used in Choquet Aggregation Method (Note: Criteria Include: (1) Basic Requirements, (2)Technical Issues, (3) Cost).

$\mu_1 = 0.52$	$\mu_2 = 0.15$	$\mu_3 = 0.15$
$\mu_{12} = 0.55$	$\mu_{13} = 0.7$	$\mu_{23} = 0.4$

(Equation 19.4); hence, 62 values should be specified. As it is a difficult task to specify these values, we use hierarchical method explained in Section 19.2.4 for this purpose. We can divide the above six criteria into three categories:

a. Meeting task requirements; Matching

b. Reliability; Controller friendliness; Efficiency

c. Cost

The division is such that the items in each category are somewhat similar to each other. For example, "Meeting task requirements" and "Matching" both have some form of veto effect. "Cost" has a negative correlation with the other items, and consequently it is isolated. "Reliability," "Controller friendliness," and "Efficiency" show the quality of the design. It is good to meet them, but a system can be acceptable even if it does not strictly meet these criteria. Now there are only three high-level criteria which can be termed "Basic requirements," "Technical issues," and "Cost," respectively. Now it is not difficult to intuitively come up with the weighting effect of each subset of these items. The fuzzy measures given in Table 19.2 are used in the present case study. Using these values, the overall importance of each of the criteria and the interaction between them are computed using Equations 19.13 and 19.15 as given in Table 19.3. It shows a negative correlation between price and two other criteria and also shows a small positive correlation between "basic requirements" and "technical issues" categories.

The next step is to repeat the same procedure for the lower level of subcriteria in each category. First category includes two subcriteria. It can be said that they have equal importance in satisfying basic requirements. Therefore, a weight of 0.5 is assigned for each of them for low-level aggregation. For the second category, the values given in Table 19.4 are assigned. This provides the results given in Table 19.5. It shows that the criteria have positive correlation with each other.

Once the partial scores of a design alternative are found, a low-level aggregation should be performed first. The score of "technical issues" is an aggregation of reliability, controller friendliness, and efficiency based on the fuzzy measures of Table 19.4. The score of "Basic requirement" is the average of "Meeting task requirements" and "Matching." Then a high-level aggregation is performed to find the global score of a design alternative based on fuzzy measures in Table 19.2. An obvious advantage of this approach is that it allows conveniently incorporating the viewpoints of the analyzer by means of weighting factors. In addition, only 14 values have to be specified instead of 62 values.

7, 8. The results of MDQ assessment for some of the design alternatives are shown in Table 19.6. The best MDQ corresponds to the choice of a roller bearing, a DC motor with encoder position sensor, and a simple PID controller.

9. It was found that some parts of the machine such as camera and all the parts involved in the positioning table are not utilized in half of each cycle, when a fish is pushed into the cutting zone. A new machine can be designed with two fish conveyor systems and with a 180° out of phase. Then these parts can be shared between them, and in this manner a design with a lower cost (per fish) can be achieved. However, reliability will be less because the number of components increases, but the overall MDQ of the machine will increase as well.

10. Final decision on the conceptual design stage is to use electrical actuation and incorporate roller bearings on the contact surfaces.

TABLE 19.3 Overall Importance and Interaction Level for High-Level Criteria

$\Psi_1 = 0.5317$	$\Psi_2 = 0.1967$	$\Psi_3 = 0.2717$
$\Theta_{12} = -0.035$	$\Theta_{13} = 0.1150$	$\Theta_{23} = 0.1850$

Note: Criteria include basic requirements, technical issues, and cost.

TABLE 19.4 Fuzzy Measures Used for Choquet Aggregation of Subcriteria of "Technical Issues"

$\mu'_1 = 0.45$	$\mu'_2 = 0.42$	$\mu'_3 = 0.35$
$\mu'_{12} = 0.8$	$\mu_{13} = 0.775$	$\mu_{23} = 0.7$

Note: Criteria include (1) reliability, (2) controller friendliness, and (3) efficiency.

19.6.2 Detailed Design

The main steps of the detail design are given below.

1. Not all components and subsystems of the machine need multicriteria and integrated design. For example, the sensory part does not have considerable interactions with other parts and also does not impose any technical limitation on the performance and the capacity of the system. Hence, the sensors can be selected separately, based on such criteria as price and to match the environmental conditions, after other parts are designed. In addition, most new actuators come with an embedded sensor [19], which facilitates the design and the production of the machine. The main objective of the detailed design stage for this machine is to select a proper motor, transmission system, bearings, and the structure of the carriage, and design an optimal controller for it, all in a concurrent manner. Other parts of the machine do not have much interaction with these essential components and can be selected or designed separately after these essential parts are designed. Therefore, the design parameters are the models of the motor, transmission, and bearings; the dimensions of the rectangular plate; and the controller gains in x and y directions.

2. Motor, transmission system, and bearings are acquired from available options in the market. The carriage has to be designed and produced separately. The controller has to be designed and then implemented in a proper digital control platform. Required technical specifications of the motor, transmission system, and bearings are roughly estimated here. In order to find the maximum speed and acceleration that may be needed for the machine motion, a simple second order system is considered. Settling time and damping ratio are taken as 0.4 and 0.7 s, respectively. Then a step input is applied to this system so that the steady state output of the system becomes 5 cm. It has been found that the maximum speed and acceleration are 0.32 m/s and 6.1 m/s², respectively.

   The weight of the cutter is approximately 10 kg, and its intersection size is approximately 10*10 cm. The size of the plate can be exactly equal to these values, and its thickness is roughly estimated to be between 2 to 20 mm. Based on these rough estimates and a reasonable safety factor, the required power for the motor is estimated to be between 100 and 200 W.

   Based on the estimated required power and by searching pertinent Web sites through the Internet, 16 different DC motors are selected. Some of them are simple DC rotary motors that need a transmission system (e.g., ball-screw unit) to convert rotary motion to linear motion. Some others are linear actuators that are formed by DC rotary motors with a ball-screw or roller-screw transmission system embedded in them. There are linear DC motors as well, that directly provide linear motion.

   Eight different transmission systems are selected that are ball-screw and roller-screw units with different values for lead rate, load capacity, and efficiency.

3. There are seven design variables including the thickness of the plate, controller gains (three gains), motor, transmission, and bearings. They are encoded as a binary string to be optimized by the genetic algorithm.

4. "Meeting task requirements" and "matching" are two criteria that have veto effects on the evaluation of a design trial, and are considered as essential criteria for the first stage of the optimization.

TABLE 19.5 Overall Importance and Interaction Level for Subcriteria of "Technical Issues"

$\Psi'_1 = 0.3842$	$\Psi'_2 = 0.3317$	$\Psi'_3 = 0.2842$
$\Theta'_{12} = -0.0975$	$\Theta'_{13} = -0.0525$	$\Theta'_{23} = -0.0975$

Note: Criteria include (1) reliability, (2) controller friendliness, and (3) efficiency.

TABLE 19.6 Evaluation of MDQ Attributes of Conceptual Design Choices for Iron Butcher

	1	2	3	4
Actuator	Electrical	Electrical	Hydraulic	Hydraulic
Contact	Bearing	Hydrostatic	Direct	Bearing
Controller	PID	PID	PID+ Friction compensation	PID
Material	Al	Al	Steel	Al
Meeting Task Requirements	1.00	1.00	1.00	1.00
Matching	0.8	0.6	0.9	0.9
Reliability	0.6	0.5	0.4	0.5
Controller friendliness	0.9	0.8	0.4	0.8
Efficiency	0.9	0.8	0.4	0.8
Cost	0.9	0.5	0.5	0.7
Overall score	0.873	0.662	0.704	0.83

5. In assessing the "meeting task requirements" criterion, a SIMULINK model is developed to analyze the performance of the controller. Rise time, settling time, overshoot, and steady state error of the response of the system are compared with ideal response specifications, and a partial score is assigned describing the degree of satisfaction of the "meeting task requirements."

   Three different issues are analyzed to assess the "matching" criterion, including:
   - Component capacities
   - Bandwidth issues
   - Stress and deflection limitations

   First, it is checked to see if the motor works near its nominal capacity. Basically, this refers to the fact that underdesigned and overdesigned components are not desirable in a good mechatronic system. For each trial design, the maximum speed and maximum force imposed on the motor are found from the response of the SIMULINK model. A partial score is then assigned to each design trial describing how the capacity of the motor has benefited from the particular design trial.

   The next issue that is analyzed to assess the matching criterion is the bandwidth. Basically, the dominant natural frequency of the system should be several times larger than the frequency content of the control action [6,19]. For each trial design, the natural frequency of the rectangular plate is computed approximately and compared with the bandwidth of the motion, and a partial score is assigned to describe the degree of satisfaction of this aspect.

   Stress and deflection limitations are also addressed in the assessment of the "matching" criterion. The stress in the plate due to its own weight, the weight of the cutter, and the vertical cutting force should be several times less than the yield stress of aluminum. The deflection of the plate under these loads should be smaller than a desired limit which is considered to be 1 mm in the present system. For each trial solution, the maximum stress and maximum displacement in the plate are computed using standard elasticity equations, and a partial score is assigned to each trial solution describing the degree of satisfaction of these issues.

   The partial score of each trial solution with respect to the "matching" criteria is considered as a linear aggregation of capacity matching, bandwidth issues, and stress-deflection limitations. The fitness evaluation of each trial solution is then a linear aggregation of the partial scores to the "meeting task requirement" and "matching" criteria.

6. The difference between two trial solutions can be computed either in genotype or phenotype representation of them. In this particular application, there are discrete design parameters that refer to the order of a part in a list of available options. The genotype representation of these

TABLE 19.7 Elite Design Trials Found by the RCS Niching Genetic Algorithm

		Elite # 1	Elite # 2	Elite # 3	Elite # 4
Motor Specifications	Motor type	Rotary DC motor	Linear DC motor	Rotary DC motor	Linear actuator
		Aerotech BMS-60	Aerotech BLMUC-111	Aerotech 1050	
	Maximum load Capacity	1.68 N.m	209 N	5.22 N.m	173 N
	Maximum speed	10000 rpm	10 m/s	5000 rpm	10 m/s
Transmission	Type	Roller screw	N/A	Ball screw	N/A
	Pitch (mm/rev)	12.7	N/A	2	N/A
	Efficiency	60%	N/A	80%	N/A
	Plate dimensions *(mm)*	14.5	11.4	48.4	14.5
	Integrator controller gain	167.96	163.49	36.07	162.7451
	Proportional gain	1213.7	1327.5	805.8824	1041.2
	Derivative controller gain	17.1373	115.8039	27.1373	94.1961
	Settling time	0.3604	0.3825	0.3457	0.3819
	Overshoot	0	0	0	0
	Steady state error	.4	0.25	0.11	0.1577
	Maximum voltage	60.8	66.37	40.3	52.1
	Maximum load	1.2345	109.31	3.98	84
	Maximum speed	0.4360	.4590	0.4	0.4439

parameters does not have a physical meaning; hence, genotype comparison is not appropriate. In the design of the machine, the degree of discrepancy between two solutions is evaluated by comparing all the parameters that are affective in the response, including even those parameters that do not appear in the genetic algorithm representation of the solutions. For example, all electrical and mechanical specifications of motors are considered in this computation.

7. A niching genetic algorithm with restricted competition selection (RCS) is used to find local optima which are adequately separated from each other to be considered as different configurations. Four elite solutions are found, which correspond to different possible configurations. Table 19.7 summarizes these four elite solutions.

8. Now that the elite solutions have been found, a full and comprehensive competition should be performed between them to find the best design. Same fuzzy measures which were identified for the conceptual design are used here.

9. Table 19.8 gives the results of evaluation of these four solutions with Choquet fuzzy integral.

10. It is clear that the first design is the best one and can be considered as the optimal design.

TABLE 19.8 MDQ Evaluation of Elite Designs

	Elite # 1	Elite # 2	Elite # 3	Elite # 4
Meeting task requirements	0.996	0.993	0.988	0.999
Matching	0.7	0.7	0.66	0.7
Reliability	.9	0.8	0.8	0.8
Controller friendliness	0.96	0.87	0.98	0.84
Efficiency	0.49	0.06	0.49	0.04
Cost	.9	0.5	.7	.8
Overall score	0.8544	0.686	0.7679	0.7852

19.7 Conclusion

In this chapter, a new mechatronic design methodology was presented based on a multicriteria design evaluation index called the MDQ. This design methodology deviates from the traditional sequential design, resulting in improved design performance. The method is not sequential because controller design issues and parameters are treated simultaneously with other issues and parameters. Controllability of the real-time parameters provides an opportunity for further improvement in the system performance after the machine is designed and built, whereas in traditional sequential design approaches it is considered an excuse to postpone the controller design to a later stage (say, after the machine is built). To utilize the opinions of human experts and to account for possible interactions between design criteria, MDQ is computed by the aggregation of different design criteria using soft computing, in particular Choquet fuzzy integral. The developed mechatronic design approach is treated as a multistage procedure, consisting of conceptual design and detailed design. The practical and straightforward steps for these stages were provided in the chapter, and illustrative examples were given, where MDQ was shown to help the designer in both stages.

References

1. Behbahani, S., Practical and Analytical Studies on Development of Formal Evaluation and Design Methodologies for Mechatronic Systems, Ph.D. thesis, Department of Mechanical Engineering, The University of British Columbia, Vancouver, Canada, 2007.
2. Behbahani, S. and de Silva, C.W., Use of mechatronic design quotient in multi-criteria design, *Proceedings of the International Symposium on Collaborative Research in Applied Science (ISOCRIAS)*, Vancouver, Canada, 2005, pp. 214–221.
3. Behbahani, S. and de Silva, C.W., A new multi-criteria mechatronic design methodology using niching genetic algorithm, *Proceedings of IEEE World Congress on Evolutionary Computation*, Vancouver, Canada, 2006, pp. 1031–1036.
4. Li, Q., Zhang, W.J., and Chen, L., Design for control — a concurrent engineering approach for mechatronic system design, *IEEE Transactions on Mechatronics*, Vol. 6, No. 2, 161–169, 2001.
5. Van Brussel, H.M.J., Mechatronics — A powerful concurrent engineering framework, *IEEE/ASME Transactions on Mechatronics*, Vol. 1, No. 2, 127–136, 1996.
6. de Silva, C.W., *Mechatronics — An Integrated Approach*, Taylor and Francis, CRC Press, Boca Raton, FL, 2005.
7. Zhang, W.J., Li, Q., and Guo, L.S., Integrated design of mechanical structure and control algorithm for a programmable four-bar linkage, *IEEE/ASME Transactions on Mechatronics*, Vol. 4, No. 4, 354–362, 1999.
8. de Silva, C.W., Sensing and information acquisition for intelligent mechatronic systems, *Proceedings of the Symposium on Information Transition*, Chinese Academy of Science, Hefei, China, November 2003, pp. 9–18.
9. Marical, J.L., An axiomatic approach of the discrete Sugeno integral as a tool to aggregate interacting criteria in a qualitative framework, *IEEE Transactions on Fuzzy Systems*, Vol. 9, No. 1, 164–172, 2001.
10. Marical, J.L., An axiomatic approach of the discrete Choquet integral as a tool to aggregate interacting criteria, *IEEE Transactions on Fuzzy Systems*, Vol. 8, No. 6, 800–807, 2000.
11. Grabisch, M., The application of fuzzy integrals in multicriteria decision making, *European Journal of Operational Research*, Vol. 89, 445–456, 1996.
12. Marical, J.L., Aggregation of interacting criteria by means of the discrete Choquet integral, *Aggregation Operators, New Trends and Applications*, Heidelberg, New York, Physica-Verlag, 2002.
13. Moulianitis, V.C., Aspragathos, N.A., and Dentsoras, A.J., A model for concept evaluation in design — an application to mechatronic design of robot gripper, *Mechatronics*, Vol. 14, No. 6, 2004, pp. 599–622.

14. Coelingh, E., de Vries, T.J.A., and Koster, R., Assessment of mechatronic system performance at an early design stage, *IEEE/ASME Transactions on Mechatronics*, Vol. 7, No. 3, 269–279, 2002.

15. Bien, Z., Bang, W.C., Kim, D.Y., and Han, J.S., Machine intelligence quotient: its measurement and applications, *Fuzzy Sets and Systems*, Vol. 127, No. 1, 3–16, 2002.

16. Cho, D.H., Jung, H.K., and Lee, C.G., Induction motor design for electric vehicle using a niching genetic algorithm, *IEEE Transactions on Industry Applications*, Vol. 37, No. 4, 994–999, July–August 2001.

17. Kim, J.K., Cho, D.H., Jung, H.K., and Lee, C.G., Niching genetic algorithm adopting restricted competition selection combined with pattern search method, *IEEE Transactions on Magnetics*, Vol. 38, No. 2, 1001–1004, 2002.

18. Himeno, M. and Himeno, R., The niching method for obtaining global optima and local optima in multimodal functions, *Systems and Computers in Japan*, Vol. 34, No. 11, 30–42, 2003.

19. de Silva, C.W., *Sensors and Actuators — Control System Instrumentation*, Taylor and Francis, CRC Press, Boca Raton, FL, 2007.

20

Kinematic Design Optimization of Acrobot

L. Yang

C.M. Chew

A.N. Poo

C.W. de Silva

Summary

A mechatronic system incorporates a combination of technologies in mechanics, electronics, and information engineering. As such, there will often be trade-offs when an overall optimum design is desired although more than one design objective may exist in the context of a multiobjective design criterion. In this chapter, a convenient and optimal design scheme that can incorporate various design criteria in a concurrent manner is presented through the use of the mechatronic design quotient (MDQ). Specifically, this design approach is adopted in the optimization of the length ratio of a robotic device known as the Acrobot. The chapter illustrates how the MDQ approach leads to a reasonably optimal decision for this mechatronic system.

20.1 Introduction

A mechatronic system consists of many different types of interconnected components and elements. A system design that is concurrent and optimally "matched," should, among other advantages, reduce system complexity, facilitate system integration and implementation, optimize system performance, reduce system cost, improve system reliability, and extend system life span. However, in reality, one design may not be the best in all aspects. MDQ provides an approach for concurrent optimal design based on a multicriteria, multidomain formulation [1,2]. It also provides for the interaction between criteria and human experience (see Chapter 19 for further details). MDQ represents the degree of satisfaction of the mechatronic design criteria in a concurrent manner. It is useful as a design evaluation index as well as a design index. The overall performance of a resulting design is very determinative. In this chapter, the Acrobot model, widely used in the study of underactuated mechanical systems, is used to illustrate a possible design approach using MDQ, which has to be formulated to fit the specific design problem. In the present application, the MDQ is formulized as a discrete function of the design solution j, as

$$MDQ(j) = \frac{\sum_{i=1}^{i-n} w_i \times m_{ij}^2}{\sum_{i=1}^{n} w_i \times 10^2} \quad j = 1,2,3,4 \tag{20.1}$$

Here, an n number of design attributes are included in MDQ. The index m_{ij}, whose maximum value is 10, denotes the degree (with respect to some design requirement) of satisfaction (level of performance) of the i-th design attribute in the j-th design solution, and w_i is the weight (level of importance) assigned to the i-th design attribute. The solution j corresponding to the largest $MDQ(j)$ gives the optimal design. The quadratic terms in expression 20.1 help strengthen the contrast between the MDQ results. Ideally, if the design is optimal for every attribute, then $m_{ij} = 10$ for all design attributes i, and the MDQ value for the corresponding design would be 1. However, there often exist contradictory requirements in design. This means that a particular design may perform well with respect to some attributes but not that well with respect to some other attributes. In the design process, one would compute the MDQ values for various designs and pick the design with the highest MDQ value.

20.2   System and Control Methodology

Acrobot is a two-link underactuated robot, which is widely used in research on legged locomotion, underactuation, and swinging-up problems. The best choice for the connection location (joint) of the two links is a design issue. With a fixed control strategy, its location will result in different response times, energy consumption, maximum torque values, overshoots, and offsets from the equilibrium configuration. The aim of the work described here is to determine the optimal ratio for the lengths of Link 1 and Link 2 (L_1/L_2) to achieve the best performance for the Acrobot. The model used for the Acrobot is shown in Figure 20.1.

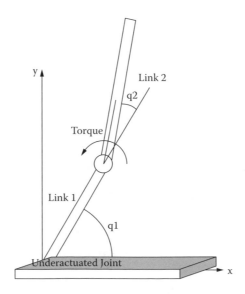

FIGURE 20.1   Model of the Acrobot.

The swinging-up control problem consists of moving the Acrobot from its stable downward position to its unstable inverted position and balancing it about the vertical. Because of the large range of motion, the swinging-up problem is highly nonlinear and challenging. Here, a somewhat simple control algorithm is adopted, which is based on the notion of partial feedback linearization [4] and a common design philosophy with the recent method of integrator back-stepping [5]. This algorithm is useful in cases where there are no limits on the rotation of the second link. The Acrobot model that is used here is a two-link planar robot with an actuator at Joint 2 but no actuator at Joint 1. Proportional derivative (PD) and linear quadratic regulator (LQR) controls are used [3]. Specifically, PD control is used to rotate Link 2 and drive Link 1 to swing up to its vertical equilibrium. LQR control is applied when both links are near their vertical equilibrium configuration where LQR control is valid with a linearized system model. PD control serves as a swinging-up controller, and LQR control performs as a balancing controller. The equations of motion of the system are [6,7]

$$d_{11}q_1 + d_{12}q_2 + h_1 + \phi_1 = 0 \tag{20.2}$$

$$d_{21}q_1 + d_{22}q_2 + h_2 + \phi_2 = \tau \tag{20.3}$$

where

$$d_{11} = m_1 l^2{}_{c1} + m_2(l^2{}_1 + l^2{}_{c2} + 2l_1 l_{c2}\cos(q_2)) + I_1 + I_2 \qquad d_{22} = m_2 l^2{}_{c2} + I_2 \qquad d_{12} = m_2(l^2{}_{c2} + l_1 l_{c2}\cos(q_2)) + I_2$$

$$d_{21} = m_2(l^2{}_{c2} + l_1 l_{c2}\cos(q_2)) + I_2 \qquad h_1 = -m_2 l_1 l_{c2}\sin(q_2)q_2 - 2m_2 l_1 l_{c2}\sin(q_2)q_2 q_1 \qquad h_2 = m_2 l_1 l_{c2}\sin(q_2)q^2{}_1$$

$$\phi_1 = (m_1 l_{c1} + m_2 l_1)g\cos(q_1) + m_2 l_{c2}g\cos(q_1 + q_2) \qquad \phi_2 = m_2 l_{c2}g\cos(q_1)$$

The Acrobot is representative of a large class of underactuated mechanical systems. A linear response may be achieved for either degree of freedom by using nonlinear feedback. First we set

$$d_{12}q_2 + h_1 + \phi_1 = -d_{11}v_1 \tag{20.4}$$

$$q_1 = v_1 \tag{20.5}$$

$$v_1 = \hat{q}_1 + k_d(\hat{q}_1 - q_1) + k_p(\hat{q}_1 - q_1) \tag{20.6}$$

where $\hat{q}_1$, $\hat{q}_1$, and $\hat{q}_1$ are the desired rotational acceleration, speed, and position, respectively, with respect to the root coordinate. A PD control is designed according to Equation (20.6). The application of Equation (20.2) in Equation (20.3) gives

$$\tau = \bar{d}_1 v_1 + \bar{h}_1 + \bar{\phi}_1 \quad \bar{d}_1 = d_{21} - d_{22}d_{11}/d_{12}$$

$$\bar{h}_1 = h_2 - d_{22}h_1/d_{12} \qquad \bar{\phi}_1 = \phi_2 - d_{22}\phi_1/d_{12} \tag{20.7}$$

After choosing suitable values for the gains K_p and K_d, Link 1 of the Acrobot is controlled to be vertical, but Link 2 will rotate freely before LQR takes over. For the balancing control, first the Acrobot model is linearized about the vertical equilibrium $q_1 - \pi/2, q_2 = 0, q_1 = 0, q_2 = 0$; then, the state–space equation of this balancing control system is

$$x = Ax + Bu \quad y = Cx + D \quad A = \left.\frac{\partial f}{\partial x_i}\right|_{x_i = x_e} \quad B = \frac{\partial f}{\partial \tau} \quad C = I \quad D = \begin{pmatrix} \dfrac{\pi}{2} & 0 & 0 & 0 \end{pmatrix} \quad Q = I \quad R = 1$$

where $f(x)$ represents the right-hand side of the general nonlinear state equation. The state vector is $x = [q_1 - \pi/2, q_2, \dot{q}_1, \dot{q}_2]^T$, and the control input is $u = \tau$. Whenever these two links swing up to a configuration close to the vertical equilibrium (about $\pm 5°$ for both links), the controller is switched from PD to LQR control. The control procedure consists of applying the PD controller first to keep Link 1 vertical, with Link 2 rotating. When these two links are in the region of the vertical equilibrium, the controller is switched to LQR control.

20.3 MDQ Design Procedure

The following five design attributes (indices) are chosen:

  I_1 response time of the entire procedure
  I_2 energy consumption of the entire procedure
  I_3 maximum torque value during the procedure
  I_4 overshoot range during the entire motion
  I_5 offset value from the desired final equilibrium configuration

Figure 20.2 illustrates the MDQ approach that is used. The first level is the index (attribute) level, which assigns weights for each attribute and computes the satisfaction values (marks) for the attributes; the second level is the design level, which computes the overall mark (MDQ value) for each coupled design.

Design aim 1—Successful swinging up and balancing: In the work here, discrete ratios $L_1/L_2 = \{0.428, 0.5, 0.58, 0.625, 1, 1.5, 2\}$ are chosen. Through computer implementation, however, it would be possible to test a range of continuous ratios and obtain a more accurate result. From Equation (20.7) we see that if $d_{12} = 0$, there will be a problem of singularity, in which case the control methodology would not be valid. It follows that the Acrobot will not be able to swing up if the chosen length ratio is unreasonable. To avoid this singularity problem, the condition $d_{12} \neq 0$ must be satisfied. The condition $L_1/L_2 < \frac{2}{3}$ was arrived at in achieving this objective in a successful design, as follows:

$$m_2 L^2_{c2} + m_2 L_1 L_{c2} \cos(q_2) + I_2 \neq 0 \quad \text{with} \quad I_2 = \frac{m_2 L^2_2}{12} \quad \text{and} \quad L_{c2} = L_2/2 \Rightarrow \cos(q_2) \neq \frac{-2}{3} \frac{L_2}{L_1} \Rightarrow L_1/L_2 < \frac{2}{3}$$

This constraint on the length ratio is not satisfied by the set $\{1, 1.5, 2\}$. Therefore, in the following design, the length ratio database $L_1/L_2 = \{0.428, 0.5, 0.58, 0.625\}$ is used. The kinematic properties of the Acrobot for the possible length ratios 0.428, 0.5, 0.58, and 0.625 are given in Table 20.1.

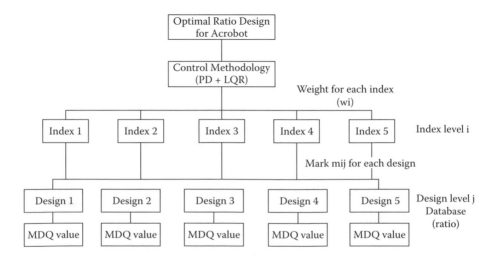

FIGURE 20.2 MDQ design procedure.

TABLE 20.1 Kinematic Properties of Acrobot for Different Length Ratios

Ratio	m_1	m_2	L_1	L_2	L_{c1}	L_{c2}	I_1	I_2	g
0.428	3.00	7.00	0.30	0.70	0.15	0.35	0.0225	0.2858	10
0.500	3.33	6.67	0.33	0.67	0.165	0.335	0.0308	0.2469	10
0.580	3.67	6.33	0.37	0.63	0.185	0.315	0.0412	0.2113	10
0.625	3.85	6.15	0.385	0.615	0.193	0.308	0.0474	0.1942	10

Design aim 2—Short overall time for the procedure: Figure 20.3 shows the responses under PD control for different length ratios. The upper profile gives the trajectory for Link 1 to swing up and remain at the vertical position, and the lower trajectory gives the motion of Link 2 before LQR control is activated. The gains used for PD control are $K_p = 21$ and $K_d = 8$. The overall time is $t = t_1 + t_2$, with t_1 being the time taken for PD control and t_2 the time that LQR control takes to balance the system. For LQR control,

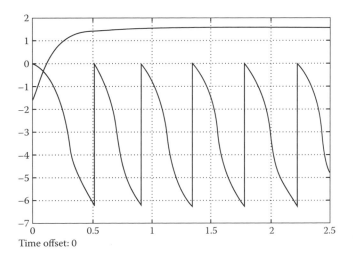

Time offset: 0

(a)

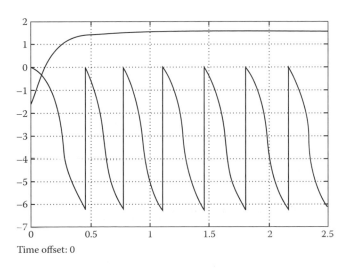

Time offset: 0

(b)

FIGURE 20.3 Response under PD control: (a) ratio = 0.428, (b) ratio = 0.5, (c) ratio = 0.58, (d) ratio = 0.625.

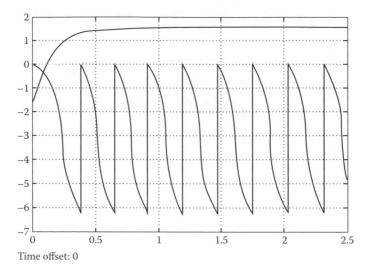

Time offset: 0

(c)

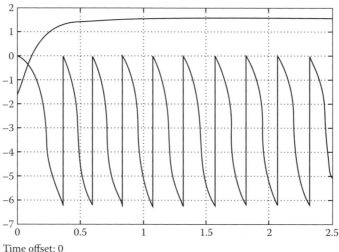

Time offset: 0

(d)

FIGURE 20.3 (Continued).

the initial condition is set at the configuration where the two links are close to the vertical equilibrium configuration or within the range of ±5° for both q_1 and q_2.

After linearizing the system about the vertical equilibrium configuration, the initial conditions for LQR control are determined, as given in Table 20.2 The corresponding responses under LQR control are

TABLE 20.2 Initial Conditions for the Designs

Ratio	Initial Condition for LQR Control
0.428	$x0 = [-0.0456, -0.0006, 0.33631, -8.6144]$
0.500	$x0 = [-0.0056, -0.0082, 0.10615, -9.9228]$
0.580	$x0 = [-0.0440, -0.0026, 0.32919, -12.164]$
0.625	$x0 = [-0.0778, -0.0105, 0.49460, -13.478]$

shown in figure 20.4. To achieve the present design aim, the four length ratios are ranked as in table 20.3, in which length ratio 0.625 corresponds to the optimal value for the time index.

Design aim 3—Low energy consumption: The principle of conservation of energy is used to determine the required energy. Note that, during the entire procedure, input energy is required to overcome gravity, generate the kinetic energy, and finally overcome the kinetic energy. The energy that is required to overcome gravity is the same for all length ratios. Hence, any variation in the energy

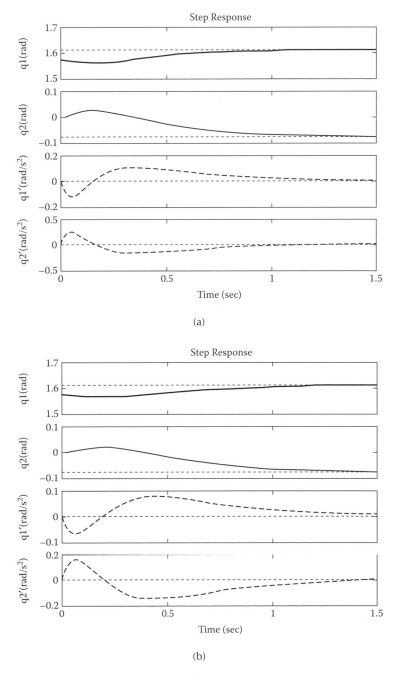

FIGURE 20.4 Response under LQR control: (a) ratio = 0.428, (b) ratio = 0.5, (c) ratio = 0.58, (d) ratio = 0.625.

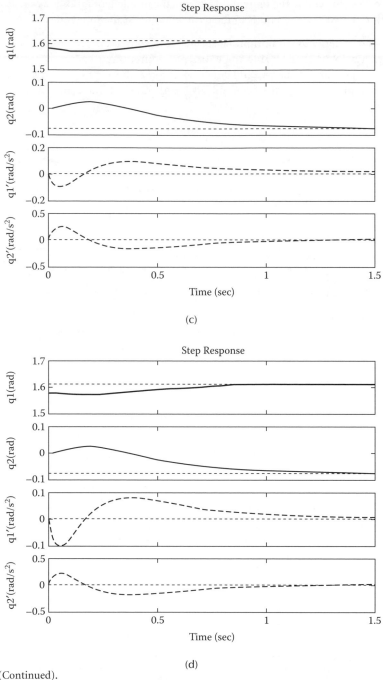

FIGURE 20.4 (Continued).

TABLE 20.3 Ranking of Response Time

Ratio	Responding Time t	Rank
0.428	2.318	3
0.500	2.496	4
0.580	2.214	2
0.625	1.901	1

TABLE 20.4 Ranking of Energy Consumption

Ratio	E_k Value	Rank
0.428	42.258	1
0.500	48.627	2
0.580	62.584	3
0.625	70.69	4

TABLE 20.5 Ranking of Maximum Torque Value

Ratio	Torque Value	Rank
0.428	1200	1
0.500	3800	2
0.580	26000	3
0.625	180000	4

consumption is governed only by the generated kinetic energy. The initial condition for LQR control is used to compute the kinetic energy E_k because, before this state, input energy is used to generate velocity as well as to overcome gravity. Thereafter the applied input energy is used to overcome the kinetic energy. Table 20.4 gives the values of E_k, which represent the input energy required during the entire procedure. From this it is clear that the longer the Link 1, the higher the energy consumption during this period.

Design aim 4—Reasonable maximum torque value: By carrying out computer simulations for sequential control, the maximum torque values of each motion are determined for each length ratio. It is seen that the value of the maximum torque increases with the length ratio. This is because the larger the length ratio, the closer it is to the value 0.667, which corresponds to the singularity condition where control becomes ineffective. Table 20.5 gives the maximum value of torque for each length ratio.

Design aim 5—Reasonable overshoot range: Overshoot occurs only under LQR control. Analysis is performed using MATLAB®, and the results are shown in Figure 20.4 and Table 20.6.

Design aim 6—Small offset value from the desired equilibrium position: Figure 20.4 and Table 20.7 give the performance index values corresponding to the offset value from the desired equilibrium position for each length ratio.

TABLE 20.6 Rank of Overshoot Range

Ratio	q_1	q_2	Rank
0.428	0.013	0.02185	4
0.500	0.0086	0.018	1
0.580	0.01	0.02188	2
0.625	0.01	0.02216	3

TABLE 20.7 Rank of Small Offset Value

Ratio	q_1	q_2	Rank
0.428	0.0385	0.078	4
0.500	0.0348	0.078	1
0.580	0.0327	0.081	2
0.625	0.0316	0.083	3

TABLE 20.8 Marks of the Coupled Designs

Ratio	I_1	I_2	I_3	I_4	I_5
0.428	6	10	10	4	4
0.500	4	8	8	10	10
0.580	8	6	6	8	8
0.625	10	4	4	6	6

TABLE 20.9 MDQ Values of Various Designs

Ratio	MDQ Value
0.428	0.516
0.500	0.772
0.580	0.528
0.625	0.260

The response time, overshoot value, energy consumption, maximum torque value, and offset value are the five main indices for the present sequence of MDQ design. Clearly, the MDQ approach facilitates interaction and incorporation of the experience of the designer and the requirements for the system. For example, among the five indices, response time is of the least concern, as the other four are more closely related with the singularity problem, control stability, and control precision. The maximum torque value and energy consumption are also somewhat related to each other. Hence, the following weights are assigned to these four indices: 0.2, 0.2, 0.25, and 0.25. A weight of 0.1 is assigned to the response time index, as this is of the least concern. Marks of the performance of various indices according to ranking are given in Table 20.8. Table 20.9 is obtained by using table 20.8 and computing the MDQ values according to Equation (20.2).

20.4 Conclusion

A mechatronic system incorporates a combination of technologies in mechanics, electronics, and information engineering. More than one design objective may exist together with a multiobjective design criterion. In this chapter, a convenient and optimal design scheme that can incorporate various design criteria in a concurrent manner was presented through the use of MDQ. The MDQ approach can be systematically applied in the integrated design of a variety of problems. When there are many alternative designs, each with its own merits and drawbacks, the MDQ approach helps to determine the best from a holistic viewpoint of overall performance. In the present chapter, this design approach was adopted in the optimization of the length ratio of a robotic device known as the Acrobot. From the analysis and design sequence presented, it was found that the maximum MDQ value occurred for the length ratio 0.5, which is the optimal length ratio. The simulation verified that, for this length ratio, the Acrobot system performed very effectively with respect to the considered design attributes (indices). The Acrobot design can be further developed and enhanced by incorporating a sub-MDQ level to choose a motor that suits the desired Acrobot model, adding a new MDQ level for comparing different control methodologies, or generalizing the approach for biped walking to judge which joint is better for control when encountering a disturbance at some specific position.

References

1. De Silva, C.W., *Mechatronics—An Integrated Approach*, Taylor and Francis, CRC Press, Boca Raton, FL, 2005.
2. De Silva, C.W., Sensing and information acquisition for intelligent mechatronic systems, *Proceedings of the Symposium on Information Transition*, Chinese Academy of Science, Hefei, China, November 2003, pp. 9–18.

3. Spong, M.W., The swing up control problem for the acrobot, *IEEE Control Systems Magazine*, Vol. 15, No. 2, 49–55, February 1995.

4. Isidori, A., *Nonlinear Control Systems*, 2nd ed., Spring-Verlag, Berlin, 1989.

5. Kokotovic, P.V., Krstic, M., and Kanellakopoulos, I., Backstepping to passivity: recursive design of adaptive systems, *IEEE Conference on Decision and Control*, Tucson, AZ, 1992, 3276–3280.

6. Murray, R.M. and Hauser, J., A case study in approximate linearization: the acrobot example, *ERL Technical Memo*, University of California, Berkeley, College of Engineering, Berkeley, California, May 1991.

7. Spong, M.W. and Vidyasagar, M., *Robot Dynamics and Control*, John Wiley & Sons, New York, 1989.

21

Evolutionary Optimization in the Design of a Heat Sink

M.R. Alrasheed

C.W. de Silva

M.S. Gadala

Summary

The approach of particle swarm optimization (PSO) is applied to design a heat sink system. This innovative approach is a robust stochastic evolutionary computation technique based on the movement and intelligence of swarms. The chapter presents the PSO algorithm in detail and investigates the rationale and means of application of the method to the optimal design of a heat sink. In the presented approach, a plate-fin heat sink design is realized for maximum dissipation of the heat generated from electronic components, as represented by the entropy generation rate. In the process, the best heat transfer efficiency is achieved.

21.1 Introduction

Mechatronic systems are multidomain systems that may involve fluid and thermal problems in addition to mechanics and electronics. The associated design problem can be quite complex and often nonanalytic. Evolutionary techniques (see Chapter 19) are one way to overcome this difficulty. In the present chapter, the approach of PSO is applied to design a heat sink system for electronic and mechatronic devices.

The recent trend in the electronic devices industry toward denser and larger heat flux densities has necessitated better thermal performance from the perspective of device cooling. For example, some 900 million computers are in use in the world today, with personal computers comprising approximately half the total. This growth is compatible with the fact that some 400 million computers were in use by the end of 2001 [1]. The rapid growth in computer systems and other digital hardware has led to the associated increase in the thermal dissipation from the constituent microelectronic devices. This has fueled the interest of engineers and researchers in controlling the maximum operating temperature and achieving long-term reliability and efficient performance in electronic components.

In electronic equipment, the temperature of each component must be maintained below an allowable upper limit, specified for each component from the viewpoint of operating performance and reliability. The power density in electronic systems is growing due to the high speeds of operation and the miniaturization of the associated components and devices. Generally, heat sinks are used to maintain the operating temperature below a specified value for reliable operation of the electronic device. Using a suitable heat sink has become crucial to the overall performance of electronic packages. The forced-air cooling technique, which is an effective method for thermal management of electronic equipment, is commonly used for cooling electronic devices [2]. The development of a systematic and rather optimal design methodology for air-cooling heat sinks is undoubtedly very important in satisfying the current thermal necessities and for successful heat removal in the future generations of critical electronic components [3].

The performance of forced-air convection heat sinks in electronic devices depends on a number of parameters, including the thermal resistance, dimensions of the cooling channels, location and concentration of heat sources, and the airflow bypass due to flow resistance through the channel. In general, an important goal of heat sink design is to reduce the overall thermal resistance [4]. An alternative and related criterion for designing a heat sink is to maximize the thermal efficiency. Both criteria would affect the maximum heat dissipation. In a practical industrial design, different criteria are chosen depending on whether the primary objective is to maximize the transmitted heat, minimize the pumping power, or obtain the minimum volume or weight under the prescribed constraints such as component size and heat transfer time [5].

21.2 Entropy Generation Minimization (EGM) of a Heat Sink

The idea of using the entropy generation rate to estimate the heat transfer enhancement was first proposed by Bejan [6] as a performance assessment criterion for thermal systems. A fin can generate the entropy associated with the external flow, and because the fin is nonisothermal, it can also generate entropy internally. The entropy generation rate that is associated with the heat transfer in a heat sink can serve as a measure of the capability of transferring heat to the surrounding cooling medium. As in all thermodynamic systems, the entropy in a heat sink is generated from the irreversibility due to the heat transfer across finite temperature differences and the friction of fluid flow. The basic thermodynamic equations for the stream channel as an open system in steady flow are

$$m_{in} = m_{out} = m \tag{21.1}$$

$$m\, h_{in} + \iint q'' dA - m\, h_{out} = 0 \tag{21.2}$$

$$S_{gen} = m\, s_{out} - m\, s_{in} - \iint \frac{q'' dA}{T_w} \tag{21.3}$$

The canonical form $dh = T\,ds + (1/\rho)dP$ may be written as

$$h_{out} - h_{in} = T_e\,(s_{out} - s_{in}) + \frac{1}{\rho}\,(P_{out} - P_{in}) \tag{21.4}$$

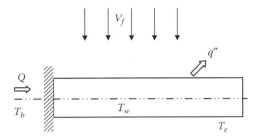

FIGURE 21.1 Schematic diagram of a general fin in convective heat transfer.

where it is assumed that the temperature and density do not change significantly between inlet and outlet. Combining Equations (21.2) through (21.4), the entropy generation rate can be written in this form

$$S_{gen} = \iint_A q'' \left(\frac{1}{T_e} - \frac{1}{T_w} \right) dA - \frac{m}{\rho_e T_e} (P_{out} - P_{in}) \tag{21.5}$$

Now, knowing that

$$m = A \rho_e V_f \tag{21.6}$$

$$F_d = A (P_{in} - P_{out}) \tag{21.7}$$

one obtains

$$S_{gen} = \frac{1}{T_e^2} \iint_A q'' (T_w - T_e) \, dA - \frac{1}{T_e} F_d V_f \tag{21.8}$$

where F_d is the drag force.

Equation (21.8) gives the entropy generation rate associated with the fin heat transfer in the external flow. As shown in Figure 21.1, a fin also generates entropy internally due to the fact that it is nonisothermal, which is given by

$$(S_{gen})_{internal} = \iint_A \frac{q''}{T_w} dA - \frac{Q}{T_b} \tag{21.9}$$

where T_w and T_b represent the local and base temperatures, respectively, and Q is the heat dissipation rate of the heat sink.

Adding equations (21.8) and (21.9), the entropy generation rate for a single fin can be written as

$$S_{gen} = \frac{Q \cdot \Delta T}{T_e^2} + \frac{F_d V_f}{T_e} \tag{21.10}$$

A uniform stream with velocity V_f and absolute temperature T_e passes through the fin as shown in Figure 21.1. Fluid friction appears in the form of drag force F_d along the direction of V_f. Equation (21.10) shows that fluid friction and inadequate thermal conductivity jointly contribute to degrading of the thermodynamic performance of the fin. Thus, the optimal thermodynamic size of the fin can be computed by minimizing the entropy generation rate given by Equation (21.10), subject to necessary design constrains.

The heat transfer rate between the fin and the stream is q''; theoretically, a heat sink is required to satisfy

$$\iint q'' \, dA \cong Q \tag{21.11}$$

For a heat sink set, the temperature drop of T is related to the overall thermal resistance of the heat sink, as given by

$$\Delta T = Q \cdot R \tag{21.12}$$

where Q is the heat dissipation rate of the heat sink and R is the overall heat sink thermal resistance. The first term of entropy generation in Equation (21.10) can be written as QT/T_e^2. The temperature difference T is represented as T_b, T_e. So, the rate of entropy change of a heat sink set can be written as

$$S_{gen} = \frac{Q^2 \, R}{T_e^2} + \frac{F_d V_f}{T_e} \tag{21.13}$$

The entropy generation rate in Equation (21.13) is a function of both heat sink resistance and viscous dissipation. The viscous dissipation term is small and may be neglected under low velocity conditions such as buoyancy-induced flow [1].

In the thermal design of a heat sink, the goal can be either to minimize the total thermal resistance or to maximize the thermal efficiency. The minimization of the entropy generation rate is equivalent to the minimization of the total thermal resistance. Therefore, the design strategy of minimizing the entropy generation rate has the same effect as maximizing the thermal efficiency, surface area, and convective coefficients. Additionally, the optimal flow velocity and viscous dissipation can be found through minimization of the entropy generation rate.

In heat sink optimization, one important implication is that, because the size parameter is linked directly to the volume and weight in a natural manner, it should be considered as one of the design constraints in the minimization of the entropy generation rate. The overall heat sink resistance is given by

$$R = \frac{1}{(N/L_{fin}) + h(N-1)sL} + \frac{b}{kLW} \tag{21.14}$$

Here, N is the number of fins, and R_{fin} is the thermal resistance of a single fin, as given by

$$R_{fin} \frac{1}{\sqrt{(h \, Pk \, A_c)} \tanh(ma)} \tag{21.15}$$

with

$$m = \sqrt{\frac{hP}{kA_c}} \tag{21.16}$$

Also, P is the perimeter of the fin, and A_c is the cross-sectional area of the fin. The total drag force on the heat sink may be obtained by considering a force balance for the heat sink. Specifically,

$$\frac{F_d}{(1/2\rho V^2 ch)} = fapp\, N\,(2aL + SL) + K_c(aW) + K_e(aW) \tag{21.17}$$

where f_{app} is the apparent friction factor for use with hydrodynamic flow. The channel velocity V_{ch} is related to the free stream velocity:

$$V_{ch} = V_f\left(1 + \frac{d}{s}\right) \tag{21.18}$$

The apparent friction factor f_{app} for a rectangular channel may be computed using a form of the model developed by Muzychka and Yovanovich [7] for developing laminar flow:

$$f_{app}\,\mathrm{Re}_{D_h} = \left[\left(\frac{3.44}{\sqrt{L^*}}\right)^2 + (f\,\mathrm{Re}_{D_h})^2\right]^{1/2} \tag{21.19}$$

where

$$L^* = \frac{L}{D_h\,\mathrm{Re}_{D_h}} \tag{21.20}$$

Also, D_h is the hydraulic diameter of the channel, and $f.\mathrm{Re}_{D_h}$ is the fully developed flow factor Reynolds number group, given by

$$f.\mathrm{Re}_{D_h} = 24 - 32.527\left(\frac{s}{a}\right) + 46.721\left(\frac{s}{a}\right)^2 - 40.829\left(\frac{s}{a}\right)^3 + 22.954\left(\frac{s}{a}\right)^4 - 6.089\left(\frac{s}{a}\right)^5 \tag{21.21}$$

The expansion and contraction loss coefficients may be computed using the standard expressions for a sudden contraction and a sudden expansion, given by

$$K_c = 0,42\left[1 - \left(1 - \frac{N.d}{W}\right)^2\right] \tag{21.22}$$

$$K_e - \left(1 - \left(1\ \frac{N.d}{W}\right)^2\right)^2 \tag{21.23}$$

The heat transfer coefficient h can be computed using the model developed by Teertstra et al. [8]:

$$N_{ub} = \left[\left(\frac{\mathrm{Re}_b^*\,\mathrm{Pr}}{2}\right)^{-3} + \left(0.664\sqrt{\mathrm{Re}_b^*}\,\mathrm{Pr}^{1/3}\sqrt{1 + \frac{3.65}{\mathrm{Re}_b}}\right)^{-3}\right]^{-1/3} \tag{21.24}$$

where

$$\text{Re}_b^* = \text{Re}_b \cdot \left(\frac{s}{L} \right) \tag{21.25}$$

$$N_{u_b} = \frac{h.s}{k_f} \tag{21.26}$$

21.3 Optimization Techniques

Optimization is the branch of computational science that searches for the best solution to problems in all areas of mathematics; physical, chemical, and biological sciences; engineering, including mechatronics; architecture; economics; and management. The range of techniques available to solve them are numerous as well. An optimization problem is made up of the following basic components:

- The quantity to be optimized (maximized or minimized), termed the *objective function*
- The parameters that may be changed in the search for the optimum, called *design parameters*
- The restrictions on allowed parameter values, known as *constraints*

The optimization process finds the values (design parameters) that minimize or maximize (optimize) the objective function while satisfying constraints. Thus, the general optimization problem may be stated mathematically as

$$\text{minimize} \quad f(x), \quad x = (x_1, x_2, \ldots, x_n)^T \tag{21.27}$$

$$\text{subject to} \quad c_i(x) = 0, \quad i = 1, 2, \ldots, m' \tag{21.28}$$

$$c_i(x) \geq 0, \quad i = m' + 1, \ldots, m \tag{21.29}$$

where $f(x)$ is the objective function, x is the column vector of n independent design parameters, and $c_i(x)$ is the set of constraint functions. Constraint equations of the form $c_i(x) = 0$ are termed *equality constraints*, and those of the form $c_i(x) \geq 0$ are *inequality constraints*.

Many engineering problems can be expressed as optimization problems, for example, process design, transportation, production scheduling, finding of optimal trajectory for a robot arm, the optimal thickness of steel in pressure vessels, heat exchanger design, and so on. The solutions to such problems are usually not easy to obtain because they have large search spaces. In practice, for a class of hard problems as just listed, it is often difficult if not impossible to guarantee convergence or find the best solution in an acceptable amount of time. Another feature of many real-life engineering problems is that they are discontinuous and noisy [9]. Furthermore, the objective function may not be analytic, and it may not be possible to determine its gradient (slope), which is needed in many optimization routines.

Generally, optimization techniques or algorithms can be broadly classified into deterministic, such as the steepest descent method, and stochastic, such as the local search method [10]. A deterministic algorithm progresses toward the solution by making deterministic decisions. On the other hand, stochastic algorithms make random decisions in their search for a solution. Therefore, deterministic algorithms produce the same solution for a given problem instance, whereas this is not the case for stochastic algorithms.

Evolutionary algorithms (EA) are search methods that take their inspiration from natural selection and survival of the fittest in the biological world (see Chapters 17–19). EA differ from more traditional optimization techniques in that they involve a search from a "population" of solutions, not from a single point. Each iteration of an EA involves a competitive selection that removes poor solutions. Evolutionary computation (EC), evolution strategies (ES), Particle Swarm Optimization (PSO), and genetic algorithms (GA) may be considered as EAs [11]. EAs and other stochastic search techniques seem to be a promising alternative to traditional or deterministic techniques. First, EA do not rely on analytic assumptions such as differentiability or continuity. Second, they are capable of handling problems with nonlinear constraints, multiple objectives, and time-varying components. Third, they have shown superior performance in a variety of real-world applications.

21.4   Particle Swarm Optimization (PSO)

This section gives the basics of the method of PSO. The structure of the method and the underlying basic components are outlined.

21.4.1   Introduction

PSO is a population-based stochastic optimization technique developed by Kennedy and Eberhart [12,13], and it has been inspired by the behavior of schools of fish and flocks of birds. Unlike other heuristic techniques of optimization, PSO has a flexible and well-balanced mechanism to enhance and adapt to the global and local exploration abilities. PSO has its roots primarily in two methodologies. Perhaps more obvious are its ties to artificial life (A-life) and the behavior of flocks of birds, schools of fish, and swarms in particular. It is also related to evolutionary computation and has ties to genetic algorithms and evolutionary strategies [14]. In general, PSO is based on a relatively simple concept and can be implemented in a few lines of computer code. It requires only simple mathematical operators and is computationally inexpensive in terms of both memory requirement and speed. It exhibits some evolutionary computation attributes; for example, it is initialized with a population of random solutions, it searches for optima by updating generations, and updating is based on the previous generations.

PSO has also been proved to perform well in test functions used in EA and may be used to solve many problems similar to those in EA. It appears to be a promising approach, and early testing has found the implementation to be effective with complex practical problems. However, PSO does not suffer from some of the difficulties of EAs. For example, a PS system has memory, which the genetic algorithms (GA) do not have. In PSO, individuals who fly past optima are pulled to return toward them, and knowledge of good solutions is retained by all particles [15]. Whereas a GA can handle combinatorial optimization problems, PSO was initially used to handle continuous optimization problems. Subsequently, PSO has been expanded to handle combinatorial optimization problems and those involving both discrete and continuous parameters as well. Efficient treatment of mixed-integer nonlinear optimization problems (MINLP) is a rather difficult issue in the optimization field. Unlike other EC techniques, PSO can be realized using only a short program in MINLP. This feature of PSO is one of its main advantages when compared with other optimization techniques [16]. In summary, compared with other methods, PSO has the following advantages [17]:

- Faster and more efficient: PSO may get results of the same quality in significantly fewer fitness and constraint evaluations.
- Better and more accurate: In demonstrations and various application results, PSO is found to give better and more accurate results than other algorithms reported in the literature.
- Less expensive and easier to implement: The algorithm is intuitive and does not need specific domain knowledge to solve the problem. There is no need for transformations or any other manipulations. Implementation in difficult optimization areas requires relatively simple and short coding.

21.4.2 The Structure of PSO

The PSO algorithm consists of the following basic components [14,15,18]:

- *Particle position vector x*: This vector contains the current location of the solution for each particle in the search space.
- *Particle velocity vector v*: It represents the amount by which vector *x* (both vectors have consistent units) will change in magnitude and direction in the next iteration. The velocity is the step size, the amount by which the changing of the *v* values changes the particle direction through the search space; that is, it causes the particle to make a turn. The velocity vector is used to control the range and resolution of the search.
- *Inertia weight w(t)*: This is a control parameter that is used to control the impact of the previous velocities on the current velocity. Hence, it influences the trade-off between the global and local exploration abilities of particles.
- *Best solution pbest*: This is the best solution of the objective function that has been discovered thus far by a particular particle.
- *Best global solution gbest*: This is the best global solution of the objective function that has been discovered by all particles of the population.

PSO can be expressed mathematically for a given problem of *D*-dimensions i.e., *D*-design parameters for each particle *i* and each channel *d* = [1,…, *D*], using

$$v_{id} = w \ v_{id} + \rho_1 \ rand_1 \ (pbest - x_{id}) + \rho_2 \ rand_2 \ (gbest - x_{id}) \qquad (21.30)$$

$$x_{id} = x_{id} + v_{id} \qquad (21.31)$$

$$w = w_{max} - \frac{w_{max} - w_{min}}{iteration_{max}} \times iteration_{current} \qquad (21.32)$$

where

- *w* is the inertia factor
- v_{id} is the value of channel *d* in the velocity vector *v* for particle *i*
- σ_1 is the cognitive learning rate
- σ_2 is the social learning rate
- $rand_1$ and $rand_2$ are random values in the range of [0–1] (accelerating), giving the current position of particle *i* along dimension *d*

21.4.3 The Trajectory of a Particle

The heart of the PSO algorithm is the process by which *v* is modified in Equation (21.30), forcing the particles to search through the most promising areas of the solution space again and again adding its velocity vector *v* to its location vector *x* to obtain a new location. Without modifying the values in *v*, the particle would simply take uniform steps in a straight line through the search space and beyond. In each iteration, the previous values of *v* constitute the momentum of a particle. The momentum is essential, as it is this feature of PSO that allows particles to escape local optima. The velocities of the particles in each dimension are clamped to a maximum velocity V_{max}, which is an important parameter. It determines the fineness or the objective function value with which the regions between the present position and the best target position thus far are searched. If V_{max} is too high, the particles might fly past good solutions. On the other hand, if V_{max} is too small, the particles might not explore sufficiently beyond locally good regions. In fact, they could become trapped in local optima, unable to move far enough to reach a better position in the problem space. The acceleration constants σ_1 and σ_2 in Equation (21.30) represent the weighting factors of the stochastic acceleration terms that direct each particle toward the *pbest* and *gbest* positions. Early experience with PSO

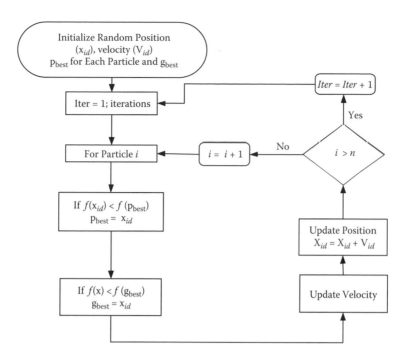

FIGURE 21.2 Flowchart of the PSO process.

has led to setting both the acceleration constants σ_1 and σ_2 to 2.0 for almost all applications. V_{max} is thus the only parameter to be adjusted by the user, and it is often set to a value of about 10 to 20% of the dynamic range of the parameter in each dimension [19]. The selected population size is problem-dependent, and a population size of 20–50 is quite common. It was found early on that smaller populations that were common for other EAs (such as GAs and evolutionary programming) were optimal for PSO in terms of minimizing the total number of evaluations (population size times the number of generations) needed to obtain a sufficient solution [20]. A flowchart for the PSO optimization process is given in Figure 21.2.

21.5 Problem Statement

Figure 21.3 shows the geometrical configuration of a plate-fin sink with horizontal inlet cooling flow. Configuration data are as follows:

- The base length L and the width W are both 50 mm.
- The total heat dissipation of 30 W is uniformly applied over the base plate of the heat sink with a base thickness b of 2 mm.
- The thermal conductivity k of the heat sink is 200 W/m.K.
- The ambient air temperature T_e is 25°C.
- The conductivity k_f of air is 0.0267 W/m.K.
- The air density ρ is 1.177 kg/m.
- The kinematical viscosity coefficient v is 1.6 (10^{-5}) m²/s.

The goal is to determine the optimal number of fins N, optimum height of fins a, optimum thickness of each fin d, and the optimum velocity of cooling flow V_f. The objective function is

$$S_{gen} = \frac{Q^2 R}{T_e^2} + \frac{F_d V_f}{T_e} \tag{21.33}$$

and the design parameters are $[x_1, x_2, x_3, x_4]^T = [\ N, a, d, V_f\ ]$.

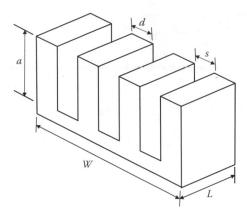

FIGURE 21.3   Geometrical configuration of a plate-fin heat sink.

The design boundaries corresponding to each design parameter are

- $2 \le x_1 \le 40$
- $25\ mm \le x_2 \le 140\ mm$
- $0.2\ mm \le x_3 \le 2.5\ mm$
- $0.5\ m/s \le x_4 \le 2\ m/s$
- The number of fins must be an integer that is restricted to the following domain:

$$2 \le N \le \text{int}\left[1 + \left(\frac{W-d}{d}\right)\right] \tag{21.34}$$

The spacing s between two fins is equal to

$$s = \left(\frac{W-d}{N-1}\right) - d \tag{21.35}$$

The first example in the paper by Shih and Liu [3] is considered here for a comparative evaluation.

21.5.1   PSO Implementation

Initially, several runs were carried out with different values of the key parameters of PSO such as the initial inertia weight and the maximum allowable velocity. In the present implementation, the initial inertia weight w is set to 0.9. Other parameters are set as number of particles $n = 35$, $c_1 = c_2 = 2$. Also, the search is terminated if the number of iterations reaches 300.

21.5.2   Numerical Results

Table 21.1 gives the results that were obtained by applying the PSO method. The last column gives the total structural volume of the heat sink, indicated as V_{oL} (mm³). The larger value of V_{oL} indicates the additional structural mass required to manufacture the heat sink.

TABLE 21.1   Results Obtained in This Work and the Paper by Shih and Liu

	N	A (mm)	d (mm)	V_f(m/s)	s (mm)	$S_{gen}(W/K)$	V_{oL} (mm³)
Current work	21	106	1.4	1.25	1.2	0.002504	155820
Shih and Liu	20	134	1.61	1.05	.9368	0.002967	220740

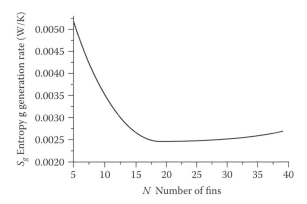

FIGURE 21.4 Variation of the optimum entropy generation rate with N.

Figure 21.4 shows that as the number of fins N increases, the entropy generation rate will decrease significantly due to increase in the surface area of the heat sink. When N reaches 22 fins and greater, the entropy generation rate will start to increase gradually and the search will start to move away from the optimal solution location. As it is clear from the figure, the optimal solution of N is between 19 and 22 fins. The optimal solution of the entropy generation rate is 0.002504 W/K. Figures 21.5, 21.6, and 21.7 show the behavior of the design parameters (V_f = flow velocity, a = height of fin, d = thickness of fin) along with the optimization process of the entropy generation rate as N varies.

A comparison has been done between PSO and GA and is shown in Figure 21.8. It shows the solutions of both PSO and GA for different values of N. It is seen that both PSO and GA have reached very close to the global solution, but PSO has generated better results than GA. Also, PSO has reached its solution with fewer numbers of iterations.

A comparison is made between the solution of PSO obtained in the present work and the solution obtained by Shih and Liu, as given in Table 21.1. The following conclusions can be made:

- The fin height a and the total structural volume V_{oL} of the heat sink are reduced by 30%. This implies that less material is needed to produce the heat sink, which may in turn reduce the cost of the cooling system.
- The entropy generation rate S_{gen} is slightly lower than the results obtained by Shih and Liu, which means that the overall thermal resistance is decreased.

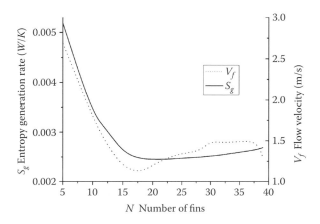

FIGURE 21.5 Variation of the optimum entropy generation rate and optimum flow velocity with N.

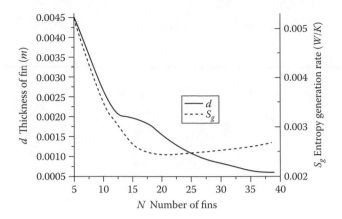

FIGURE 21.6 Variation of the optimum entropy generation rate and optimum thickness of fin with N.

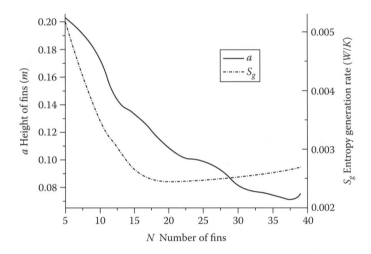

FIGURE 21.7 Variation of the optimum entropy generation rate and optimum height of fin with N.

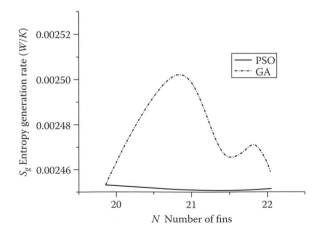

FIGURE 21.8 Variation of the optimum entropy generation rate and optimum flow velocity with N (PSO and GA solutions).

- The thickness d of each fin and the spacing s between two fins are slightly lower than what has been reported by Shih and Liu.
- The free stream velocity V_f is 20% higher, which means more pumping power is required. This will increase the running cost of the cooling system

21.6 Conclusion

Mechatronic systems are multidomain systems, which may involve fluid and thermal problems in addition to mechanics and electronics. The associated design problem is complex and often nonanalytic. They may require gradient-free optimization techniques. As one such approach, the PSO algorithm was presented in this chapter. Next, the applicability of the PSO algorithm to the optimal heat sink design was investigated. In particular, the PSO process was presented for the design of a plate-fin heat sink with the objective of maximizing the removal of the heat generated by electronic components. The entropy generation rate was used as the fitness function to realize the highest heat transfer efficiency. A practical application was presented as an illustrative example. A comparative evaluation was made with respect to design results available in the literature. The following comments may be made regarding the developed approach:

- PSO uses objective function information to guide the search in the problem space. Therefore, PSO can easily deal with nondifferentiable and nonconvex objective functions. Additionally, this property relieves PSO of assumptions and approximations, which are often required in the traditional optimization methods.
- PSO uses probabilistic rules for particle movement in the search space, not deterministic rules. Hence, PSO is a type of stochastic optimization algorithm that can search a complicated and uncertain problem domain. This makes PSO more flexible and robust than conventional methods.

References

1. Bar-Choen, A., Least-energy optimization of air-cooled heat sinks for sustainable development, *IEEE Transactions on Components and Packaging Technologies*, Vol. 26, No. 1, 16–25, March 2003.
2. Ogiso, K., Assessment of overall cooling performance in thermal design of electronics based on thermodynamics, *Journal of Heat Transfer*, Vol. 123, 999–1005, October 2001.
3. Shih, C.J. and Liu, G.C., Optimal design methodology of plate-fine heat sinks for electronic cooling using entropy generation strategy, *IEEE Transactions on Components and Packaging Technologies*, Vol. 27, No. 3, 551–559, September 2004.
4. Lyergar, M.K., Resource Constrained Heat Sink Optimization, Ph.D. thesis, University of Minnesota, Minneapolis, MN, 2002.
5. Wei, X. and Joshi, Y., Optimization study of stacked micro-channel heat sinks for micro-electronic cooling, *IEEE Inter Society Conference on Thermal Phenomena*, San Diego, CA, May 2002, pp. 441–448.
6. Bejan, A., *Energy Generation Minimization*, CC Press, Orlando, FL, 1995.
7. Muzychka, Y.S. and Yovanovich, M.M., Modeling friction factors in noncircular ducts for developing laminar flow, *Proceedings 2nd AIAA Theoretical Fluid Mechanics Meetings*, Albuquerque, NM, June 1998, pp. 15–18.
8. Teertstra, P., Yovanovich, M.M., Culham, J.R., and Lemczyk, T.F., Analytical forced convection modeling of plate fin heat sink, *Proceedings 15th IEEE SEMI-THERM Symposium*, San Diego, CA, 1999, pp. 34–41.
9. Onwubolu, G. and Babu, B. (Eds.), New optimization techniques in engineering, *Studies in Fuzziness and Soft Computing*, Vol. 141, Springer-Verlag, New York, 2004, pp. 2–15.
10. Sait, S. and Youssef, H., *Iterative Computer Algorithms with Applications in Engineering–Solving Combinatorial Optimization Problems*, Wiley-IEEE Computer Society Press, Los Alamitos, CA, 1999.

11. Lee, K.Y. and El-Sharkawi, M.A. (Eds.), *Modern Heuristic Optimization Techniques with Applications to Power Systems, IEEE Power Engineering Society*, New York, 2002.

12. Kennedy, J. and Eberhart, R.C., Particle swarm optimization, *Proceedings of IEEE International Conference on Neural Networks*, Piscataway, NJ, 1995, pp. 1942–1948.

13. Eberhart, R.C. and Kennedy, J., A new optimizer using particle swarm theory, *Proceedings of the Sixth International Symposium on Micromachine and Human Science*, Nagoya, Japan, 1995, pp. 39–43.

14. Abido, M.A., Optimal power flow using particle swarm optimization, *International Journal of Electric Power and Energy Systems*, Vol. 24, 563–571, 2002.

15. Eberhart, R. and Kennedy, J., A new optimizer using particle swarm theory, *Proceedings of the Sixth International Symposium on Micro Machine and Human Science*, Nagoya, Japan, Vol. 4–6, October 1995, pp. 39–43.

16. Fukuyama, Y., Fundamentals of particle swarm techniques, in Lee, K.Y. and El-Sharkawi, M.A. (Eds.), *Modern Heuristic Optimization techniques with Applications to Power Systems*, IEEE Power Engineering Society, New York, 2003, pp. 45–51.

17. Xiaohui, H., Eberhart, R.C., and Yuhui, S., Engineering optimization with particle swarm, *Proceedings of the 2003 IEEE Swarm Intelligence Symposium*, Indianapolis, IN, April 2003, pp. 53–57.

18. Carlisle, A.J., Applying the Particle Swarm Optimizer to Non-Stationary Environments, Ph.D. thesis, Auburn University, Auburn, AL, 2002.

19. Eberhart, R.C. and Yuhui, Shi, Particle swarm optimization: developments, applications and resources, *Proceedings of the 2001 Congress on Evolutionary Computation*, Seoul, South Korea, Vol. 1, May 2001, pp. 81–86.

20. Karray, F.O. and de Silva, C.W., *Soft Computing and Intelligent Systems Design—Theory, Tools, and Applications*, Addison Wesley, New York, 2004.

22

Actuator Optimization and Fuzzy Control in Mechatronics

H. Marzi

Summary

In this chapter, the characteristics of multi-input–output mechatronic systems are described, and application of fuzzy controllers to stabilize nonlinear multivariant systems is examined. The ability of different types of fuzzy controllers with respect to the number of inputs is investigated. The problem of balancing an inverted pendulum on a moving cart is simulated to demonstrate the performance of fuzzy control. Actuator optimization is an important step in the design optimization of a mechatronic system. As an exercise toward this goal, a mathematical model of a DC motor actuator is incorporated to optimize the selection of the actuator for improving the dynamic performance.

22.1 Introduction

Actuators are constituent components of mechatronic systems. Consequently, actuator optimization is an important step in the design optimization of a mechatronic system. This chapter considers the characteristics and performance of actuators using fuzzy logic control. The performance of a mechatronic system depends heavily on the quality and operation of its actuators. The operation and functionality of actuators depend on their physical characteristics and design as well as the control schemes employed in their operation. In this chapter, a brief description of fuzzy logic is given, followed by examining the performance of different types of fuzzy controllers in balancing an inverted pendulum mounted on a cart that is moving in a short track. A direct current (DC) electrical motor is used as the actuator, which changes the position of the cart to balance the pendulum in an upright orientation. A simple mathematical model for the DC motor is described and incorporated in the control scheme of the mechanism. The effects of various parameters of the motor in selecting the optimum actuator are discussed.

22.2 Actuator

An actuator refers to a device or component that provides the driving action to a mechanism or a plant to perform an intended task [23]. In robotics, it may be used, for example, to move the joints or in the grasping mechanism (hand) for picking parts. In a more general case of mechatronic systems, actuator may refer to a driving device that receives an input signal that represents a motion command and produces a corresponding motion. Examples of common types of actuators include hydraulic pistons, electromagnetic valve actuators, electrical relays, and electrical motors [23]. Electrical motors by far are the most popular type of actuators. In this chapter, a DC electrical motor is used as an actuator to mobilize the cart carrying an inverted pendulum in order to balance the pendulum in an upright orientation.

22.3 Inverted Pendulum

The problem of balancing an inverted pendulum mounted on a cart is a classical exercise and a stimulating example that can demonstrate the quality of a control scheme. This section describes a mathematical model for the inverted pendulum problem and develops conditions for the stability of the pendulum. It is assumed that the pendulum is hinged to the cart and has an angular movement in a vertical plane. The cart must be moved back and forth along the horizontal axis to keep the inverted pendulum in an upright balanced position. Figure 22.1 shows the diagram of the pendulum with the direction of motion and the forces acting on it. Parameters describing the kinematics of the cart and pendulum are as follows:

m_c	Mass of the cart
m_p	Mass of the pendulum
2L	Length of the pendulum
q	Angle of deflection with vertical line
x	Cart position
F	Force applied to the cart
N	Horizontal reaction force from the pendulum
P	Vertical reaction force from the pendulum
bx	Friction of the cart

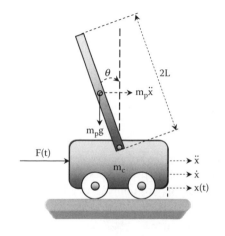

FIGURE 22.1 An inverted pendulum on a cart (the pendulum has motion in the vertical plane).

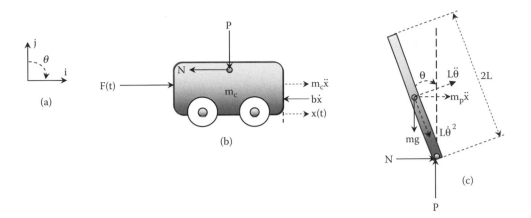

FIGURE 22.2 Cart and pendulum: (a) direction of coordinates, (b) free-body diagram of the cart, (c) free-body diagram of the pendulum.

To derive equations of motion for the cart-and-pendulum problem, Newton's second law is applied. Newton's law for rotational systems (i.e., torque/moment = inertia * angular acceleration) is used to generate the relation for rotational movement of the pendulum about the hinged point. The free-body diagram of Figure 22.2(a) is used to drive the equation of motion for the cart in the horizontal direction:

$$F - bx - N = m_c x \qquad (22.1)$$

where N is the reaction force from the pendulum, and bx is the friction force on the cart. Figure 22.2(b) shows the free-body diagram of the pendulum. The pendulum has three components of acceleration. The first one is centripetal, which is proportional to $L\theta^2$ and is along the pendulum pointing toward the axis of rotation. The second component is due to the acceleration of the cart and is proportional to x. The third is due to the angular acceleration of the pendulum, and it is perpendicular to the pendulum and proportional to $L\theta$. Applying Newton's law to the pendulum in the horizontal direction yields

$$N = m_p x + m_p L\theta \cos\theta + m_p L\theta^2 \sin\theta \qquad (22.2)$$

Equation (22.2), in the direction perpendicular to the pendulum, will take the form

$$P \sin\theta + N \cos\theta - m_p g \sin\theta = m_p L\theta + m_p x \cos\theta \qquad (22.3)$$

The application of Newton's law to the rotational motion of the pendulum, where the moments are summed about the centre of mass, and I is the moment of inertia about the center of mass, yields

$$-pL \sin\theta - NL \cos\theta = I\theta \qquad (22.4)$$

Substituting N from equation (22.2) into equation (22.1),

$$F = (m_c + m_p)x + bx + m_p L\theta \cos\theta + m_p L\theta^2 \sin\theta \qquad (22.5)$$

Combining equations (22.3) and (22.4) to eliminate reaction forces N and P results in

$$(I + m_p L^2)\theta + m_p gL \sin\theta = - m_p Lx \cos\theta \qquad (22.6)$$

Equations (22.5) and (22.6) describe the motion of the pendulum.

The equations of motion can be linearized by assuming small motions for the pendulum along the vertical axis. For small motions, θ is small. Then, $\cos\theta \cong 1$, $\sin\theta \cong \theta$, and $\theta \cong 0$. Consequently, equations of motion can be approximated by

$$F = (m_c + m_p)x + bx + m_p L\theta \tag{22.7}$$

$$(I + m_p L^2)\theta + m_p gL\theta + m_p Lx = 0 \tag{22.8}$$

At this stage, in solving the problem using conventional control approaches, the model is transformed into the Laplace domain using Laplace transformation: $x(t) \rightarrow X(s)$, $x \rightarrow X(s)s$, and $x \rightarrow X(s)s^2$. The Laplace transforms of equations (22.7) and (22.8) are

$$F(s) = (m_c + m_p)X(s)s^2 + bX(s)s + m_p L\theta(s)s^2 \tag{22.9}$$

$$(I + m_p L^2)\theta(s)s^2 + m_p gL\theta(s) + m_p LX(s)s^2 = 0 \tag{22.10}$$

Substituting $X(s)$ from equation (22.10) into (22.9) and rearranging the equation will result in a relation between output $\theta(s)$ and input $F(s)$, referred to as the transfer function of the system:

$$\frac{\theta(s)}{F(s)} = \frac{\dfrac{m_p L}{C}s}{s^3 + \dfrac{b(I + m_p L^2)}{C}s^2 + \dfrac{(m_c + m_p)(m_p Lg)}{C}s + \dfrac{m_p Lgb}{C}} \tag{22.11}$$

where $C = (m_p L)^2 - (I + m_p L^2)(m_c + m_p)$. Alternatively, the system can be represented by a state–space model [1].

22.4   Proportional-Integral-Derivative (PID) Controllers

Designing a control system using conventional approaches requires knowledge of the mathematical model of the plant and controller. Plants are physical systems, and their mathematical models can be inaccurate due to a number of factors such as approximations, change of parameters, unmodeled dynamics, unmodeled time delays, change in operating points, sensor noise, and unpredictable disturbance [2]. Therefore, designing accurate control systems is challenging, in particular, designing high-dynamic systems in the presence of uncertainties and approximations. Figure 22.3 shows the block diagram of a typical control system with uncertainties due to external unpredictable disturbances and random or uncontrolled noise.

Standard control schemes and the use of compensators can achieve the desired performance. However, in the presence of uncertainties, a robust control scheme is needed to assure system performance. Classical control design techniques may achieve reliable performance to a certain degree; however, there are a number of robust control design methods such as root locus, frequency response, and PID control that can provide desirable performance in the absence of large disturbances and errors. Among these methods, the PID control scheme provides a popular robust control design approach that has functional simplicity. In designing a PID controller, three control parameters—proportional gain K_p, integral gain K_I, and derivative gain K_D have to be determined, because the PID control law in the time domain is given by

$$f(t) = K_p.e(t) + k_I \int_{t_0}^{t} e(t).dt + k_D.T_D.\frac{d}{dt}e(t) \tag{22.12}$$

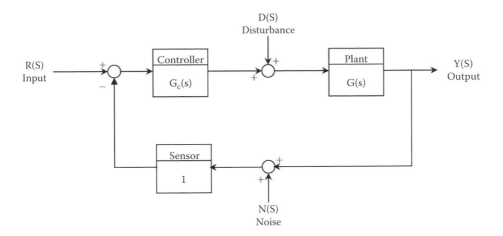

FIGURE 22.3 Block diagram of a closed-loop system with uncertainties.

The corresponding transfer function representation is

$$G_c(s) = K_p + \frac{K_I}{s} + k_D s \qquad (22.13)$$

22.5 Fuzzy Logic

The conventional approach in controlling the inverted pendulum system is to use a PID controller, which has the proportional (*P*), integral (*I*), and derivative (*D*) actions. To model the system, the model developer has to know its technical details and be able to express such details mathematically. Fuzzy logic control (FLC) challenges this traditional approach by using educated guesses about the system to control it [3]. Passino states that differential equations are the language of conventional control (PID), whereas "rules" about how the system works is the language of fuzzy control [4].

Fuzzy logic has found its way into the everyday life of people since Lotfi Zedah first introduced it in 1962. In Japan, the use of fuzzy logic in household appliances is common. Fuzzy logic can be found in such common household products as video cameras, rice cookers, and washing machines [5]. From the weight of the clothes, fuzzy logic is able to determine how much water as well as the time needed to effectively wash the clothes. Japan developed one of the largest fuzzy logic projects when they opened the Sendai Subway in 1987 [6]. In this subway system, the trains are controlled using fuzzy logic.

In a general sense, fuzzy logic is an application area of fuzzy set theory and utilizes its concepts, principles, and methods [7]. More specifically, fuzzy logic is a subset of traditional Boolean logic. Boolean logic states that something is either true or false, on or off, 0 or 1, and so on. Fuzzy logic extends this into saying that something is somewhat true, or not completely false. In fuzzy logic there is no clear definition as to what is exactly true or false. It uses a degree of membership (DOM) to generalize the inputs and outputs of the system [8]. The DOM ranges from 0 to 1 and can lie anywhere in between. In using fuzzy set theory for approximate reasoning, a connection between degrees of membership in fuzzy sets and degrees of truth of fuzzy propositions must be established.

22.6 Membership Functions

A membership function quantifies the meaning of a linguistic statement [4]. In other words, a membership function defines or maps an input value to a linguistic quantity. Therefore, the membership function is the method of quantification of knowledge in fuzzy logic [9,10]. It quantitatively determines the level

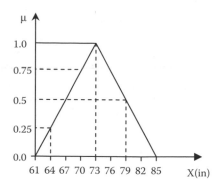

FIGURE 22.4 Membership function of tall people in a sports club.

of possibility that one instance satisfies a criterion or the possibility of belonging to a descriptive set [24]. For example, Figure 22.4 shows membership function of tall people in a certain sports club. In this graph, x-axis indicates the person's height measured in inches. Let us assume that, in this club, the interpretation of tallness, that is, the presumed height for a tall member, is 73 in. Therefore, club members who are 73 in. tall, certainly satisfy the condition. These members have a membership value of 1 (or $\mu = 1$). Any other member taller or shorter than this height is tall to a certain degree. The triangular shape of figure 22.4 is the map or the membership function of the tall members of the club. Then, in this club, a 79-in. person is tall with a level of possibility of 0.5 or the membership function of 0.5 (i.e., $\mu = 0.5$). Also, a height of 64 in. has $\mu = 0.25$, but a height of 87 in. or 59 in. in this club has a membership of 0 (i.e., $\mu = 0$). The shape of a membership function is decided by the designer of the fuzzy system, and it may be any shape such as trapezoid, Gaussian, and so on.

Membership function is used in fuzzy set theory to address the level possibility of belonging to a set that is not precisely defined. There have been long debates and discussion on similarities and confusions between fuzzy set theory and the probability theory. As probability governs uncertainty arising from randomness, fuzzy set theory governs uncertainty arising from vagueness and imprecision [24]. A number of articles [10–15] address this debate, including the seminal article of 1965 by Lotfi Zadeh [11], in which he explains that the fuzzy set is a natural way of dealing with the problems in which the source of imprecision is the absence of sharply defined criteria and class membership, rather than the presence of random variables.

22.7 Fuzzy Logic Controller (FLC)

The main focus of fuzzy set has been to solve problems of imprecision and vagueness. The early target areas were pattern classification and information processing, as indicated by Lotfi Zadeh in his 1965 paper on fuzzy sets. In the 1970s, when the practical applications of fuzzy sets started, the most widely used application area of fuzzy logic was fuzzy control. A wide range of control problems is tackled by FLCs. These may contain complex tasks requiring a number of synchronized actions or a simple task such as maintaining a single parameter. As the number of coordinated actions or the input space of fuzzy logic increases, so does the complexity of the required FLCs. Fuzzy control solutions are mostly based on human knowledge and expertise in operating the control system. This knowledge is expressed in the form of a set of *if and then* rules.

The majority of inverted-pendulum systems developed using fuzzy logic use a two-dimensional approach, where only the angle and angular velocity of the pendulum's arm are measured. The work presented in this chapter will show why this method is insufficient for the development of an inverted pendulum on a track of limited size. To have an efficient fuzzy controller for an inverted pendulum, the system must also include inputs for the position of the cart that the pendulum is balanced on and the velocity of the cart. Two-dimensional fuzzy controllers are simple examples of fuzzy control. They will balance the inverted pendulum but are not in control of the cart's position on the track. Nafis [16] proposed a two-dimensional fuzzy controller to balance an inverted pendulum on a track. Tests showed that the controller would balance the pendulum but neglected to control the position of the cart and, eventually, the cart's position would exceed the length of the track. Another FLC was proposed by Passino and Yurkovich [4]. Again, this cart had the same adverse result as the previous FLC.

Control of the system considered here requires the cart holding the pendulum to be moved by a specific mechanism. For simulation purposes, in the experiment presented here, an armature-controlled DC motor [2,23] is used.

22.8 Multi-Input System

As discussed in the preceding section, FLCs can have more than one input. It is simple to implement a two-input FLC. Layne and Passino [3] modeled a fuzzy controller that had good performance in balancing the pendulum, but the cart's positioning was unstable, making it an impractical rule set for real-life implementation. Two-input FLCs are the most commonly used ones for inverted-pendulum systems.

The two-input system receives angle θ (theta) and angular velocity $\dot{\theta} = \omega$ (theta-dot) as its inputs. The system uses five membership functions for each input and another five for the outputs (force). The system consists of 25 (5^2) rules. Table 22.1 shows the rule base for the two-input inverted pendulum system.

For the rule-base structure with five rules, 2 represents a negative large value, 1 a negative small value, 1 a positive small value, and 2 a positive large value. For example, a situation is selected in Table 22.1 where, if angle θ is positive large (i.e., 2) and angular velocity $\dot{\theta}$ or ω is negative small (i.e., −1), then, in the rule at the cross section of the two input values will be fired.

Figure 22.5 shows a simulation that is run over a time period of 1 s. The pendulum has an initial angle of 0.2 rad (dashed line). When the simulation is run, the angle of the pendulum balances quickly in about 1 s, but the position of the cart is not controlled (continuous line); so the cart's position will eventually drift off into the end of the track even though the pendulum arm is balanced. In this case, fuzzy control rules do not control the position of the pendulum, and it is only the angle and the angular velocity that are under the control of the rule base. This approach is not acceptable for linear tracks; however, it is acceptable in cases where the track of the pendulum is circular.

Adding two more inputs into the system to control the X-position and the velocity of the cart will greatly benefit the stability of the system. There is a cost, however, for better stability, which includes a greater computation time and greater complexity in the model. The cost of adding more inputs increases exponentially with the number of inputs added. The two-input system described in the preceding text uses five membership functions for each input; this results in a 25 ($= 5^2$) rules in the rule base. By adding two more inputs to the system, the system's rule base would grow to 625 ($= 5^4$) rules. The development and computational time for a rule base of this size can be excessive. Bush proposed the use of an equation to calculate the rules rather than developing the rules individually [17]. The system had 5^4 rules with 17 output membership functions (OMF). The following equation was used:

$$I + (J - 1) + (-K + 5) + (L + 5) \tag{22.14}$$

TABLE 22.1 Rule-Base for a Two-Input Inverted Pendulum

$\theta/\dot{\theta}(\omega)$	−2	−1	0	1	2
−2	2	2	2	1	0
−1	2	2	1	0	−1
0	2	1	0	−1	−2
1	1	0	−1	−2	−2
2	0	(−1)	−2	−2	−2

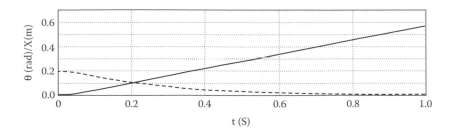

FIGURE 22.5 Variation of angle (θ/rad: dashed) and position (X/m: continuous) of pendulum versus time (t/s) in a two-input FLC with 25 rules.

where I, J, K, and L represent values of membership functions for each of the inputs. This equation results in values ranging between 1 and 17 for the five membership functions. Assume that parameter I in this equation stands for angle θ, J for angular velocity θ or ω, K for position X, and L for velocity X or V. Assume further the following membership functions for each input: θ = *large positive* = 2; ω = *large positive* = 2; X = *large negative* = 2; L = *large positive* = 2. Inserting these values for membership functions will result in the rule 17 to be fired, which is the highest rule number. This corresponds to the OMF that is to be used in the calculation of the output. The performance of the system using this approach is not consistent with that of the original simulation given by equation (22.14) [17]. The force given to the cart holding the pendulum was found to be insufficient to balance the pendulum, and the system failed quickly. It can be concluded that this system is a good starting point for developing a large rule set on but needs tweaking of the rules and the membership functions to obtain satisfactory balancing of the pendulum.

The final FLC that was modeled for simulation was a Takagi–Sugeno (T–S) type fuzzy controller, whereas all the previous FLCs were of the Mamdani type. The T–S type controller was first proposed in 1985 by the joint work of T. Takagi and M. Sugeno [18]. Since that time, it has been the subject of other research [19–22]. The T–S type fuzzy controller deviates from the traditional Mamdani type controller by using linear or constant OMFs instead of the likes of triangular, trapezoidal, and Gaussian functions. The present system uses four inputs with only two input membership functions for each. This resulted in 2^4 = sixteen rules. The linear output membership functions are calculated using equation

$$y = c_0 + (c_1 * x_1) + (c_2 * x_2) + (c_3 * x_3) + (c_4 * x_4) \tag{22.15}$$

where c_n are the parameters of the OMF, and x_n are the values of θ, θ, X and X, respectively. The system modeled here uses the fuzzy logic toolbox of MATLAB® [9].

The control of all four parameters with only two membership functions causes the system to operate very fast. The downside to this quick response is that it takes more time for the system to stabilize because there are so few membership functions. The system will overshoot the targeted position and eventually come to rest. The settling time of this system is greater than that for the other systems.

Figure 22.6 shows the result of the simulation. The pendulum is started with an initial disturbance of 0.2 rad, as in the previous case. As shown, the fuzzy controller overcompensates for this initial disturbance and sends the pendulum angle (dashed line) in an opposite direction in an attempt to balance it, creating an overshoot. It takes approximately 5 s for the pendulum arm to balance.

22.9 Modeling the Actuator

The actuator that operates the inverted pendulum mechanism is a DC electrical motor. DC motors are versatile and are widely used due to their special characteristics such as high speed and highly controllable speed and torque [23]. There are different types and classifications of DC motors. For example, wound-field DC motors include shunt wound, series wound, and compound wound. Another classification is

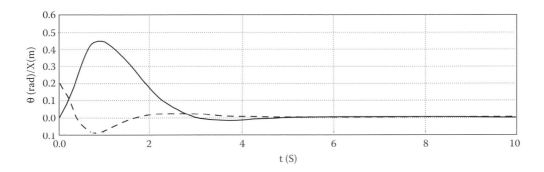

FIGURE 22.6 Variation of angle (θ/rad: dashed) and position (X/m: continuous) of pendulum versus time (t/s) using a FS type FLC with 16 rules.

permanent magnet and brushless types DC motors. DC motors are also classified based on their controllability of speed or torque: field-current-controlled DC motors and armature-controlled DC motors [23]. In DC motors, flux generated in the air gap (air-gap flux) is proportional to the field current i_f; that is, $\phi = K_f i_f$, where K_f is a constant gain of field and i_f is the field current. Motor torque T_m is proportional to the flux ϕ and armature (rotor) current i_a; that is, $T_m = K_1 \phi i_a$, or

$$T_m = K_1\, K_f\, i_a(t)\, i_f(t) \tag{22.16}$$

For linear control of DC motors, one of the currents in Equation (22.16) is kept constant, and the other current is changed in order to control the speed or torque of the motor. For field-current-controlled DC motors, armature current i_a is kept constant, and i_f controls the DC motor. The Laplace transform of the motor torque is given by

$$T_m(s) = K_m\, I_f(s) \tag{22.17}$$

In armature-controlled DC motors, i_f is kept constant, and i_a is used to control the DC motor. The Laplace transform equation of the motor torque in armature-controlled DC motors is given by

$$T_m\,(s) = K_m\, I_a(s) \tag{22.18}$$

The DC motor chosen for the present simulation is an armature-controlled DC motor. The motor is modeled using Equation (22.18). By combining Equation (22.18), inserting values of I_a and V_b from Equations (22.19) and (22.20), and using Equation (22.21) for load, an output/input relation (transfer function) for the actuator can be obtained. Specifically,

$$V_a(s) - V_b(s) = (R_a + L_a s)I_a(s) \tag{22.19}$$

$$V_b(s) = K_b \omega(s) \tag{22.20}$$

$$T_L(s) = (Js + b)s\theta(s) = T_m(s) - T_d(s) \tag{22.21}$$

The transfer function of the armature-controlled DC motors [2] is then obtained as

$$T(s) = \frac{\theta(s)}{V_a(s)} = \frac{K_m}{[(R_a + L_a s)(Js + b) + (K_b K_m)]s} \tag{22.22}$$

Figure 22.7 displays a block diagram for armature-controlled DC motors.

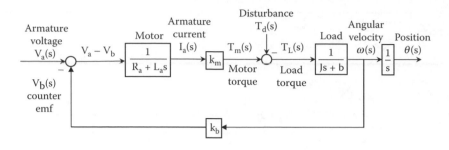

FIGURE 22.7 Block diagram of armature controlled DC motor.

The parameters used for the motor are as follows:

K_m = *Motor constant* [(N.m)/A]
K_b = *Back emf constant* [(N.m)/A] or [V.s/rad]
J = *Rotor inertia* [(N.m.s²)/rad] or [kg.m²]
R_a = *Electric resistance* [Ω]
L_a = *Electric inductance* [H]
b = *Damping ratio* [N.m.s]
V_a = *Armature voltage* [V]
V_b = *Counter or back emf* [V]
I_a = *Armature current* [Amp]
T_m = *Motor torque* [kg.m]
T_L = *Load torque* [kg.m]
T_d = *Disturbance torque* [kg.m]
θ = *Angular position* [rad]
ω = *Angular velocity* [rad/s]

In the present simulation, it is assumed that $K_m = K_b$ in the steady state.

The transfer function of the DC motor yields position (angle θ). This output is then converted into the torque or force required as the input to the pendulum. The motor responds well, reaching its maximum force exerted on the cart in less than 0.5 s.

22.10 Results of Simulation

The simulation consists of four main components: the fuzzy controller, DC motor, cart, and the inverted pendulum, as shown in Figure 22.8. The parameters θ, θ, X, and X are provided by the cart to the fuzzy controller. Based on these four parameters, the fuzzy controller generates a voltage to the motor. The motor, in turn, produces the force that is exerted on the cart. The system then calculates the new values for the parameters θ, θ, X, and X, and the cycle is repeated.

The fuzzy controller used in the simulation, with the DC motor included, is a 2^4 FLC, as described earlier. The system runs identical to the 2^4 system. Only the settling time for the simulation, with the motor included, is greater. Figure 22.9 shows the results of the simulation, which was run using the same fuzzy controller as in [19], with the DC motor included in the simulation.

The DC motor has a delay where it takes the motor a given time to reach the maximum force. This, in turn, causes the simulation to take longer to reach the steady state. Table 22.2 lists values of the parameters used in the simulation.

Figure 22.9 shows that it takes approximately 8 s for the pendulum angle to become steady, and even longer for the cart position to stabilize. The difference in the response time of this system with that shown

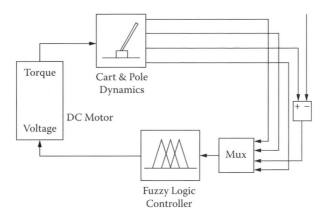

FIGURE 22.8 Block diagram of the fuzzy controller for inverted pendulum.

in Figure 22.6 is attributed to the motor dynamics. The motor has a time constant that delays the response time of the motor to an the input voltage. A typical armature-controlled DC motor has a time constant of about 100 *ms*. The shorter the time constant of the motor, the quicker the system response. The armature time constant is negligible; that is, $\tau_a = \frac{L_a}{R_a} \cong 0$. With this assumption, Equation (22.22) will change to

$$T(s) = \frac{\theta(s)}{V_a(s)} = \frac{K_m}{s\left[R_a\left(1 + \frac{L_a}{R_a}s\right)(Js + b) + K_b K_m\right]} = \frac{K_m}{s[R_a(Js + b) + K_b K_m]} = \frac{K_m/(R_a b + K_b K_m)}{s(\tau_l s + 1)} \quad (22.23)$$

where $\tau_l = R_a J/(R_a b + K_b K_m)$ is the equivalent time constant.

The simulation shows that the system responds well even with a motor attached and its dynamics included in the system. The cost of implementing a motor into the simulation is the added response time for the pendulum to stabilize. Simulations done without the addition of the DC motor cannot be considered representative of real-life implementations because the motor is needed in real life.

In the present approach, the actuator dynamics is separated from the fuzzy logic control mechanism. In this way the stability of the physical system is examined with different types of actuators. Then, the optimum actuator is selected that best suits the particular operation.

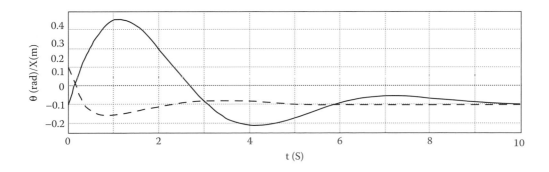

FIGURE 22.9 Variation of angle (θ/rad: dashed) and position (X/m: continuous) of pendulum versus time (t/s) after the DC motor dynamics included with FLC.

TABLE 22.2 Parameters Used for the Armature-Controlled
DC Motor Simulation

K_m/K_b	$= 50 \times 10^{-3}$ N.m/A		L_a	$= 0.5$ H
J	$= 1 \times 10^{-3}$ N.m.s^2/rad		b	$= 0.1$ N.m.s
R_a	$= 1.00$ Ω		r (radius)	$= 0.03$ m

22.11 Conclusion

Actuators are constituent components of mechatronic systems. Consequently, actuator optimization is an important step in the design optimization of a mechatronic system. This chapter presented the design of a fuzzy controller for a multi-input-output system consisting of an actuator. It demonstrates a trade-off between precision, which requires complex design, and simplification, which achieves less precise system. There is no one best approach in developing FLCs. The designer of an FLC system must consider whether precision will be sacrificed for performance and simplicity. The system with 5^2 rules, developed in this work, was rather simple and computationally efficient. The drawback of this initial design was that precision was compromised. The 2^4 system was also very simple and ran quickly, but the performance of the system was not satisfactory. The 5^4 system was very complex and the performance was slow, but if tuned correctly, a system of this size would be very precise.

Implementation of the considered system requires a high performance DC motor. Simulation results showed that this type of motor would be satisfactory. Having a smaller time constant in the DC motor would result in a shorter response time of the system. The FLC would need to be fine-tuned for other types of motors. With the DC motor implemented in the simulation model, the system did not react as well to high disturbances as it did when the motor was neglected in the simulation. This indicates that the present system can react well to small disturbances and is able to recover from them quickly. As the results indicate, for this system to handle large disturbances, a motor with high performance dynamics and a very small time constant has to be used. By the present approach, the optimum actuator may be selected for optimal performance of the system, as the actuator dynamics are separated from the fuzzy logic controller of the physical mechanism.

Acknowledgment

Material has been extracted, with permission, from the author's article: "Multi-input fuzzy control of an inverted pendulum using an armature controlled DC motor," *Robotica* (2005), 23:6, 785–788, Cambridge University Press.

References

1. Franklin, G., Powell, J.D., and Emami-Naeini, A., *Feedback Control of Dynamic Systems*, 5th ed., Prentice Hall, Englewood Cliffs, NJ, 2006.
2. Dorf, R.C., Bishop, and Robert, H., *Modern Control Systems*, 10th ed., Prentice Hall, Englewood Cliffs, NJ, 2005.
3. Layne, R.L. and Passino, K.M., A fuzzy dynamic model based state estimator, *Fuzzy Sets and Systems*, Vol. 122, No. 1, 45–72, August 2001.
4. Passino, K.M. and Yurkovich, S., *Fuzzy Control*, Addison-Wesley, Reading, MA, 1998.
5. Jenson, J., *Design of Fuzzy Controllers*, European Network for Fuzzy Logic and Uncertainty Modeling in Information Technology, http://www.iau.dtu.dk/~jj/pubs/design.pdf, last accessed August 31, 2006.
6. Kahaner, D.K., Fuzzy systems activities in Japan, 1993, http://www.atip.org/public/atip.reports.93/fuzzy 5.93.html, last accessed August 31, 2006.

7. Klir, G.J., Clair, U.S., and Yuan, B., *Fuzzy Set Theory: Foundations and Applications*, Prentice Hall, Englewood Cliffs, NJ, 1997.
8. Lin, C.T. and Lee, G., *Neural Fuzzy Systems*, Prentice Hall, Englewood Cliffs, NJ, 1996.
9. Cox, E., *The Fuzzy Systems Handbook*, 2nd ed., AP Professional, San Diego, CA, 1999.
10. Ross, T., *Fuzzy Logic with Engineering Applications*, 2nd ed., John Wiley, New York, 2004.
11. Zadeh, L., Fuzzy sets, *Information and Control*, Vol. 8, 338–353, 1965.
12. Zadeh, L., Outline of a new approach to the analysis of complex systems and decision processes, *IEEE Transactions on Systems, Man, Cybernetics*, Vol. 8, 338–353, 1973.
13. Zadeh, L., Discussion: probability theory and fuzzy logic are complementary rather than competitive, *Technometrics*, Vol. 37, 271–276, 1995.
14. Zadeh, L., Forward to fuzzy logic and probability applications: bridging the gap, *Society for Industrial and Applied Mathematics*, Philadelphia, PA, 2002.
15. Gaines, B., Fuzzy and probability uncertainty logics, *Information and Control*, Vol. 38, pp. 154–169, 1978.
16. Nafis, A., Fuzzy control of inverted pendulum on a cart, *The Institute of Electrical and Electronics Engineers (IEEE), Karachi Section Conference on* Technology Extravaganza, August 2001, http://ewh. ieee.org/sb/iiee/new/misc/pfcip.pdf, last accessed August 31, 2006.
17. Bush, L., *Fuzzy Logic Control for the Inverted Pendulum Problem*, Technical Report, Computer Science Department, Rensselaer Polytechnic Institute, 2001.
18. Takagi, T. and Sugeno, M., Fuzzy identification of systems and its applications to modeling and control, *IEEE Transactions on Systems, Man and Cybernetics*, Vol. 15, No. 1, 116–132, 1985.
19. Sugeno-Type Fuzzy Inference, in Chapter 2 of *Fuzzy Logic Toolbox for use with* MATLAB, The Mathworks, Natick, MA, July 2002.
20. Liang, W. and Langari, R., Building sugeno-type models using fuzzy discretization and orthogonal parameter estimation techniques, *IEEE Transactions on Fuzzy Systems*, Vol. 3, No. 4, 454–458, November 1995.
21. Johansen, T.A., Slupphaug, O., Lo, J.C., and Chen, Y.M., Comment on stability issues on Takagi-Sugeno fuzzy model-parametric approach [and reply], *IEEE Transactions on Fuzzy Systems*, Vol. 8, No. 3, 345–346, June 2000.
22. Tanaka, K., Hori, T., and Wang, H.O., Multiple Lyapunov function approach to stabilization of fuzzy control systems, *IEEE Transactions on Fuzzy Systems*, Vol. 11, No. 4, 582–589, August 2003.
23. De Silva, C.W., *Sensors and Actuators—Control System Instrumentation*, CRC Press, Boca Raton, FL, 2007.
23. Karray, F. and de Silva, C.W., *Soft Computing and Intelligent Systems Design*, Addison-Wesley, New York, 2004.

V

Monitoring and Diagnosis

23

Health Monitoring of a Mechatronic System

F. Pirmoradi

F. Sassani

C.W. de Silva

Summary

In this chapter, first the need for health monitoring in mechatronic systems is explained. Next, the nature of failures and the way they affect the system are outlined. The detailed health monitoring process and how failures are characterized are presented. An overview of the existing approaches for fault detection and diagnosis (FDD) is discussed. Next, an example of a mechatronic system is given where schemes of health monitoring and FDD are crucial. In concluding the chapter, some results from the developed algorithm are presented and discussed.

23.1 Introduction

Mechatronic systems use control systems for their operation. Malfunction and failure of sensors are critical in an automatic control system where the human operator is removed from the control process. A good example of such systems is spacecraft. Failures in space missions have resulted in the loss of human lives, interruption in research activities, and enormous economic losses. However, the increasing demand for conducting experiments in space and deploying satellites into orbits for observation, communication, and commercial purposes have necessitated reliable, durable, and safe control systems and instrumentation, which contribute directly to the intended mission of the spacecraft. This demand has motivated the development and design of sophisticated technologies, that improve the performance of the spacecraft control system and instruments installed on-board. These technologies not only aim to increase the vehicle operating life and reduce costs but are also intended for advanced control systems that counteract problems such as the presence of external disturbances in uncertain environments and the detection and compensation of system failures.

Spacecraft operations mostly rely on ground-based support and communication through radio commands. Large amounts of data are continuously collected and downlinked to the ground for processing, computation, and analysis of the current status of the spacecraft by human operators and, in anomalous situations, for decision making by ground experts. Subsequently, the necessary control commands are transmitted to the spacecraft. Exception to this arises when the resulting levels of communication time delay between the spacecraft and the ground station cannot be tolerated in the specific mission. Delay-free communication (or delay levels consistent with real-time communication) becomes particularly important when some faults or failures occur in the system, and their early detection, identification, and diagnosis become crucial for taking the necessary subsequent actions (e.g., control, parameter adjustment, reconfiguration, and emergency maneuvering). Furthermore, any loss of ground control and monitoring results in the failure of the spacecraft mission. In view of these, modern spacecraft need greater autonomy on-board to effectively handle their operation in the presence of faults and failures in sensors, actuators, and other components. In addition, the introduction of increased intelligence and autonomy on-board the satellites will reduce the amount of data collected and downlinked to the ground. These demands necessitate the development of health monitoring schemes that can result in increasing the ability of spacecraft to detect and diagnose faults and failure in a timely manner, with minimal human intervention from a ground station.

The focus of this chapter is the health monitoring of mechatronic systems. In particular, health monitoring and associated fault categorization are explained, and an overview of the relevant major approaches is given. As an illustrative example, a health monitoring concept of the attitude determination system of a spacecraft is presented and its advantages are explained.

23.2 Nature of Failures

In general, faults represent any type of malfunction in an actual dynamic (mechatronic) system that leads to an unacceptable anomaly in the overall performance of the system. In other words, according to the literature, faults are unpermitted deviations of a characteristic property of the system that lead to its inability to fulfill the intended purpose [1]. Such malfunctions can occur either in the plant of the system or in its measurement and control instruments. The way the faults impact the system in each case is different; so, the faults may be categorized into one of the following groups:

1. *Additive measurement faults:* These are discrepancies between the measured and the actual values of the system output or input variables. This category of faults can mostly describe sensor biases. Actuator malfunctions can also be considered as additive faults, as they are discrepancies between the input command of the actuator and its actual output. These faults cause offsets in the input and output vectors [2], as given by

$$
\begin{aligned}
y(t) + f_M(t) \\
u(t) + f_L(t)
\end{aligned}
\tag{23.1}
$$

where $y(t)$ and $u(t)$ are the output and the input of the system, and $f_M(t)$ and $f_L(t)$ are the output and input offset faults, respectively.

2. *Additive process faults:* These are disturbances (including unmeasured inputs) acting on the system that cause a shift in the outputs. These faults best describe system component leaks, load variations, and so on. In the context of fault detection, these disturbances are the ones that we wish to detect and isolate.

3. *Multiplicative process faults:* These are changes (abrupt or incipient) of some system parameters that also depend on the magnitude of the known inputs. Such faults best describe the deterioration of system equipment; for example, surface contamination or clogging. It can be expressed in the state–space form as [2]

$$
\begin{aligned}
x(t) &= (\mathbf{A} + \Delta\mathbf{A})x(t) + (\mathbf{B} + \Delta\mathbf{B})u(t) \\
y(t) &= (\mathbf{C} + \Delta\mathbf{C})x(t)
\end{aligned}
\tag{23.2}
$$

where $\mathbf{A}$, $\mathbf{B}$, and $\mathbf{C}$ are system matrices, and delta (Δ) denotes the change in their parameters in the presence of fault.

It should be noted that this classification is primarily analytical. From a practical point of view, some faults may be considered both additive and multiplicative. Furthermore, the classification of failures depends on the approach of fault detection and diagnosis that is used. The foregoing classification is correct for model-based approaches, which will be discussed later in this chapter.

23.3 Failure Characterization

Fault characterization and condition monitoring involve the following tasks:

- *Fault detection*: identification of the presence of an unknown fault.
- *Fault diagnosis*: identification of the nature and location of a detected fault. It identifies which component is faulty and determines the nature of the fault. Fault signature or fault model may be used for this purpose. This task also includes determination of the magnitude of the fault. Often, the term *isolation* is used as a synonym for diagnosis.
- *Fault accommodation/resolution*: correction of the fault by activating new controllers, adjusting control parameters, or scheduling gains. If a fault cannot be removed, the task is to accomplish the original or modified operation in the presence of the fault by either taking hardware actions (e.g., activating backup systems) or software actions (e.g., reconfiguring the system for the intended task).

The performance of detection of a diagnostic technique is characterized by some quantifiable parameters and is described by

- *Fault sensitivity*: the ability of the technique to distinguish reasonably small faults.
- *Reaction speed*: the ability of the techniques to detect faults with reasonably small time delay after their occurrence.
- *Robustness*: the ability of the technique to detect faults in the presence of noise, disturbances, and modeling errors with minimal presence of false alarms.

Achieving a required level of performance in these properties depends on the system to be monitored. In addition, due to the presence of noise, disturbances, and modeling errors, there are trade-offs between various properties in designing an algorithm for FDD.

23.4 Approaches to Fault Detection and Diagnosis

Generally, FDD methods are designed based on physical or analytical redundancies in the system. Figure 23.1 gives a general overview of the methods and the way they are categorized, which are briefly discussed in this section.

In physical redundancy—a traditional engineering approach—multiple sensors (instrument sets and hardware elements) are installed to measure the same physical quantity. A fault signal is generated when a serious discrepancy occurs in the output of the sensors. In this method, because two parallel sensors are not generally adequate for fault isolation, three or more redundant hardware units are necessary to isolate the faulty elements [3]. This method has several disadvantages: it incurs extra hardware cost and requires extra space for the installation of hardware elements and instrument sets. It might be necessary to have one-to-one component redundancy in some critical subsystems. However, the concepts of triple module redundancy with voting [4] and quadruple redundancy with parity check [5] are not efficient because of the weight and space limitations in many engineering systems, which is a particularly serious concern in aerospace applications. Besides, if faults affect all the elements in the same way, this method cannot help in the detection of the faults.

Most FDD methods are based on analytical redundancy and can be grouped into two major approaches:

1. Model-free (signal-based): methods that do not use a system model
2. Model-based: methods that use a system model

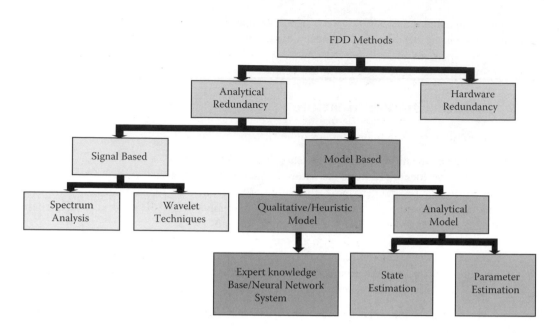

FIGURE 23.1   Major approaches of fault detection and diagnosis (FDD).

23.4.1   Model-Free Approaches

In model-free approaches, a mathematical model of the system is not used. Because physical redundancy eliminates the need for a model, it can also be categorized in this group [6]. Model-free approaches are briefly reviewed here.

1. *Limit checking and installing special limit sensors:* In this method, system measurements are compared to preset limits. A fault is declared if those measurements exceed the limits. Many existing engineering systems use this so-called "redline system" for fault detection and diagnosis, where redlines are defined as the limits at which the system is inoperable. These limits correspond to the values at which some measurements that are selected, or some core parameters that are derived, exceed their nominal operating values. By installing limit sensors, the physical properties in the system hardware (e.g., limit temperature or pressure) or some specific variables (e.g., sound or vibration) are measured. Although this approach is effective in avoiding catastrophic failures caused by a specific component, it has many drawbacks. For example, because of possible input variations, the limits should be set conservatively to avoid false alarms, which usually result in the premature removal of a properly operating component from the loop. Another limitation is inconsistent redlined limits that vary from one instrument to another because the components of each unit may be produced by different manufacturers. In many systems, two levels of limits are set. The first level is only for warning, whereas the second triggers an emergency action. However, if the fault in one component propagates to other parts of the system, this method causes confusion in diagnosis. Therefore, more sophisticated methods need to be developed to achieve efficient and desirable health monitoring systems.
2. *Frequency analysis:* In this method, a typical frequency spectrum of the system measurements is obtained under normal operating conditions. Any deviation from this spectrum indicates an abnormality. Certain types of failure may have a characteristic signature in the spectrum, which can be used for failure isolation. This method is widely used in such situations as the spectral analysis for feature extraction using wavelet technique in a diesel engine [7]. However, it is limited to dynamic systems involving vibrations. Also, because closed-loop control systems generally compensate for changes in measured output by changing the input, the deviations caused by faults cannot be recognized by range checking alone.

Complementary methods together with the previously mentioned techniques are used for evaluating the detected symptoms. They consist of simple logical rules to reach a conclusion. They work on the information presented to them by the detection hardware or software, or interact with the human operator for information through the entire logical process.

23.4.2 Model-Based Approaches

In this method, a quantitative (analytical model) or qualitative (knowledge-based) model of the physical system is used for fault detection and diagnosis. Here, sensory measurements are compared to analytically computed values of the particular variable, and subsequently the resulting differences—the residuals—are generated. These residuals are evaluated to indicate the presence of a fault. A fault signal is declared if the residuals exceed a certain threshold value. Due to noise in the system, these residuals are never zero even when a fault does not exist, which leads to the need for testing the residuals against the thresholds.

The residuals generated to indicate faults may also react to the noise and disturbances that can exist in the process and the measurements of the system. Desensitizing the residuals to these sources is an important aspect in the design of an FDD algorithm. To deal with the effect of noise, the residuals can be filtered, and statistical techniques can be applied for their evaluation. For this purpose, the knowledge of the statistical properties of noise and the noise-transfer dynamics of the system should be available.

23.4.2.1 Analytical Models

The quality of the analytical model of a system will directly affect the FDD scheme. Therefore, a realistic system specification should be used. For a realistic representation, it is important to model all effects that can lead to alarms or false alarms. Such effects are

- Faults in the actuators, or in the components of the plant dynamics, or in the sensors
- Modeling errors between the actual system and its mathematical model
- System noise and measurement noise

The general procedure of analytical model-based fault detection and isolation can be roughly divided into two steps [8]:

1. Generation of residuals (functions that are accentuated by the fault vector *f*)
2. Detection and isolation of the faults (time, location, sometimes also type, size, source, and cause)

If a fault occurs, the redundancy relations are no longer satisfied, and the respective residual becomes distinctly nonzero. This residual is used to form the appropriate decision functions. A fault signature—a signal which is obtained from a faulty system model defining the effects associated with a fault—is the basis for the decision on the occurrence of a fault. They are evaluated in the fault decision logic to monitor both the time of occurrence and the location of the fault. There are, of course, some problems with generating residuals. For instance, the mathematical model of the nominal system is uncertain, and system noise and measurement noise are present. Consequently, for the existence of the residuals, the knowledge of the "normal" behavior and definition of "faulty behavior" should be available. Basically, there are two different ways of generating residuals and fault isolation. The residuals are generated by either *state estimation* techniques including the Kalman filter, parity relations, and diagnostic observer schemes, or by *parameter estimation* techniques. These techniques are briefly introduced next. Figure 23.2 shows a general fault characterization process based on the analytical model.

23.4.2.1.1 State Estimation Approach

In the *Kalman filter* method, the innovation of the Kalman filter is used as the fault detection residual. This residual is white and has a zero mean under nonfaulty conditions and it becomes nonzero in the presence of a fault. The information about the failure types should be available for fault isolation [9–11].

The key idea in the *parity relation* approach is to check the consistency of the mathematical equations of the system (analytical redundancy relations) by using actual measurements. In this method, residuals

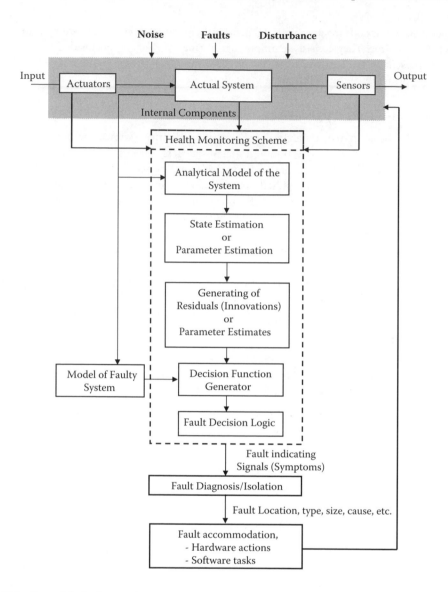

FIGURE 23.2   General fault characterization process based on analytical redundancy.

are colored, and disturbance decoupling becomes better. A fault is declared once preassigned error bounds are exceeded [8].

The basic idea of the *diagnostic observer* approach is to reconstruct the outputs of the system from the measurements or subsets of measurements with the aid of observers and use the estimation error as a residual to detect and isolate the faults. The residual sequence is colored, which makes statistical testing somewhat complicated but enhances the isolation of faults. In the case of a nonfaulty system, residuals are still influenced by unknown inputs, whereas they would increase in magnitude in the case of a fault. Therefore, to detect faults in the system, the increment of the residual caused by the faults should be checked. One way to do this is to compare them with a threshold.

The state estimation methods have been implemented for nuclear reactors [12], aerospace systems [13,14], and automobiles [15].

23.4.2.1.2 *Parameter Estimation Approach*

This approach is used for the detection and isolation of parametric (multiplicative) faults. The involved residuals are generated using a parameter identification technique. It makes use of the fact that faults of a dynamic system are reflected in the physical parameters. Therefore, estimation of the parameters of the mathematical model is used for fault detection. In physical systems, coefficients that indicate process faults may not be directly measurable but may be estimated using other measurements. Therefore, determination of their changes may be done by determining the changes in model parameters. A fault event and the time of its occurrence are determined by using a statistical decision-making method. This approach has been used in hydraulic systems [16], spacecraft main engine [17], and others.

After a fault event has been detected, the patterns of parameter changes are submitted to a classification procedure with the aim of determining the fault type, size, location, and cause. This includes whiteness, chi-square, generalized likelihood, and different types of multiple hypotheses tests. A detailed account of these techniques can be found in the literature [18,19,13,4].

23.4.2.2 Qualitative Model Approach

Expert systems (knowledge-based methods) can be used as an alternative means to detect and diagnose faults in a system. In this approach, artificial intelligence (AI) techniques are used to achieve human-like reasoning [20]. This method is mostly used for fault detection and diagnosis of the systems or processes for which an accurate model is either not available or very complex to develop. Whereas the algorithmic methods use quantitative analytical models, the expert system approach makes use of qualitative models based on available knowledge of the system.

Knowledge-based methods can complement the model-based methods of fault detection in many applications. The combination of both strategies allows the evaluation of all available information and knowledge of the system for fault detection. The ability of an expert system to integrate both strategies has made them ideal for fault diagnosis purposes. The main task can be divided into fault detection by analytic and heuristic symptom generation and fault diagnosis [2,21].

In the context of analytical symptom generation, analytical knowledge of the system is used to produce residuals. The characteristics of the system can be obtained by data processing and feature extraction, which were described previously. In heuristic symptom generation, qualitative information is used from human operators. Through human observation and inspection, heuristic characteristic values in the form of special noise, color, smell, vibration, wear and tear, and so on are obtained. Process history in the form of performed maintenance, repair, former faults, lifetime, and load measures represents a further form of heuristic knowledge. Also, statistical data from experience with the same or a similar process can be included. In this manner, heuristic symptoms may be generated as linguistic variables.

It is advantageous to use a unified representation of the two types of symptoms. One possibility is to present both types of symptoms with confidence numbers between 0 and 1 and use probabilistic approaches. Another possibility is the representation as membership functions of fuzzy sets. By using these fuzzy sets and corresponding membership functions, all analytic and heuristic symptoms can be represented in a unified way within the range $0 \leq \mu(S_i) \leq 1$, where S_is are the generated symptoms. These integrated symptom representations are then provided as inputs for the inference mechanism. For establishing a heuristic knowledge base, logical interactions between observed symptoms (effects) and unknown faults (causes) are used to generate the rule base. Finally, different diagnostic strategies involving forward and backward reasoning may be applied. The final results of this scheme are diagnosed faults [22].

Sometimes, in formulating an FDD problem, a model may not exist either qualitatively or quantitatively that relates the symptoms to the system faults. This, in turn, might be due to a lack of information about how faults occur and are propagated in the system. In the neural-network approach, instead of having a mathematical model or expert knowledge, the data collected from the system are used to generate a neural network, which replaces the model. This method is particularly useful when no prior information about the relationship between the faults and the symptoms is available, which is mostly the case in complex processes [23,24]. Commonly, the neural-network approach is used as a fault classifier for fault diagnosis.

It should be noted that the growing demand for higher operational safety together with the increase in the complexity in automation has resulted in the integration of different FDD approaches in different applications [25,26].

23.5 Case Study—Health Monitoring of a Spacecraft System

In this section, an example of a mechatronic system together with a health monitoring scheme is presented. Finally, the advantages of the proposed scheme are illustrated with some sample results. The health monitoring system provides information about the condition of the system state variables, which will help detection and diagnosis of failures. This information allows the main controller to execute appropriate actions for system reconfiguration, such as gain adjustment, activation of alternative system, and so on. These actions minimize damage after a failure is detected.

The main objective of this example is to develop a health monitoring scheme that addresses the problem of FDD in a spacecraft control system, specifically for the spacecraft attitude determination (AD) system. The AD system provides information of the current status of the spacecraft for use by controllers to command the necessary actions to the actuators. The AD system is a key component of the spacecraft attitude control system, and improvements to its accuracy and reliability contribute directly to the success of a spacecraft mission.

The objective of AD is to generate estimates of the angular rates and the attitudes of the spacecraft. Multiple sensors are used to obtain the measurement data in the attitude determination and control system (ADCS), depending on its architecture. In different phases of the mission, different types of sensors are selected, depending on the attitude determination scheme, and are configured to provide the necessary data. The failure detection and diagnosis of the AD system of a spacecraft involves the detection and isolation of faults and failures in these on-board sensors.

The solution offered in this example is a health monitoring scheme for the AD system in which, combined with the AD algorithms, an FDD design introduces a significant degree of autonomy. In particular, by this approach, a fault can be detected and identified without ground interaction and intervention.

The model-based state estimation approach is utilized for health monitoring. Specifically, Kalman filters are utilized as optimal state estimators, based on the kinematics and dynamics of the system and sensor measurements, to predict the states of the system and form the residuals using the observed measurement. Systems in aerospace applications are safety-critical, where security and reliability play an important role. Because, in an aerospace application, it is not difficult to establish a mathematical model for the system, specifically for the control system, the model-based estimation approach is more robust and can outperform other methods. This approach also handles noise in the system, which is one of the main concerns in attitude determination. The approach is based on monitoring the innovation sequence of the designed Kalman filters followed by statistical testing of residuals for fault detection and hypothesis testing for fault isolation. However, in the design of the FDD scheme, the isolation of fault is performed in two stages.

In this example, an integrated AD system is devised for data collection, which includes rate gyros, Sun sensors, and magnetometers. Figure 23.3 shows how the FDD module is incorporated into the ADCS of a spacecraft. Measurements from sensors are processed for AD and monitored for FDD. The FDD module provides the information about the sensors that have failed. This information is processed and utilized for recovery actions and resetting the controller parameters.

Because the equations of motion of a spacecraft are nonlinear, the AD problem involves a nonlinear filtering problem. Therefore, nonlinear equations are used to design four Kalman filters. The first Kalman filter is employed in AD (AD-KF) to provide estimates of the states of the system. It uses a linearized model about a nominal trajectory obtained from the mission requirements. The remaining three Kalman filters, which are employed for FDD (FDD-KF-1, FDD-KF-2, and FDD-KF-3), use nonlinear state equations,

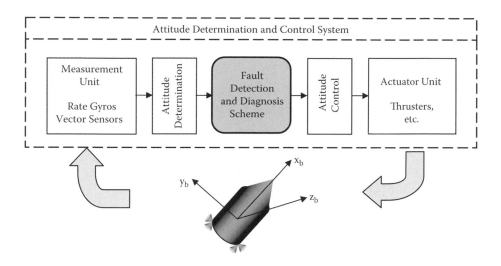

FIGURE 23.3 Incorporation of the FDD module in the attitude determination and control system (ADCS) of a spacecraft.

but the state and the measurement nonlinearities are linearized around the latest state estimates for covariance propagation.

A significant advantage of using a Kalman filter in AD arises from the fusion of measurement data from different types of sensors, which results in getting more accurate estimates of the attitude of the spacecraft out of the observed measurements. Due to the presence of unbounded errors in the output of the rate gyros, the data fusion is important for the removal of the erroneous gyro measurements in the AD system.

Figure 23.4 shows the designed overall structure of AD and FDD. The FDD scheme consists of three main phases: fault detection, primary isolation, and fault diagnosis/isolation.

For the purpose of FDD the difference is generated between the predicted outputs, given by the Kalman filters and the observed measurements, referred to as the residuals. In the absence of failure, the residuals are unbiased (of zero mean), demonstrating agreement between the estimates and the observed measurements. In contrast, biased residual are indicative of abnormal behavior or failures.

In the first phase of FDD, the presence of an unknown fault is detected. Therefore, statistical threshold tests are continuously used to monitor the residuals and to detect faults. If the test statistic exceeds the defined threshold, then a failure is declared. However, our knowledge of rate gyro outputs indicates that they consist of several biases, some of which grow with time. This causes the triggering of fault signals by gyros during long-term usage. To detect sensor faults that do not originate from gyro biases, and also to obtain accurate angular rate measurements, these biases are estimated by the AD-KF as well as the attitude angles and are fed back to the system to remove erroneous gyro measurements. In summary, AD-KF fuses data from different sensors and generates estimates of the angular position of the spacecraft and the bias-free angular rates from gyros. Subsequently, the bias free measurement data are used in the FDD algorithm.

The next step in the FDD scheme is designed to partly isolate the failure. The primary isolation phase aims to identify whether the source of failure is from the rate gyros, vector sensors, or both. This is performed by designing two filters running in parallel and using statistical testing. The statistical tests employed are *t*-test and chi-square test, where their thresholds are obtained empirically by considering the accuracy of the sensors and using statistical tables.

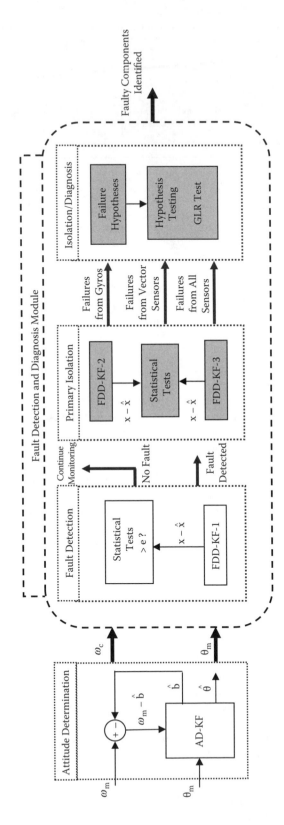

FIGURE 23.4

The information obtained from the primary isolation phase is utilized in the next phase of the FDD scheme (fault isolation/diagnosis). In this phase, the failed components are identified using the approach of hypotheses testing. The isolation phase consists of multiple hypotheses for different failure scenarios. The best match for the present system state from the hypothesized states is found using the technique of generalized likelihood ratio (GLR) test.

To illustrate the performance of the developed FDD scheme, the simulation results of a failure scenario are presented. In this failure scenario, a jump of $0.3°/s$ occurs in the pitch rate gyro output at $t = 40$ s. The spacecraft moment of inertia matrix is given by $I = \text{Diag}\{10,12,2\}$ kg.m^2. This specification closely corresponds to the one used in Mehra et al., 1998 [27]. In all simulations, the Kalman filter sampling time is set to 0.1 s. The initial conditions of the spacecraft angular positions and angular rates, as used in the simulations, are

Angular positions	$\psi = 10°$, $\theta = 10°$, $\phi = 10°$
Angular velocities	$p = 0.005$, $q = 0.005$, $r = 0.005$ [rad/sec]
Moments [28]	Disturbance torques ($\sigma_M = 0.0001$) N.m

The error in the initial states is set to be equal to the standard deviations of the measurements. The sensor outputs are modeled to include bias and white noise of Gaussian type with zero mean and standard deviations as given below [14]:

Bias time constant	$\tau = 300$ [s]
Gyros noise standard deviation	$\sigma_w = 0.05$ [deg/s]
Vector sensor standard deviation	$\sigma_\theta = 0.5$ [deg]
Bias standard deviation	$\sigma_b = 0.3$ [deg/s]

For brevity, only the simulation results from FDD-KF-1 are presented here. Figures 23.5 through 23.8 demonstrate the behavior of the system in the healthy mode.

It is observed that the filter performs properly because the errors are within the standard deviation bounds that have been estimated by KF. Note that when a fault is not present, all the residuals are zero-mean and white. Also, note that the rate gyro standard deviations of $0.05°/s$ each have decreased to the steady state values of $0.003°/s$, $0.0025°/s$, and $0.007°/s$, for the roll, pitch, and yaw rates, respectively. This is also true for the angle measurement standard deviations. They decrease from $0.5°$ to approximately $0.09°$.

Under the failure scenario, the rate and angle residuals from FDD-KF-1 behave as illustrated in figures 23.9 and 23.10. The mean of the pitch rate residual has changed significantly, whereas the mean of the pitch and yaw angle residuals have changed slightly for this scenario. Figure 23.11 shows the t-statistic calculation for each residual and its threshold.

At $t = 41.5$ s, the statistic crosses the threshold for the first time and the algorithm declares a fault. Figure 23.12 shows the results for the chi-square test.

It shows that the value of the statistic β jumps and crosses the threshold at $t - 40.4$ s. By comparing the graphs in figures 23.11 and 23.12, it is noted that the chi-square test can detect the presence of a fault faster than the t-test.

As described previously, it is decided in primary isolation whether the failure is from the vector sensors, the rate gyros, or both. Figure 23.13 presents the result of the chi-square tests for making this decision. It shows that the β-statistic for the rate residuals reacts to the fault, thus triggering the flag showing that the fault source is from the gyros.

Under the failure scenario 1, only the β-statistic for the rate residuals reacts to the fault, thus triggering the flag showing that the fault source is from the gyros.

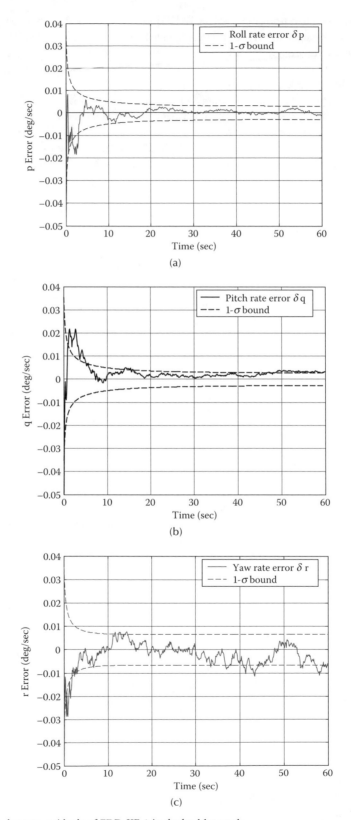

FIGURE 23.5 Angular rate residuals of FDD-KF-1 in the healthy mode.

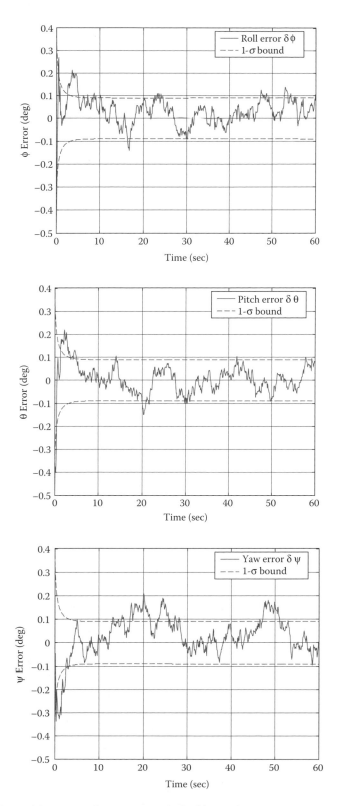

FIGURE 23.6 Angular position errors of FDD-KF-1 in the healthy mode.

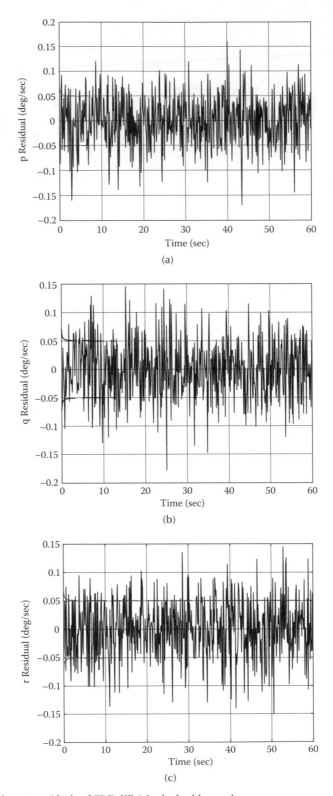

FIGURE 23.7 Angular rate residuals of FDD-KF-1 in the healthy mode.

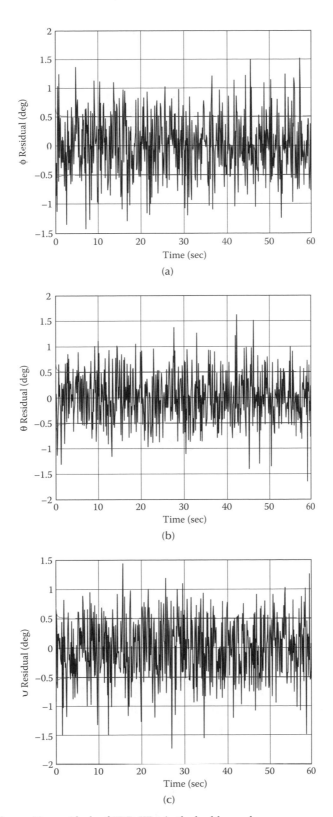

FIGURE 23.8 Angular position residuals of FDD-KF-1 in the healthy mode.

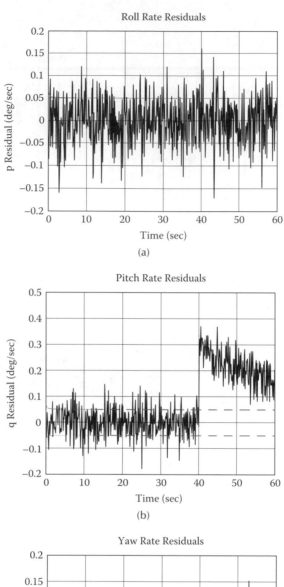

(a)

(b)

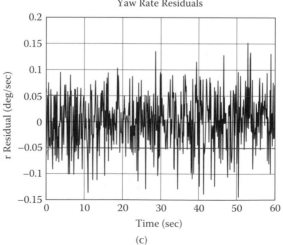

(c)

FIGURE 23.9 Angular rate residuals of FDD-KF-1 in the failure mode.

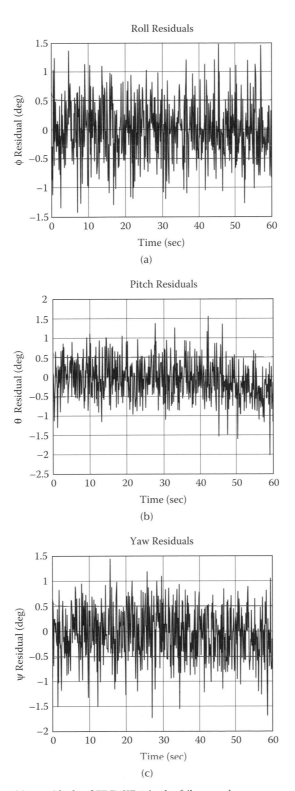

FIGURE 23.10 Angular position residuals of FDD-KF-1 in the failure mode.

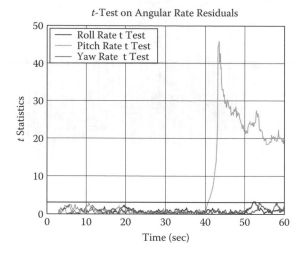

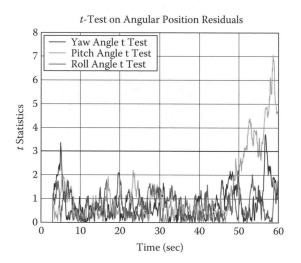

FIGURE 23.11   *t*-Test on FDD-KF-1 residuals in fault detection phase.

Figure 23.14 shows the likelihood functions for all activated hypotheses computed for this failure scenario. It shows that only the rate angle hypotheses have been activated because, based on the information from the primary isolation module, the required set of hypotheses is activated. Here, the second hypothesis, which is the pitch rate gyro failure, has the highest value and thereby is declared as a faulty sensor.

Figure 23.15 illustrates the time duration of each phase for the fault detection and isolation scheme, the time in which a fault is detected, primarily isolated, and completely diagnosed.

According to the simulation results, it is seen that the developed health monitoring and FDD scheme for the AD system of a spacecraft is able to reliably detect and isolate the faulty components on line. The main advantage of having primary isolation in the proposed scheme is the reduction in the number of hypotheses to a quarter of what would be normally needed. This contributes directly to fast isolation of the faulty components. In addition, it leads to a higher distinguishing ability of faults. This FDD algorithm is readily expandable to include monitoring of a greater number of sensors in the system.

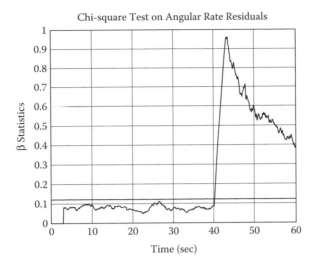

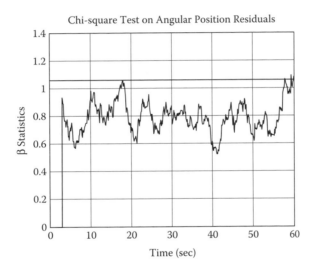

FIGURE 23.12 Chi-square test on FDD-KF-1 residuals in the fault detection phase.

23.6 Conclusion

Health monitoring and the detection and diagnosis of faults are important for proper operation of a mechatronic system. A spacecraft is a good example of a complex mechatronic system. Sensor failure can lead to serious problems in its control system, with potentially disastrous results. In this chapter, the need for health monitoring in mechatronic systems was emphasized. The nature of failures and the way they affect a mechatronic system were outlined. An overview of the existing approaches for fault detection and diagnosis (FDD) was discussed. The detailed health monitoring processes and how failures are characterized were presented. Next, an illustrative example, the spacecraft control system, was given where schemes of health monitoring and FDD are crucial. Results from the application of a developed algorithm for FDD to this system were presented and discussed.

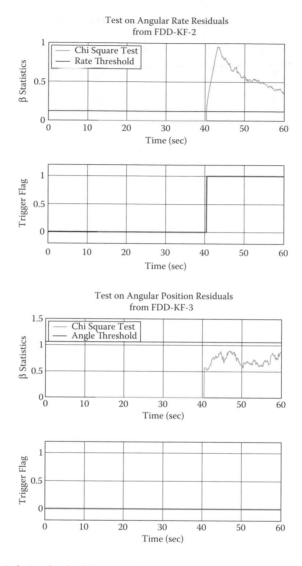

FIGURE 23.13   Primary isolation for the failure scenario.

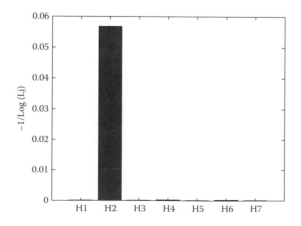

FIGURE 23.14   Likelihood functions for hypotheses testing.

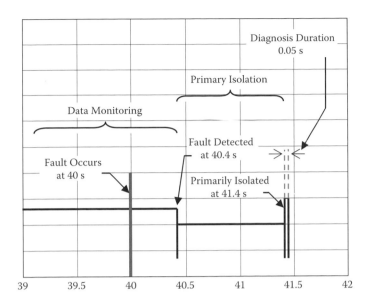

FIGURE 23.15 Timing of FDD phases.

References

1. Iserman, R., Process fault detection based on modeling and estimation methods—a survey, *Automatica*, Vol. 20, No. 4, 387–404, 1984.
2. Iserman, R., Model based fault detection and diagnosis methods, *Proceedings of the American Control Conference*, Seattle, Washington, D.C., Vol. 3, June 1995, pp. 1605–1609.
3. Betta, G. and Pietrosanto, A., Instrument fault detection and isolation: state of the art and new research trends, *IEEE Transactions on Instrumentation and Measurement*, Vol. 49, No. 1, 100–107, February 2000.
4. Willsky, A.S., A survey of design methods for failure detection in dynamic systems, *Automatica*, Vol. 12, No. 6, 601–611, 1976.
5. Satin, A.L. and Gates, R.L., Evaluation of parity equations for gyro failure detection and isolation, *Journal of Guidance and Control*, Vol. 1, 14–20, 1978.
6. Gertler, J.J., Survey of model-based failure detection and isolation in complex plants, *IEEE Control Systems Magazine*, Vol. 8, No. 6, 3–11, 1988.
7. Tafreshi, R., Sassani, F., Ahmadi, H., and Dumont, G., Local discriminant bases in machine fault diagnosis using vibration signals, *Journal of Integrated Computer-aided Engineering*, Vol. 12, No. 2, 147–158, 2005.
8. Frank, P.M., Fault diagnosis in dynamic systems using analytical and *knowledge*-based redundancy, *Automatica*, Vol. 26, No. 3, 459–474, 1990.
9. Da, R. and Lin, C.F., Sensor failure detection with a bank of Kalman filters, *Proceedings of the American Control Conference*, Seattle, WA, Vol. 2, June 1995, pp. 1122–1126.
10. Hajiyev, C. and Caliskan, F., Sensor/actuator fault diagnosis based on statistical analysis of innovation sequence and robust Kalman filtering, *Aerospace Science and Technology*, Vol. 4, No. 6, 415–422, 2000.
11. Mehra, R., Seereeram, S., Bayard, D., and Hadaegh, F., Adaptive Kalman filtering, failure detection and identification for spacecraft attitude estimation, *Proceedings of the 4th IEEE Conference on Control Applications*, Albany, New York, September 1995, pp. 176–181.
12. Elmadbouly, E. and Frank, P.M., Robust instrument failure detection via luenberger observers, *CIGRE Symposium on Control applications for Power System Sensitivity*, Florence, Italy, 1983, pp. 1–6.

13. Patton, R.J., Fault detection and diagnosis in aerospace systems using analytical redundancy, *Computing and Control Engineering Journal*, Vol. 2, No. 3, 127–136, May 1991.

14. Venkateswaran, N., Siva, M.S., and Goel P.S., Analytical redundancy based fault detection of gyroscopes in spacecraft applications, *Acta Astronautica*, Vol. 50, No. 9, 535–545, 2002.

15. Ribbens, W.B., A mathematical model based method for diagnosisng failures in automotive electronic systems, *SAE Transactions*, Vol. 100, 12–21, 1991.

16. Khoshzaban-zavarehi, M., "Online condition monitoring and fault diagnosis in hydraulic system components using parameter estimation and pattern classification", Ph.D. dissertation, The University of British Columbia, Vancouver, BC, Canada, October 1997.

17. Cikanek, H.A., Space shuttle main engine failure detection, *IEEE Control Systems Magazine*, Vol. 6, No. 3, 13–18, June 1986.

18. Brumback, B.D. and Srinath, M.D., A chi-square test for fault detection in Kalman filters, *IEEE Transactions on Automatic Control*, Vol. AC-32, No. 6, 552–554, June 1987.

19. Da, R., Failure detection of dynamical systems with state chi-square test, *AIAA Journal of Guidance, Control, and Dynamics*, Vol. 17, No. 2, 271–277, 1994.

20. Patton, R.J., Lopez-Toribio, C.J., and Uppal, F.J., Artificial intelligence approaches to fault diagnosis, *IEE Colloquium on Condition Monitoring: Machinery, External Structures and Health*, Birmingham, England, April 1999, pp. 511–518.

21. Iserman, R. and Ulieru, M., Integrated fault detection and diagnosis, *Proceedings of IEEE/SMC Conference on Systems Engineering in the Service of Humans*, Le Touquet, France, Vol. 1, October 1993, pp. 743–748.

22. Iserman, R., On fuzzy logic applications for automatic control, supervision, and fault diagnosis, *IEEE Transactions on Systems, Man and Cybernetics*, Part A, Vol. 28, No. 2, 221–235, March 1998.

23. Leonard, J.A. and Kramer, M.A., Diagnosing dynamic faults using modular neural nets, *IEEE Expert*, Vol. 8, No. 2, 44–53, April 1993.

24. Sobhani-Tehrani, E., Khorasani, K., and Tafazoli, S., Dynamic neural network-based estimator for fault diagnosis in reaction wheel actuator of satellite attitude control system, *International Joint Conference on Neural Networks*, Montréal, Québec, Canada, 2005, pp. 2347–2352.

25. Da, R. and Lin, C.F., Failure diagnosis using the state chi-square tests and the ARTMAP neural networks, *Proceedings of the American Control Conference*, Seattle, Washington, Vol. 5, June 1995, pp. 3279–3283.

26. Palma, L., Coito, F., and Silva, R., A combined approach to fault diagnosis in dynamic systems, application to the three-tank benchmark, *Intelligent Control Systems and Optimization*, Vol. 1, August 2004, pp. 163–171.

27. Mehra, R., Rago, C., and Seereeam, S., Autonomous failure detection, identification and fault-tolerant estimation with aerospace applications, *Proceedings of IEEE Aerospace Conference*, Aspen, CO, Vol. 2, March 1998, pp. 133–138.

28. Kim, Y.H., Park, Y., Bang, H., and Tahk, M.J., Covariance analysis of spacecraft attitude control system, *Proceedings of 2003 IEEE Conference on Control Applications*, Vol. 1, June 2003, pp. 480–485.

24

Defect Detection of Patterned Objects

H.Y.T. Ngan

G.K.H. Pang

Summary

Mechatronic systems are used in the production and inspection of fabrics. Monitoring these systems and detecting possible defects in the products are important in their proper operation. This chapter presents a comparison of the defect detection methods recently developed for a popular kind of patterned objects Jacquard fabric. These methods include golden image subtraction (GIS), wavelet-based defect detection, and Bollinger Bands (BB). The GIS method makes use of a golden image, which is larger than a repetitive unit, to perform a convolution filter on the fabric image based on the golden image. Not all types of defects can be successfully detected by the basic GIS method. Hence, a wavelet-preprocessed golden image subtraction (WGIS) method has been developed. The detection success rate of this method is as high as 96.7%, based on 60 images. A wavelet-based method, called Direct Thresholding on Detailed Subimages (DT), has been developed as well. The DT method has a detection success rate of 88.3% on 60 images. However, the detection method based on the use of BB is simple and very effective. In an extensive evaluation of 230 images, the detection success rate for this method is an impressive 99.57%.

24.1 Introduction

Mechatronic systems are utilized in the production and inspection of various products. Monitoring of these systems, and accurate and automatic detection of possible defects in their products, are crucial for the proper operation of the systems. This chapter introduces the technologies of defect detection on patterned objects. It is organized as follows. First, the background of the inspection of patterned objects is outlined. Then, a comparison of the existing methods for detection of the defects on patterned fabric is given. A brief description of the traditional methods—the image subtraction and Hash function—for patterned fabric inspection are presented. Next, three newly designed methods are compared for defect detection: WGIS, DT, and BB. The key concepts, detection procedures, detection success rate, and a comparison of the inspection results from these three methods are presented. Discussion and Conclusion are given at the end of the chapter.

24.2   Background of Patterned Objects Inspection

Automated inspection is typically a mechatronic process and is the dream of many manufacturers in various industries to reduce the labor cost [1,2], enhance the quality check results [2], and shorten the production time [3]. Patterned objects such as wallpaper, ceramics, and fabric are suitable products for automated inspection. Fabric flaw detection will be treated as an example in this chapter. In the textile industry, traditional visual inspection of fabric conducted by humans is still popular, and suitably trained workers are needed for this repetitive task. Many researchers have tried to use automated inspection, but mainly on unpatterned fabric such as plain and twill fabrics. An unpatterned fabric has no designed pattern on the fabric surface. Many methods have been developed for unpatterned fabric inspection, including neural networks [4], Fourier transform [5], Gabor filters [6], and wavelet transform [7], which can provide a detection success rate of over 95%. However, the research on patterned fabric inspection is rather limited. Patterned fabric is defined as fabric with repetitive patterned units in its design. Under the class of patterned fabric, there are many further categories. The repetitive units of the pattern can range from the simplest charter boxes and dots to the most complicated multiple flowers, animals, and designed patterns.

There are several reasons why researchers have had difficulties in the research of patterned fabric inspection. First, the texture complexity in patterned fabric is much more sophisticated than that in unpatterned fabric. Second, the categories of patterned texture are numerous. Third, the similarity in shape between defects and background texture (Figure 24.5(c)) is another obstacle for patterned fabric inspection. Fourth, the acquired image of patterned fabric may have slight shift in the patterns, which could cause alignment problems in traditional texture analysis approaches.

Figure 24.1(a) shows a sample of dot-patterned Jacquard fabric, which is our target object for defect detection in the present chapter. Every input dot-patterned Jacquard fabric sample would be processed by histogram equalization (Figure 24.1(b)) to have a better contrast for defect detection. All images used in the present evaluations have 256×256 pixels in gray level scale, and the computation is carried out using MATLAB (version 6.5).

In our evaluations, sixty images including thirty defect-free images and thirty defective images of dot-patterned Jacquard fabric are inspected using the methods presented in this chapter. Within the set of sixty images, there are six types of defects—broken end, holes, knots, netting multiple, thick bar, and thin bar—as shown in Figure 24.4(a–f), which represent the most popular types of defects in fabric

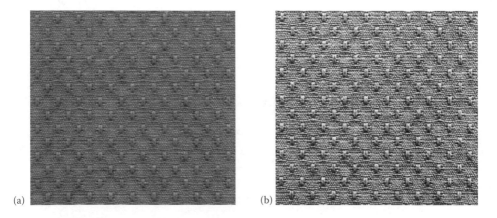

(a)   (b)

FIGURE 24.1   (a) Dot-patterned designed Jacquard fabric, (b) histogram equalization of (a).

inspection. The BB method is found to provide the best result among all the methods in the evaluation of the sixty images. Consequently, four extra types of defect samples—oil warp, oil weft, loose pick, and missing pick—as shown in Figure 24.5, are added for an extensive evaluation.

24.3 Patterned Fabric Defect Detection

In this section, various approaches for defect detection in patterned fabrics are outlined. Their key concepts are given. A comparative evaluation of a newly developed technique for pattern detection in fabrics is given with respect to other popular methods. Their performance is critically examined.

24.3.1 Traditional Methods for Defect Detection in Patterned Fabric

In the early days, a popular approach to the inspection of patterned fabric was the traditional image subtraction (TIS) method, developed by Sandy [8]. A recent method that can be applied to patterned fabric inspection is based on the Hash function, developed by Baykal [9]. The disadvantages of the TIS and Hash function methods are their sensitivity to noise, alignment problems, and inability in outlining the shape of the defect after detection.

24.3.2 Comparison of the WGIS, DT, and BB Methods

The three new methods for patterned fabric inspection—WGIS [10], DT of detailed subimages [11], and BB [12]—developed recently, are based on different spatial properties of the input images. The aim of this chapter is to give an overview and a comparison of these three methods. Underlying key concepts, detection procedures, and the detection results are presented and compared.

24.3.3 The Key Concepts

The basic GIS method [10] has been developed for detecting defects on patterned fabric, and the WGIS is an improved version of it. The GIS method developed by us is different from that of Sandy [8] because our golden template can contain several repetitive units taken from a defect-free image. In addition, the golden image used in GIS performing like a convolution filter, slides on a test image, and it is not a static comparison between the golden template and the test image. The GIS method can trigger out the defective region after thresholding the matrix of energies of GIS (Figure 24.2d) for every input image. The definition of energy of GIS [10,11] is the main formula that is used in the GIS method, as given by

$$R = (r_{xy}) = \frac{1}{m \cdot n} \sum_{i=1}^{m} \sum_{j=1}^{n} |g_{ij} - h_{ij}| \qquad (24.1)$$

where $x = 1, \quad , M - m + 1$ and $y = 1, \quad , N - n + 1$; $i = 1, \quad , m$ and $j = 1, \quad , n$ ($0 < n \le N, 0 < m \le M$). The golden image $G = (g_{ij})$ has a size of $m \times n$ pixels, and the subtracted image $H_{xy} = H_{xy}(i,j) = (h_{ij})$ has a size of $m \times n$ pixels. The size of every input image is $M \times N$ pixels.

In the DT method [11], the Haar wavelet fourth-level detailed subimages are extracted from each input image. In the Haar wavelet transform, the length of decomposition is two, and it is sufficient to obtain good detailed subimages that would enhance the defective regions in a defective image. (Figure 24.2e) There are four functions measuring different issues of the original image at level m size of $2^{-m} \times 2^{-m}$. Suppose that $\Phi(x,y)$ gives an approximation image at a resolution of $2^{-(m+1)} \times 2^{-(m+1)}$, $\Psi^H(x,y)$ gives the subtle change of intensity in horizontal view, $\Psi^V(x,y)$ gives the subtle change of intensity in vertical view, and $\Psi^D(x,y)$ gives the subtle change of intensity in diagonal view. Note that ψ and ϕ are one-dimensional wavelet and scaling functions, respectively. The subscripts L and H denote that the function

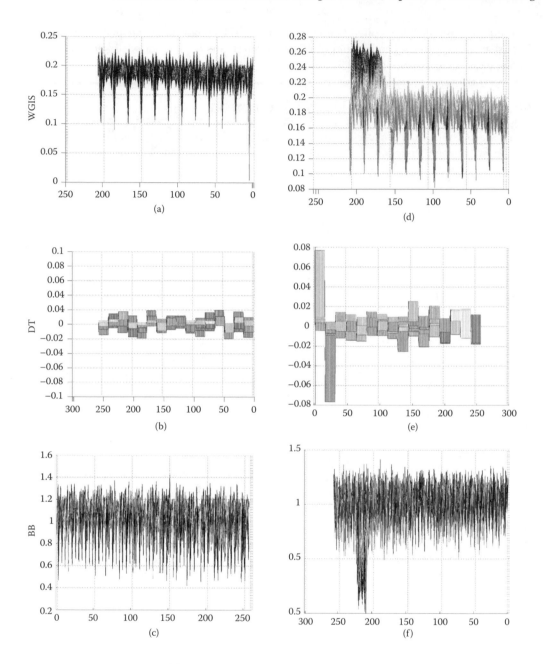

FIGURE 24.2 Horizontal views of the energies of the transformed reference image by (a) WGIS, (b) DT, (c) BB; horizontal views of the energies of the transformed defective image by (d) WGIS, (e) DT, (f) BB.

acts as a low-pass or a high-pass filter, respectively. The key formulas are as follows:

$$\Phi(x,y) = \Phi_{LL}(x,y) = \phi(x)\phi(y) \tag{24.2}$$

$$\Psi^{H}(x,y) = \Psi_{LH}(x,y) = \phi(x)\psi(y) \tag{24.3}$$

$$\Psi^{V}(x,y) = \Psi_{HL}(x,y) = \psi(x)\phi(y) \tag{24.4}$$

$$\Psi^{D}(x,y) = \Psi_{HH}(x,y) = \psi(x)\psi(y) \tag{24.5}$$

Defects usually appear as high-frequency elements at the boundaries between defective and defect-free regions in the defective image. The wavelet transform of the DT method is used to enhance the defective information in the horizontal and vertical directions, respectively.

The application of BB on patterned Jacquard fabric inspection is based on three constituted bands—middle, upper, and lower. Its original application was in the financial area, by Bollinger [13]. Ngan and Pang [12] extended BB from a one-dimensional approach for stock market analysis into a two-dimensional approach for patterned Jacquard fabric image processing. Robust defect detection has been achieved in this manner. The formulas of the BB method [13] are as follows.

For a particular row in an image of size $p \times q$,

$$\text{Middle band: } M_{r_n} = \left(\sum_{j=r_1}^{r_n} x_j \right) \Big/ n \tag{24.6}$$

$$\text{Upper band: } U_{r_n} = M_{r_n} + \left\{ d \cdot \sqrt{\left\{ \sum_{j=r_1}^{r_n} (x_j - M_{r_n})^2 \right\} \Big/ n} \right\} \tag{24.7}$$

$$\text{Lower band: } L_{r_n} = M_{r_n} - \left\{ d \cdot \sqrt{\left\{ \sum_{j=r_1}^{r_n} (x_j - M_{r_n})^2 \right\} \Big/ n} \right\} \tag{24.8}$$

where n is an integer value denoting the length of period on the row dimension, and x_j is the pixel value. The summation is from the r_1-th pixel to the r_n-th pixel with $1 \le r_1 \le r_n \le q$, and d is a value that denotes the number of standard deviation. There is a similar procedure for the columns.

24.3.4 The Detection Procedures

A comparison of the detection procedures of the three methods is shown in Figure 24.3.

The similarities of the three methods are discussed next.

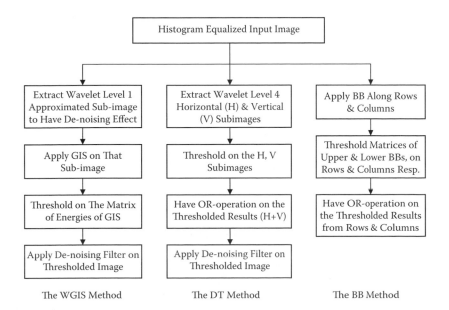

FIGURE 24.3 Procedures of the three different inspection methods.

24.3.4.1 Smoothing Techniques

In the WGIS method, a wavelet-preprocessed stage is used for smoothing the input image to remove excessive noise. However, in the DT method, wavelet transform is applied for a different purpose—extracting the localization property in the detailed subimages. In the methods of WGIS and DT, it is necessary to use a smoothing filter to remove excessive noise at the end of the procedures. Smooth filtering should be used, and the detection results will be significantly improved after removing excessive Gaussian noise.

24.3.4.2 Thresholding

Because all three methods have been designed to enhance the defective regions, thresholding is used in each method to segment out the defective regions. For the WGIS method, a matrix of energies that would enhance the defective regions is obtained. A defect would show a "pop-up" effect as shown in Figure 24.2(d). So, a threshold value is defined as the maximum of the peak values in the matrix of GIS of a reference image. For the DT method, from the detailed subimages of the input image, the defective regions would appear as two-sided pop-ups, as shown in Figure 24.2(e). Therefore, a bilevel thresholding can be implemented. The method of BB would take advantage of the standard deviation to push the defective regions downward (Figure 24.2(f)), which is similar in appearance to the pop-up region in the matrix of energies of GIS (Figure 24.2(d)). In all cases, the upper bound and the lower bound in the transformed matrix from the reference images (Figure 24.2 (a) from WGIS, (b) from DT, and (c) from BB) could be used as the threshold boundaries.

24.3.4.3 Time Complexity of Different Methods

To assess the effectiveness and the feasibility of implementation, the time complexity has been measured by MATLAB® with the same set of images. The computer used for defect detection is a Pentium 4 1.8 GHz, 256 DDR RAM. The WGIS, DT, and BB methods have their own individual time complexities for running all procedures (Figure 24.3) for one input image in detection. The execution times are 9.813 s, 18.782 s, and 28.656 s for the WGIS, DT, and BB methods, respectively. Accordingly, the WGIS method is the fastest among all three methods that have been implemented.

24.3.5 The Detection Success Rate

To compare the three methods, they are evaluated using a set of 60 test images (of size 256×256 pixels) that have six popular types of defects (Figure 24.4).

24.3.5.1 The WGIS Method

For the WGIS method, a 36×26 golden image and a Haar wavelet first-level approximated image are used. After the WGIS process, the white pixels that have appeared in the defect-free images is in the range from 0 to 100, so that 100 white pixels representing of defective images is set as the threshold for a defective image. Then, 28 of 30 defect-free images and all 30 defective images could be correctly detected after smooth filtering. Detection results of six typical defective samples are shown in Figure 24.4. The detection success rates are 93% for defect-free images and 100% for defective images. Therefore, the overall detection success rate is 96.7%.

24.3.5.2 The DT Method

The Haar wavelet type is chosen for its simplicity. By studying the detection result of defect-free images, all of them were found to have less than 8 white pixels at the end of detection. Hence, for a 16×16 final image from the fourth-level Haar wavelet decomposition, 8 white pixels out of 256 are set as a threshold for defective image. Of the 30 defect-free images, 26 of them had less than 8 pixels after filtering, and 27 of 30 defective images could be correctly detected. The detection success rates are 86.77% for defect-free images and 90% for defective images (Figure 24.4). Accordingly, the overall detection success rate is 88.3%.

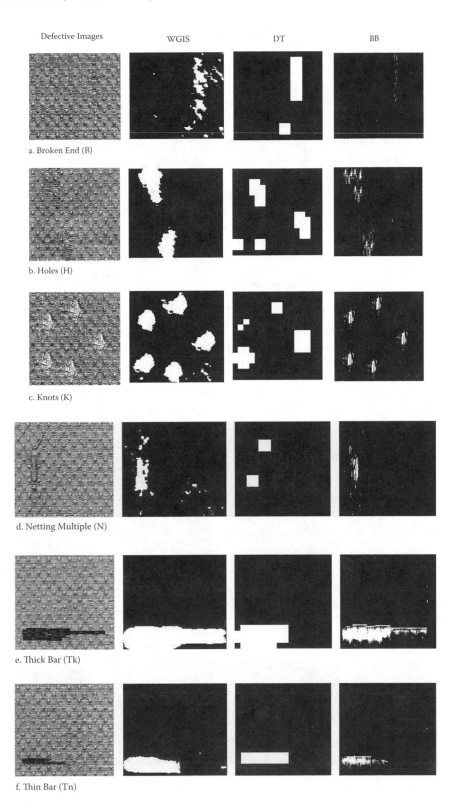

FIGURE 24.4 Final results of methods of WGIS, DT, and BB on different defective images: (a) broken end, (b) holes, (c) knots, (d) netting multiple, (e) thick bar, (f) thin bar.

24.3.5.3   The BB Method Using 60-Image Set

The detection success rate for the 60-image set is 100%. In the training stage, all defect-free images would show less than 50 white pixels at the end of detection. So, for evaluation purposes, a defective image is defined as one exceeding 50 white pixels after thresholding. The defect detection results are found to be excellent (Figure 24.4), but there is a need to demonstrate the potential of this method using a more extensive evaluation.

24.3.5.4   The BB Method Using 230 Images

The BB method has been extensively evaluated with 110 defect-free images and 120 defective images with 10 types of defects: broken end, holes, knots, netting multiple, thick bar, thin bar, oil warp, oil weft, loose pick, and missing pick, as shown in Figures 24.4(a)–(f) and 24.5(a)–(d). For evaluation, the threshold of the number of white pixels is still set as 50. Then, it was found that all 110 defect-free images and 119 defective images could be correctly detected. Four examples of four extra types of detection results are shown in Figure 24.5. The detection success rate is 100% for defect-free images, and 99.17% for the defective images. Therefore, the overall detection success rate is 99.57%. These results are quite satisfactory and encouraging for the detection of defects in this kind of dot-patterned fabric.

24.3.6   Comparison of Inspection Results for WGIS, DT, and BB Methods

For the 60-image database, the WGIS method showed a high overall detection success rate, i.e., 96.7%. Only two defect-free images were misclassified. Each final image (the second left column of Figure 24.4) in the six types of defects could be outlined by the WGIS method. However, the final images of broken end and netting multiple in Figure 24.5 showed excessive noise surrounding the defective regions. Besides, the whitened areas of the final images would show an enlarged version of actual defective patterns in the original defective images. The enlarged portion was mainly due to the effect of GIS, because the golden image was larger than one repetitive-unit-size window acting like a convolution filter on the input image so that it would lead the borders of the subtracted image to have a padding effect (similar to the effect of actual convolution) with extended defective areas.

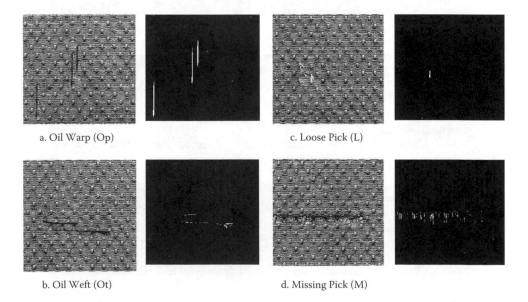

a. Oil Warp (Op)   c. Loose Pick (L)

b. Oil Weft (Ot)   d. Missing Pick (M)

FIGURE 24.5 Examples of four extra types of defective images in the testing set ($N1 = 15$, $d1 = 2$ and $N2 = 25$, $d2 = 2$, in BB where N1 and N2 are the period lengths on row and column dimensions, respectively).

TABLE 24.1 Comparison of WGIS, DT, and BB Methods

Method	Dimensionality	Ability to Outline the Defective Shape	Detection Success Rate on 60 Images	Strength	Weakness
WGIS	2	Good	96.7%	1. Easy to choose the size of golden image 2. Shortest time complexity among all methods in implementation	Cannot detect defects near the borders
DT	2	Poor	88.3%	1. Haar wavelet transformed subimages have lower resolution and lower computational power in detection	Coarse in detection results
BB	1	Outstanding	100%[a]	1. Simplicity: BB uses moving average and standard deviation 2. Clear final image 3. Applicable in on-loom machine	Cannot detect defects near the borders

[a] Application of BB was evaluated with 230 images, generating 99.57% detection success rate.

Compared with the WGIS and BB methods, the DT method generated coarser results. The final images of six types of defects (second right column in Figure 24.4) generated rectangular coarse results due to the low resolution of the detailed subimages (16×16 in fourth-level Haar wavelet decomposition) through the multiresolution property of WT. Excessive noise was commonly found in the final images such as those of broken end and holes. The coarser results and excessive noise actually would be quite confusing to some extent if the classification of defects had been implemented as a whole.

By using the concepts of moving average and standard deviation, the BB method has generated much better results than the previous two methods. Not only were the final images of the six types of defects clean and sharp, especially for the final images of thick bar and thin bar, but the overall detection success rate of 100% was also the highest among the three methods (correctly detected on all 60 images). This result is robust and satisfactory for the defect detection the on the dot-patterned Jacquard fabric. With a more careful and extensive evaluation of 230 defective images, the overall detection success rate was found to remain high at 99.57%, and only one failed among all defective images. From Figures 24.4 and 24.5 it is seen that the BB method gives a good outline of the defective regions.

A comparison of the detection procedures of the three methods is given in Table 24.1, giving their strengths and weaknesses.

24.4 Discussion

The regularity property of patterned fabric is the common principle utilized by the three methods of defect detection evaluated in this chapter. The intensities of any one row or column, which is called a row signal or a column signal, would generally generate a periodic signal regardless of the patterned texture. On applying specific types of transform (GIS, DT, or BB) on these periodic intensities in row or column, the resultant signal will be periodic signal as well. The aim of the transformation from one periodic signal to a new periodic one in this manner is to enhance the irregularities—defect information—in the new signal. For examples, the WGIS method uses subtractions to enhance the defective regions. The DT method applies wavelet transform to generate horizontal and vertical details for localizing the defective information. The BB method enhances the defective regions through the computation of moving average and standard deviation. The enhancement of defective regions is in the appearance of subtle changes compared to the normal path of the transformed signal. Therefore, after transformation of the signals, applying thresholding on the new signal can segment out the defects. Extension of the usage from one-dimensional signal to two-dimensional image enables the detection of defects on fabric.

24.5 Conclusion

Mechatronic systems are used in the production and inspection of fabrics. Monitoring these systems and detecting possible defects in the products are important in their proper operation. This chapter presented a comparison of the defect detection methods recently developed for a popular kind of patterned objects—Jacquard fabric. These methods include GIS, wavelet-based defect detection, and BB. By evaluation of the three methods, it was found that the BB method is the best among them. In summary, the strengths of BB are as follows: The method is effective in segmenting out and outlining the shape of the defective regions for patterned fabric. It is sensitive to small defects and clean in the final thresholded image. The method is easy to use and implement. It is a one-dimensional approach for defect detection. The image alignment or distortion problems do not seem to weaken the method. Although the time needed for detection of an image was 28.656 s when implemented using MATLAB®, the time taken using a C routine was only 0.47 s. It follows that the BB method is fast enough for real-time on-loom defect detection.

References

1. Chin, R.T. and Harlow, C.A., Automated visual inspection: a survey, *IEEE Transactions on Pattern Analysis and Machine Intelligence*, Vol. PAMI-4, No. 6, 557–573, 1982.
2. Moganti, M. and Ercal, F., Automatic PCB inspection systems, *IEEE Potentials*, Vol. 14, No. 3, 6–10, 1995.
3. Smith, M.L. and Stamp, R.J., Automated inspection of textured ceramic tiles, *Computers in Industry*, Vol. 43, 73–82, 2000.
4. Hoffer, L.M., Francini, F., Tiribilli, B., and Longobardi, G., Neural network for the optical recognition of defects in cloth, *Optical Engineering*, Vol. 35, No. 11, 3183–3190, November 1996.
5. Chan, C.H. and Pang, G.K.H., Fabric defect detection by Fourier analysis, *IEEE Transactions on Industry Applications*, Vol. 36, No. 5, 1267–1276, September–October 2000.
6. Kumar, A. and Pang, G.K.H., Defect detection in textured materials using Gabor filters, *IEEE Transactions on Industry Applications*, Vol. 38, No. 2, 425–440, March–April 2002.
7. Yang, X.Z., Pang, G.K.H., and Yung, N., Discriminative training approaches to fabric defect classification based on wavelet transform, *Pattern Recognition*, Vol. 37, No. 5, 889–899, 2004.
8. Sandy, C., Norton-Wayne, L., and Harwood, R., The automated inspection of lace using machine vision, *Mechatronics Journal*, Vol. 5, No. 2–3, 215–231, 1995.
9. Baykal, I.C., Muscedere, R., and Jullien, G.A., On the use of hash functions for defect detection in textures for in-camera web inspection systems, *IEEE ISCAS*, Vol. 5, 665–668, 2002.
10. Ngan, H.Y.T., Pang, G.K.H., Yung, S.P., and Ng, M.K., Defect detection on patterned jacquard fabric, *IEEE International Workshop on 32nd Applied Imagery Pattern Recognition*, Washington D.C., pp. 163–168, 2003.
11. Ngan, H.Y.T., Pang, G.K.H., Yung, S.P., and Ng, M.K., Wavelet based methods on patterned fabric defect detection, *Pattern Recognition*, Vol. 38, No. 4, 559–576, 2005.
12. Ngan, H.Y.T. and Pang, G.K.H., Novel method for patterned fabric inspection using Bollinger bands, *Optical Engineering*, Vol. 45, 087202, 2006.
13. Bollinger, J., *Bollinger on Bollinger Bonds*, McGraw-Hill, New York, 2001.

25

Fault Detection and Diagnosis in Mechatronic Systems

K.K. Tan

S. Huang

T.H. Lee

A.S. Putra

C.S. Teo

C.W. de Silva

Summary

Computer-numerical-control (CNC) machines used in automated manufacturing are mechatronic systems. In this chapter, a fault detection method is developed based on a state observer model for a milling machine in a CNC machining center. The CNC machining center is treated as an uncertain linear system. To obtain more information, a robust observer is designed based on the uncertain linear model. Subsequently, this model is used as a state (tool wear) estimator, and fault diagnosis is carried out using two-variable information. The approach can be used for the detection of faults arising from the malfunction of a sensor or an actuator.

25.1 Introduction

Computer-numerical-control (CNC) machines are commonly used in automated factories for producing machined parts. A CNC machine, which consists of mechanical components, actuators, sensors, controllers, and interface hardware and software, is a mechatronic system. In a metal-cutting process using a CNC milling machine, it is possible that a fault occurs during operation even though the process parameters have been set properly. Faults in the cutting tools, which are frequently caused by tool wear, can potentially damage the workpiece. Prevention of faults is therefore important to minimize possible loss in manufacturing.

Before a fault occurs, parameters of the cutting process change beyond their normal values. By detecting the unusual change of the parameters, it is possible to anticipate faults and take preventive or corrective action. This possibility of preventing faults has made continuous monitoring of metal-cutting processes and detection of the changes of parameters an important topic in the area of manufacturing automation and mechatronics. Successful monitoring systems can properly maintain the machine tools and delay the occurrence of tool wear.

Various methods have been developed to detect the tool wear state. The application of statistical algorithms to associate patterns in measurable signals with wear states is found in Reference [1]. Weck [2] and Byrne et al. [3] use the signal of cutting force to monitor tool wear. By using inexpensive current sensors, several intelligent tool-wear-monitoring systems have been developed [4–6]. Coker and Shin [7] developed an in-process monitoring and control system for surface roughness during machining via ultrasonic sensing. In Reference [8] an acoustic emission sensor and accelerometers are used to monitor progressive stages of flank wear on carbide tool tips. An alternative approach to tool-wear monitoring is to apply system-theoretic ideas to estimate the wear states during the cutting process. In Reference [9] a linear model is built to detect the tool wear and breakage in the drilling process. In References [10,11] another linearization model is used to design an adaptive observer for online tool-wear estimation in a turning operation.

The aim of tool-wear detection is to find the loss of the original functioning or capability of the tool to detect an abnormal state. The methods of tool-wear diagnosis have focused on the development of signal processing techniques on the measurements such as cutting force, vibration, and spindle motor current. However, a tool signal from a single measurement may make a misjudgment because of the complicated dynamic characteristics of the cutting process and sensor noise. To prevent this, a multisensor approach has been presented in References [12,13]. This requires a higher hardware device supplied for simultaneously treating increased amounts of information.

This chapter presents a method of model-based process supervision with fault detection and diagnosis. The model is built based on the data collection from a practical manufacturing plant. Unlike the results of References [9–11], the method developed in this chapter is focused on a milling machine (see Section 25.2). Note that in general the dynamic model of a milling operation is different from drilling or turning operations, as used in References [9–11]. The method is also different from what is presented in References [12,13], in that the sufficient observer information (software) is used to make decisions so as to enhance the reliability of tool wear, as opposed to using a multisensor (hardware) technique.

The approach presented in this chapter can be summarized in the following steps. First, multiple linear models are identified based on different working conditions, and a dominant model is obtained from the models. The used model is the dominant model plus an uncertainty model with bounded signals. Second, an observer is built based on the identified model. Third, tool-wear signatures are detected by using two signal processing methods: the estimated wear rate based on the observer and the error between the observer and actual cutting force.

25.2 Model of the Metal-Cutting Process

A milling machining process is considered in this chapter. Milling is the process of cutting away material of a workpiece by feeding a material stock against a rotating tool/cutter. The workpiece to be machined may have several combinations of shape, such as flat, angular, curved, or tubular surfaces. The process of milling is executed by a milling machine—a mechatronic system—whose construction and working mechanism allow it to perform a variety of operations, including machining processes that are normally performed by specifically designed machines (e.g., drilling, turning, and shaping). This makes the milling machine among the most versatile machines in manufacturing.

The typical feed system of a milling machine consists of the following basic components: cutting tool and tool post, table, saddle, bearings, ball screw, feed box, and feed motor. Figure 25.1 illustrates the typical feed-drive system of a horizontal milling machine.

Models for the cutting process have been studied in References [14–16]. For example, Lauderbaugh and Ulsoy [14] have proposed the following model:

$$F + \varsigma \omega_n F + \omega_n^2 F = K_s f_s, \tag{25.1}$$

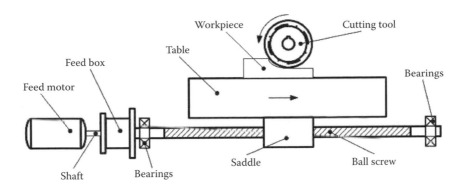

FIGURE 25.1 A feed-drive system of a horizontal milling machine.

where F is the cutting force and is the output of the system, f_s is the feed rate, and the parameters ς, ω_n, and K_s depend on the depth of cut d, spindle speed v, and feed rate f_s. This equation can be rewritten in the state–space form as

$$x = Ax + bu,$$
$$y = C^T x,$$

(25.2)

with $x = [F\ F]^T$, $u = f_s$, and

$$A = \begin{bmatrix} 0 & 1 \\ -\omega_n^2 & -\varsigma\omega_n \end{bmatrix}, \quad b = \begin{bmatrix} 0 \\ K_s \end{bmatrix}, \quad C = [1 \quad 0]^T$$

(25.3)

The model is weakly nonlinear and has significant process parameter variations [14]. Now, consider an uncertain linear model given by

$$x = (A + \Delta A)x + (b + \Delta b)u + d,$$
$$y = C^T x,$$

(25.4)

where **A** and **b** are the nominal matrices (system matrix and the input gain matrix, respectively) of the system, and

$$\Delta A = \begin{bmatrix} 0 & 1 \\ \Delta a_1(t) & -\Delta a_2(t) \end{bmatrix}, \quad \Delta b = \begin{bmatrix} 0 \\ \Delta b_1(t) \end{bmatrix},$$

(25.5)

where $\Delta a_1(t)$, $\Delta a_2(t)$, and $\Delta b_1(t)$ are the perturbation parameters of ω_n^2, $\varsigma\omega_n$, and K_s, respectively, and d is the bounded disturbance. It is seen that this model can include more classes than that of Reference [14].

In tool-wear detection, it is well known that the tool life can be divided into three phases characterized by three different wear processes: (1) break-in, (2) normal wear, and (3) abnormal or catastrophic wear. The present objective is to detect the rise in the tool wear and to diagnose the fault types so that a tool replacement decision could be made. Because fault accommodation is not addressed in this chapter, we can make the standard assumption that the control u and the state vector x remain bounded prior to and after the occurrence of a fault.

Assumption 1

There exist compact sets $\Omega_x \subset R^2$, $\Omega_u \subset R$, such that $x \in \Omega_x$ and Ω_u for all $t \geq 0$.

25.3   Fault Detection

Having obtained the model for the cutting process, the proposed method of fault detection and its constituent components are now discussed in detail according to the following steps. First, an observer model is presented for estimating the states. Second, the stability of the process is discussed based on the Lyapunov theory. Third, the fault detection strategy is proposed based on the observer.

25.3.1   Observer Model and Stability Analysis

Consider the uncertain plant (25.4). The dominant model in (25.4) can be identified in an offline or online manner. For the model represented in (25.4), the cutting force can be measured, whereas the other state is not available (one cannot use the derivative of F to represent x_2 because F has strong noise). However, the state variable x can be estimated by an observer.

An observer for the estimation of the states in (25.4) are given by

$$\dot{\hat{x}} = A\hat{x} + bu + K(y - C^T \hat{x}),$$
$$\hat{y} = C^T \hat{x},$$

(25.6)

where $\hat{x}$ denotes the estimate of the state x and $K = [k_1 \quad k_2 \quad \quad k_n]^T$ is the observer gain vector. Only the output y is assumed to be measurable.

Define the state and output estimate errors as $x = x - \hat{x}$ and $y = y - \hat{y}$, respectively. Thus, the error dynamics is given by

$$\dot{x} = (A - KC^T)x + \Delta Ax + \Delta bu + d,$$

(25.7)

$$y = C^T x.$$

(25.8)

Theorem 1

Consider the nonlinear system described by (25.4) and the observer by (25.6). If Assumption 1 holds, $x_0 \in \Omega_x$. Then, all of the signals are bounded and the state estimate $\hat{x}$ still remains in the compact set $\Omega_{\hat{x}} = \{\hat{x} \mid \|x - \hat{x}\| \leq B_Q, x \in \Omega_x\}$. In addition, a small error of $\|x\|$ may be achieved by selecting gain K.

25.3.2   Model-Based Fault Detection

For a practical cutting process, the tool wear can be formulated by

$$\omega = \omega_0 + \omega t,$$

(25.9)

where ω is the tool wear level, ω_0 is the initial tool wear level, ω is the wear rate, and t is the cutting time. In the normal phase, the wear rate is constant. However, a sudden rise in the wear rate can be observed in an abnormal phase. Our objective is to monitor the rise in the wear rate to give a warning to the operators so that they can determine whether to replace the tool or to take some other action. It is observed that the tool wear is related to the cutting force [5]. One may represent this by

$$F = F_0 + L\omega,$$

(25.10)

where F_0 is the cutting force arising under identical cutting conditions, but with an unworn cutting tool, and L is a parameter dependent on the cutting speed, feed rate, and depth of cut. Substituting (25.9) into (25.10) yields

$$F = F_0 + L\omega_0 + L\omega t. \qquad (25.11)$$

The following wear rate is derived in Reference [5]

$$\omega = \frac{\Delta F}{L\Delta t}, \qquad (25.12)$$

where ΔF and Δt are the differences of F and t, respectively. Unfortunately, the measured cutting force is noisy and this causes difficulty in the calculation of (25.12) in reality.

On the other hand, from (25.10) it follows that

$$F = L\omega. \qquad (25.13)$$

This implies that F can be used to estimate the wear rate. Although F cannot be computed due to noise, its observer $\hat{x}_2$ is available without the need of differentiation. When the observer is designed to satisfy the stability requirement, one can use $\hat{x}_2$ in place of F. In order to monitor the tool wear, a time interval is defined as $[t_0, t_f]$. This interval can be computed by

$$t_f - t_0 = \frac{l_f}{f_r / 60}, \qquad (25.14)$$

where l_f is the reference distance that is determined by the user, f_r is the feed rate (given in unit length per minutes, which is a machining parameter), and 60 is one minute in seconds. The sampling points can be calculated by $N = \frac{t_f - t_0}{T}$, where T is the sampling time. Thus, an estimate of $\hat{x}_2$ during the interval is given by

$$\bar{\hat{x}}_2 = \frac{\sum_{i=1}^{N} |\hat{x}_2(i)|}{N}. \qquad (25.15)$$

The threshold value is given by

$$F_T = C_1 \bar{\hat{x}}_2, \qquad (25.16)$$

where C_1 is a constant that is determined by experiments.

Another variable to monitor is the error between the cutting force and its estimated value:

$$e = y - C^T \hat{x}. \qquad (25.17)$$

Similarly, the threshold value is given by

$$E_T = C_2 \frac{\sum_{i=1}^{N} |e(i)|}{N}. \qquad (25.18)$$

The fault detection problem consists of checking whether the following conditions hold:

$$|e| > E_T, |\hat{x}_2| > F_T. \tag{25.19}$$

The fault detection based on multiple variables strengthens the reliability of the method.

25.4 Experimental Results

The model of the cutting process and the state observer for tool-wear detection as discussed in Section 25.2 and Section 25.3 have been implemented in an actual milling machine. The results of the implementation of the developed method are discussed now.

The machine used in the experiment was a horizontal milling machine, designed and manufactured by MAKINO. The actual cutting force was measured by a force dynamometer at a sampling rate of 2000 Hz. In order to estimate the tool wear, a camera system was mounted on the machining center.

For model identification, several cutting tests were conducted under various cutting conditions, as shown in Figure 25.2.

Utilizing the identification technique, the following models were obtained:

$$G_1(s) = \frac{1.633 \times 10^3}{s^2 + 3.9488 \times 10^3 s + 1.2247 \times 10^3} \quad \text{for test 1,} \tag{25.20}$$

$$G_2(s) = \frac{1.2034 \times 10^3}{s^2 + 1.7258 \times 10^3 s + 1.2584 \times 10^3} \quad \text{for test 2,} \tag{25.21}$$

$$G_3(s) = \frac{433.5270}{s^2 + 1.2310 \times 10^3 s + 580.9918} \quad \text{for test 3,} \tag{25.22}$$

$$G_4(s) = \frac{699.6232}{s^2 + 1.7190 \times 10^3 s + 996.5475} \quad \text{for test 4,} \tag{25.23}$$

The nominal model was then constructed as

$$A = \begin{bmatrix} 0 & 1 \\ -1.0152 \times 10^3 & -2.1562 \times 10^3 \end{bmatrix}, \tag{25.24}$$

$$|\Delta a_1(t)| \le 435, \quad |\Delta a_2(t)| \le 1793$$

$$b = \begin{bmatrix} 0 \\ 992.3876 \end{bmatrix}, \quad |\Delta b_1(t)| \le 642. \tag{25.25}$$

The observer gain K is chosen as $[100 \quad 10]^T$ so that $A - KC$ is stable (where the eigenvalues are -100.4987 and -2155.7). By choosing $Q = 2I$, the Lyapunov equation can be computed, providing an error estimation.

The cutting processes according to the aforementioned four tests were used to compare the estimated results and the actual measurement. Figure 25.3 presents the comparisons. The estimated error was found to be within 12 μm, which validates the accuracy of the proposed method. It follows that the established observer is suitable for use as a monitoring method to detect faults.

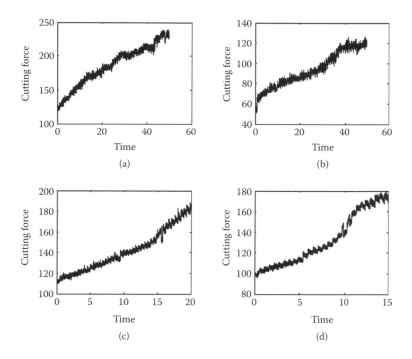

FIGURE 25.2 Cutting force obtained from CNC milling center: (a) SS = 800 rpm, f_r = 150 mm/min, depth of cut = 1 mm; (b) SS = 1000 rpm, f_r = 100 mm/min, depth of cut = 1 mm; (c) SS = 1000 rpm, f_r = 200 mm/min, depth of cut = 1 mm; (d) SS = 1200 rpm, and f_r = 200 mm/min, depth of cut = 1 mm.

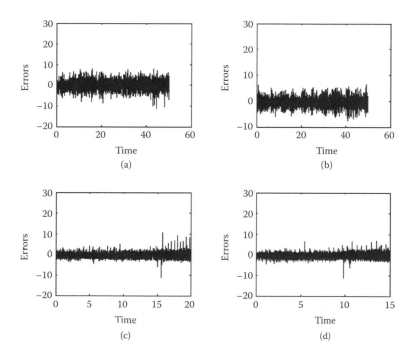

FIGURE 25.3 Comparison of actual and estimated cutting forces: (a) SS = 800 rpm, f_r = 150 mm/min, depth of cut = 1 mm; (b) SS = 1000 rpm, f_r = 100 mm/min, depth of cut = 1 mm; (c) SS = 1000 rpm, f_r = 200 mm/min, depth of cut = 1 mm; (d) SS = 1200 rpm, and f_r = 200 mm/min, depth of cut = 1 mm.

25.5   Conclusion

CNC machines used in automated manufacturing are mechatronic systems. In this chapter, a fault detection method was developed based on a state-observer model for a milling machine in a CNC machining center. Specifically, a state-observer model of the cutting force was used to detect tool wear in milling operations, with satisfactory results. In this method, only the cutting force was used for monitoring the automated machining. The inexpensive technique based on the observer model was applied to a CNC milling center. Experimental results showed that the proposed method would provide robust performance and could be easily used to monitor tool wear.

Acknowledgment

The authors thank Dr. Wang Wenhui for his helpful suggestions and the effort in data collection.

References

1. Rangwala, S., Liang, S., and Dornfeld, D., Pattern recognition of acoustic emission signals during punch stretching, *Mechanical Systems and Signal Processing*, Vol. 1, 321–332, 1987.
2. Weck, M., Machine diagnostics in automated production, *Journal of Manufacturing Systems*, Vol. 2, 101–106, 1983.
3. Byrne, G., Dornfled, D., Inasaki, I., Ketteler, G., Konig, W., and Teti, R., Tool condition monitoring (TCM)—the statue of research and industrial application, *Annals of the CIRP*, Vol. 44, 541–567, 1995.
4. Altintas, Y., Prediction of cutting forces and tool breakage in milling from feed drive current measurements, *ASME Journal of Engineering for Industry*, Vol. 114, 386–392, 1992.
5. Li, X., Djordjevich, A., and Venuvinod, P.K., Current-sensor-based feed cutting force intelligent estimation and tool wear condition monitoring, *IEEE Transactions on Industrial Electronics*, Vol. 47, 697–702, 2000.
6. Li, X., Li, H.X., Guan, X.P., and Du, R., Fuzzy estimation of feed-cutting force from current measurement—a case study on intelligent tool wear condition monitoring, *IEEE Transactions on Systems, Man, and Cybernetics*, Part C, Vol. 34, 506–512, 2004.
7. Coker, S.A. and Shin, Y.C., In-process control of surface roughness with tool wear via ultrasonic sensing, *Proceedings of American Control Conference, Seattle*, WA, 1995, pp. 1717–1721.
8. Prateepasen, A., Au, Y.H.J., and Jones, B.E., Acoustic emission and vibration for tool wear monitoring in single-point machining using belief network, *Proceedings of IEEE Instrumentation and Measurement Technology Conference*, Budapest, Hungary, 2001, pp. 1541–1546.
9. Isermann, R., Ayoubi, M., Konrad, H., and Rei, T., Model based detection of tool wear and breakage for machine tools, *International Conference on Systems, Man and Cybernetics*, Le Touquet, France, Vol. 3, 1993, pp. 72–77.
10. Danai, K. and Ulsoy, A.G., An adaptive observer for on-line tool wear estimation in turning, Part I: Theory, *Mechanical Systems and Signal Processing*, Vol. 1, 211–225, 1987.
11. Danai, K. and Ulsoy, A.G., An adaptive observer for on-line tool wear estimation in turning, Part II: Results, *Mechanical Systems and Signal Processing*, Vol. 1, 227–240, 1987.
12. Luo, R.C. and Kay, M.G., Multi-sensor integration and fusion in intelligent systems, *IEEE Transactions on Systems Man and Cybernetics*, Vol. 19, 901–931, 1989.
13. Noori-Khajavi, A. and Komanduri, R., On multisensor approach to drill wear monitoring, *Annals of the CIRP*, Vol. 42, 71–74, 1993.
14. Lauderbaugh, L.K. and Ulsoy, A.G., Model reference adaptive force control in milling, *ASME Journal of Engineering for Industry*, Vol. 3, 13–21, 1989.
15. Park, J. and Ulsoy, A., On-line tool wear estimation using force measurement and a nonlinear observer, *ASME Journal of Dynamic Systems, Measurement, and Control*, Vol. 14, 666–672, 1992.
16. Kim, T.Y. and Kim, J., Adaptive cutting force control for a machining center by using indirect cutting force measurements, *International Journal of Machine Tools and Manufacture*, Vol. 36, 925–937, 1996.

26

Sensor Fusion for Online Tool Condition Monitoring

W.H. Wang

Y.S. Wong

G.S. Hong

Summary

Sensors are indispensable in the condition monitoring of mechatronic systems. A milling machine is a mechatronic system. This chapter presents an intelligent integrated approach combining a direct sensor (vision) and an indirect sensor (force) for online monitoring of flank wear and breakage of the cutting tool in milling. For flank wear, images of the tool are captured and processed in-cycle using successive moving image analysis. Two features of the cutting force are extracted in-process and appropriately preprocessed. A self-organizing map (SOM) network is trained in a batch mode after each pass, using the two features derived, and measured wear values obtained by interpolating the vision-based measurement. The trained SOM network is applied to the succeeding machining pass to estimate the flank wear in-process. The in-cycle and in-process procedures are employed alternatively for the online monitoring of the flank wear. To detect breakage, two other features in the time domain, as derived from cutting force, are used. Vision is used to verify whether this breakage has actually occurred. Experimental results show that this sensor fusion scheme is independent of cutting conditions, and is feasible and effective for the implementation of online tool condition monitoring (TCM) in milling.

26.1 Introduction

Automated manufacturing processes are good examples of mechatronic systems. In manufacturing, it is desirable to reduce labor cost, minimize operator errors, and enhance the productivity and quality of products [1]. Online tool condition monitoring (TCM) is an important technique [2] that is used in achieving this objective. It helps operate the machine tool at its maximum efficiency by detecting and measuring tool conditions such as flank/crater wear, chipping, breakage, and so on. A successful TCM system can increase productivity, and hence competitiveness, by maximizing the tool life, minimizing the machine downtime, reducing scrap, and preventing workpiece damage [3].

Much of TCM research has been dedicated to the monitoring of tool conditions online. However, most available TCM techniques are intended for single-point cutting processes such as turning. Their results may not be directly suitable for multitooth milling processes [2]. Although milling is an important machining process in manufacturing, much less effort has been made to monitor it [4]. The systems developed for milling need to be more reliable, robust, and responsive for use in truly automated manufacturing [5]. Obviously, there is still much to understand and do before online TCM systems in milling can be used in industry.

For decades, researchers have proposed various methods based on sensors to monitor tool conditions in milling online. In the early years, only a single sensor was used, but it was found to be inadequate. A single sensor is less likely to provide reliable, robust, and accurate condition monitoring in a tough industrial environment. To overcome the deficiency of the single-sensor method, researchers have looked into using sensor fusion for TCM by combining two or more sensors together, especially using a vision and other indirect sensors [6–9]. In this chapter, a vision-based sensor and a force sensor are integrated to implement an online TCM system that can monitor progressive flank wear and detect breakage in milling.

26.2   Overview of the Method

The overall scheme is shown in Figure 26.1. For the flank wear estimation, accurate vision measurement is obtained by processing the tool images captured in-cycle. Two features of the cutting force extracted online and the flank wear increment in the current machining pass are used to train an unsupervised

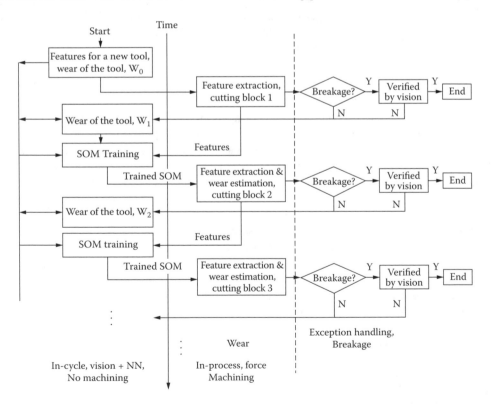

FIGURE 26.1 Scheme of the sensor fusion method. (From Wang, W.H., Wong, Y.S., Hong, G.S., and Zhu, K.P., *International Journal of Production Res.*, 2007, in press. With permission.)

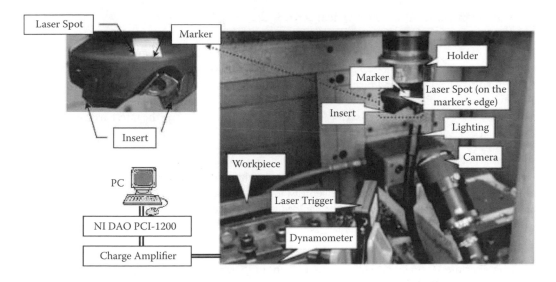

FIGURE 26.2 Experimental setup. (From Wang, W.H., Wong, Y.S., Hong, G.S., and Zhu, K.P., *International Journal of Production Res.*, 2007, in press. With permission.)

Kohonenís SOM neural network, which in turn serves to estimate the wear increment in the next machining pass. The two individual sensors play complementary roles.

To handle exceptions, i.e., breakage events, two force features extracted in-process are first checked with preset thresholds. If the thresholds are reached, the image of the tool is then captured and processed to verify if a breakage has actually occurred. Alternatively, breakage can be detected solely by vision in-cycle.

26.3 Experimental Setup and Vision Subsystem

The experimental setup is shown in Figure 26.2. The system consists of two sensor subsystems, specifically, vision and force.

26.3.1 Flank Wear Measurement Based on Individual Still Images [11]

The terms are defined and shown in Figure 26.3.

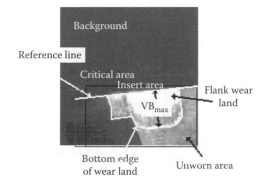

FIGURE 26.3 Terms in image processing. (From Wang, W.H., Hong, G.S., and Wong, Y.S., *International Journal of Machine Tools Manufacturing*, Vol. 46, No. 2, 199–207, 2006. With permission.)

26.3.1.1 Identification of the Critical Area

The identification of the critical area consists of four parts: preprocessing, histogram stretch, thresholding, and segmentation. In preprocessing, a median filter [12] is employed to remove noise. In histogram stretch, the original gray level interval is mapped into a fixed interval of [0, 255] with a linear transform [12]. In thresholding, the background pixels are assigned a value of 0, other pixels 1, with a suitable threshold *T*. In the extraction of the critical area, a segmentation method of "line segment encoding" [13] is used to pick out the insert area. Then, the critical area is obtained by magnifying the insert area appropriately.

26.3.1.2 Identification of the Reference Line

This consists of the sequential processing of edge detection, edge enhancement, thresholding, edge line extraction, and morphology operation. For the edge detection operation, the Sobel operator is applied. A local window function, called adaptive contrast enhancement at edges [14], is employed to improve the edge contrast. The resultant image is thresholded with an optimal threshold based on the Otsu method [15]. The reference line can be obtained using the Hough transform [12–14].

26.3.1.3 Flank Wear Measurement

The flank wear measurement is based on the resultant histogram stretch image (gray-level image) $S(x, y)$ and the binary image after morphology $M(x, y)$. The processing is confined to an enclosed area centered around the worn region of the insert area, called the *critical area*, which is the region of interest (ROI).

As illustrated in Figure 26.4, the measurement procedure, referred to as *orthogonal scanning*, involves repetitive scanning along lines perpendicular to the reference line. This determines points on the bottom edge of wear land and further, provides the wear value along each line and the entire wear land.

A *scan line* is the perpendicular line that starts from a point on the reference line and ends at a point on the boundary of the critical area. For the i-th scan line $L(i)$, the start point is denoted as $P_A(i)$, and the end point as $P_B(i)$, with respect to the origin at the upper-left corner of the critical area.

In wear detection, first the edge point, denoted as $P_E(i)$, is determined in the i-th scan line, using a windowing technique. The detected $P_E(i)$ is used as a reference point to more precisely detect the point on the bottom edge of wear land, denoted as $P_{RB}(i)$. The distance between $P_{RB}(i)$ and $P_A(i)$ gives a single scan line of the flank wear region, and is denoted by $W(i)$.

26.3.1.3.1 Determination of $P_E(i)$

The binary edge image data $M(x, y)$ is used to perform this task. The procedure involves scanning from $P_B(i)$ to $P_A(i)$, by which the first white ("1") pixel may be basically regarded as $P_E(i)$. To make the method more robust to noise, a windowing technique is applied, which improves the performance.

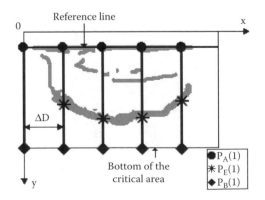

FIGURE 26.4 Orthogonal scanning. (From Wang, W.H., Hong, G.S., and Wong, Y.S., *International Journal of Machine Tools Manufacturing*, Vol. 46, No. 2, 199–207, 2006. With permission.)

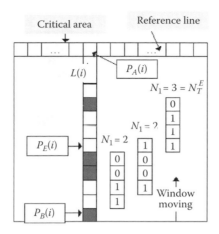

FIGURE 26.5 Moving window. (From Wang, W.H., Hong, G.S., and Wong, Y.S., *International Journal of Machine Tools Manufacturing*, Vol. 46, No. 2, 199–207, 2006. With permission.)

After finding a candidate edge point whose binary value is 1, a given number of adjacent pixels (a w_1-pixel length window) is considered. If the number of "1" pixels (N_1) in the window exceeds a preset threshold N_T^E, this candidate edge point is taken as $P_E(i)$; otherwise, another candidate edge point is found, and the previous operation is repeated. This process thus uses a moving window, as illustrated in Figure 26.5, where the window length is taken as 4 pixels ($w_1 = 4$) and $N_T^E = 3$.

26.3.1.3.2 Determination of $P_{RB}(i)$ and the Flank Wear Area

The difference in the gray level between the wear land and the unworn area of the scan line $L(i)$ serves as a criterion in determining $P_{RB}(i)$. The average gray level in the wear land, $\bar{G}_w(i)$, and the average gray level in the unworn area, $\bar{G}_{\bar{w}}(i)$, can be easily computed because $P_A(i)$, $P_E(i)$, and $P_B(i)$ are available for scan line $L(i)$.

26.3.1.3.3 Determination of $P_{RB}(i)$

The average gray-level difference, $\Delta\bar{G}(i)$, between $\bar{G}_w(i)$ and $\bar{G}_{\bar{w}}(i)$ is given by:

$$\Delta\bar{G}(i) = \bar{G}_w(i) - \bar{G}_{\bar{w}}(i) \tag{26.1}$$

A scan line $L(i)$ with wear (that is, $P_{RB}(i)$ on this line) has to meet the following condition:

$$\Delta\bar{G}(i) \geq G_{w-\bar{w}} \tag{26.2}$$

where $G_{w-\bar{w}}$ represents the minimum gray-level difference between the wear land and unworn area. This can be obtained experimentally.

For scan line $L(i)$ satisfying Equation (26.2), we assume that wear occurs in the vicinity of Q pixels of $P_E(i)$. To locate the wear bottom point, a threshold independent method [16], based on moment invariance, is used.

A scan line across a step edge in the absence of noise is characterized by a set xi ($i = 0, 1, 2, \ldots, n = 1$), which are either monotonically nondecreasing or nonincreasing. An ideal edge is a sequence of one brightness value h_1, followed by a sequence of another brightness value h_2, as shown in Figure 26.6a, where k denotes the edge location and n is the number of input data.

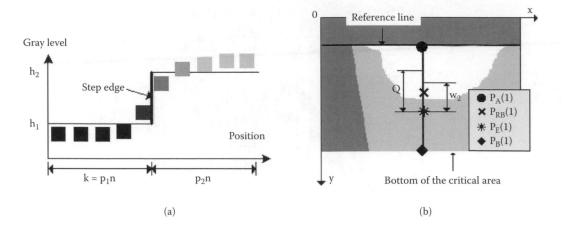

FIGURE 26.6 (a) Moment-invariance-based edge detection; (b) Searching for wear land bottom edge. (From Wang, W.H., Hong, G.S., and Wong, Y.S., *International Journal of Machine Tools Manufacturing*, Vol. 46, No. 2, 199–207, 2006. With permission.)

The first three moments of the input data can be calculated by:

$$\bar{m}_i = \frac{1}{n} \sum_{j=0}^{n-1} (x_j)^i, \quad i = 1, 2, 3 \tag{26.3}$$

The solutions of the edge are given by:

$$h_1 = \bar{m}_1 - \sigma \sqrt{\frac{p_2}{p_1}} \tag{26.4}$$

$$h_2 = \bar{m}_1 + \sigma \sqrt{\frac{p_1}{p_2}} \tag{26.5}$$

$$p_1 = \frac{1}{2} \left[1 + s \sqrt{\frac{1}{4 + s^2}} \right] \tag{26.6}$$

where

$$s = \frac{\bar{m}_3 + 2\bar{m}_1^3 - 3\bar{m}_1\bar{m}_2}{\sigma^3}, \quad \sigma = \sqrt{\bar{m}_2 - \bar{m}_1^2}, \, p_1 + p_2 = 1$$

Thus, the edge location is given with a subpixel precision by

$$k = p_1 n \tag{26.7}$$

The preceding edge detection method is repeated on gray-level data with increasing length from w_2 to Q, with the input data starting each time from $P_E(i)$ with length increased by "1" pixel. For clarity, postfix form $h_1(n)$ is defined as h_1 calculated with data length n, and the other parameters follow the same rule.

The distance K from $P_E(i)$ to $P_{RB}(i)$ is then given by

$$K = k(j) = p_1(j)j \tag{26.8}$$

where $j = \arg \max\{h_2(n) - h_1(n)\}$, $n = w_2, w_2+1, w_2+2, \ldots, Q$.

Correspondingly, $P_{RB}(i)$ can be determined by offsetting $P_E(i)$ with the value of K. The illustration of $P_{RB}(i)$ search procedure is shown in Figure 26.6b.

The wear land on line $L(i)$ can be calculated, i.e., the single scan line of the flank wear region, as $W(i)$ given by:

$$W(i) = |P_{RB}(i) - P_A(i)| \tag{26.9}$$

26.3.2 Flank Wear Measurement Based on Successive Moving Images

Based on the analysis of successive moving images, instead of individual images, a flank wear measurement system has been developed [17], which is more suitable for unmanned, automated manufacturing.

From a study of the successive moving images, it was observed that there was close correlation between these images. First, the reference line remains unchanged in the successive images. The same is true for the critical area. This indicates that these two features can be extracted from the image of the fresh insert and directly be used for the subsequent images, instead of calculating them again. In this case, the right border of the critical area is not necessarily fixed. The wear value in the previous image is used to expand the critical area dynamically, as shown in Figure 26.7d.

For a new image, the width of the critical area can be calculated by

$$w = (d_0 + V + M)/\cos\alpha_0 \tag{26.10}$$

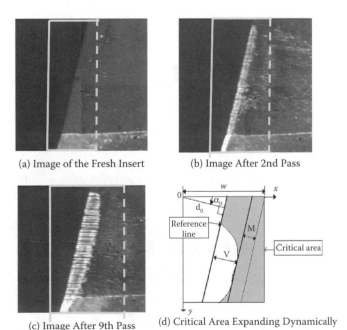

(a) Image of the Fresh Insert   (b) Image After 2nd Pass

(c) Image After 9th Pass   (d) Critical Area Expanding Dynamically

FIGURE 26.7 Critical area and reference line reusable for successive images. (From Wang, W.H., Wong, Y.S., and Hong, G.S., *Computers in Industry*, Vol. 56(8–9), 816–830, 2005. With permission.)

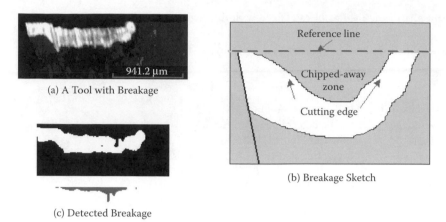

(a) A Tool with Breakage

(b) Breakage Sketch

(c) Detected Breakage

FIGURE 26.8 Breakage detection by vision.

where, w is the width of the critical area; d_0 is the distance from the origin to the reference line; α_0 is the angle from the x-axis to the normal to the reference line; V is the maximum width of the wear land of previous image; and M is the margin set for the critical area.

Obviously, V is a time-varying parameter. In particular, $V = 0$ when the insert is fresh. After each pass, V is increased by some marginal amount M as wear progresses. Hence, a dynamic critical area can be used. A dynamic critical area with different width as wear progresses is shown in Figures 26.7a, 26.7b, and 26.7c.

The value of V also serves *a priori* knowledge regarding the wear search scope for the current image. Therefore, this method can provide accurate, robust, and consistent wear measurements.

26.3.3 Breakage Detection

Breakage occurs as a substantial amount or chunk of material is chipped away from the cutting edge. Figure 26.8a shows a sample image of a broken tool. It can be observed that the chipped-away zone actually has the same gray level as the background. Therefore, one may just use the binary image to detect this chipped-away zone, as shown in Figure 26.8b.

The processing is confined to the critical area, where the image can be binarized using the optimal threshold calculated by the Otsu method [15]. The small regions (isolated small blocks or noise) of the resultant binary image are removed first, so that there is no noise within the area circumvented by the reference line and the cutting edge. After this, the chipped-away zone can be easily extracted because the location of the pixels on the cutting edge can be determined in the binary image (scanned from the pixel on the reference line downward to the first white-point pixel on the cutting edge). Figure 26.8c shows the binary image (upper) and the extracted chipped-away zone (lower).

26.4 SOM-Based In-Process Wear Estimation

In this section in-process wear estimation of a cutting tool in a milling process with a self-organizing map (SOM) is presented.

26.4.1 Self-Organizing Map

The basic SOM defines a mapping from the input data space $\Re_n$ onto an array of one- or two-dimensional units [18]. A weighted vector $mi \in \Re_n$ is associated with each unit i. A metric function is used to compare an input vector $x \in \Re_n$ with each mi. The closest match is defined as the response, and hence the input is mapped onto this location. A common metric function used is the Euclidean norm $\|x - m_i\|_2$, and the winning neuron c is defined as

$$\|x - m_c\|_2 = \min_i\{\|x - m_i\|_2\}, \quad \text{or} \quad c = \arg\min_i\{\|x - m_i\|_2\} \tag{26.11}$$

For the purpose of topological ordering, during learning, the weighted vector of those units that are topographically closer to the winner c will also be updated from the same input. The learning rule is given by

$$m_i(t + 1) = m_i(t) + h_{ci}(t)[x(t) - m_i(t)] \tag{26.12}$$

where t is the discrete-time coordinate and $h_{ci}(t)$ is a neighborhood kernel defined over the lattice points.

26.4.1.1 Batch Training Algorithm

This algorithm proceeds as follows:

1. Calculate the sum of the vectors in each Voronoi set:

$$s_i(t) = \sum_{j=1}^{N_{Vi}} x_j \tag{26.13}$$

where N_{Vi} is the number of input samples in the Voronoi set of unit i.

2. Update the weight vectors:

$$m_i(t+1) = \frac{\sum_{j=1}^{u} h_{ij}(t)s_j(t)}{\sum_{j=1}^{u} N_{Vj}h_{ij}(t)} \tag{26.14}$$

where u is the number of map units.

26.4.2 SOM as Estimator

A two-phase learning strategy with asymmetric mapping [18] is employed here. We refer to as *labeled data* the input data consisting of force features and corresponding flank wear output (wear increment value); otherwise, the data consisting of only force features are referred to as *unlabeled*. The unlabeled data set is utilized first for a preliminary unsupervised learning, which readily approximates the density function of input data. Then, the labeled data set is used for refinement of the density function.

26.4.2.1 Phase One

This involves the presentation of all the available unlabeled input data to a normal SOM, where the input vectors are of the form: $x = [in \; \phi]^T$, with $in = [f_1, f_2, ..., f_m]^T$. This is a feature vector at one instant of time, and its corresponding output vector is denoted by $out - [o_1, o_2, ..., on]^T$ (in this case, $n = 1$, which is the flank wear increment). The symbol ϕ denotes a "don't care" condition, i.e., when searching for the winner, only the *in* part of x is compared with the corresponding components of the weight vectors, and no output (*out*) part is presented in the learning algorithm.

26.4.2.2 Phase Two

After the convergence of phase one, the labeled data are applied to the SOM, and training is continued. During this second phase, the winner is determined only based on the *in* part of x, while the weight vectors corresponding to the *in* part of x are no longer changed. Instead, the SOM learning algorithm is now applied to change the weight vectors of the out part only.

Thus, during winner search, $x = [in\ \phi]^T$, and during learning, $x = [\phi\ out]^T$. Note that the *in* and *out* in phase two are labeled, whereas in phase one they are unlabeled.

26.4.3 Estimation by SOM

26.4.3.1 Feature Extraction

The average force and the standard deviation over one rotation have been found to have a strong correlation with flank wear [19,20].

First, the two features are filtered by a median algorithm. For scaling, the filtered force in the *k*-th pass is used as a reference. The force at the beginning (force at 1st rotation) of the *k*-th pass is denoted by $L(0)$. Then, any force at *n*-th rotation in this pass $L(n)$ is scaled by

$$F'(n) = (L(n) - L(0))/L(0); \qquad (26.15)$$

The scaled value is then accumulated to form a stable feature by

$$F(n) = \sum_{i=0}^{n} F'(i) \qquad (26.16)$$

26.4.3.2 Working with SOM

A one-dimensional SOM is constructed with 15 units. In training, the majority of the data in the *k*-th pass are unlabeled (only the two force features) and used in phase one. A small portion of the data are labeled (the two force features and the corresponding wear increment value) and used in phase two. The trained SOM is then applied on each data set in the $(k + 1)$-th pass to estimate flank wear increment at different rotations.

The output of the SOM is flank wear increment with respect to the wear value at the beginning of the $(k + 1)$-th pass. This wear value has been made available with the successive moving image analysis. Thus, the absolute wear obtained each time is the summation of the wear value plus the wear increment value estimated by SOM.

26.5   Breakage Detection by Force

It has been shown that the residual error (denoted by E_r) of the average force is able to detect the breakage [21]. A residual error exceeding a preset threshold (i.e., $E_r > T_r$, where T_r is the threshold) indicates breakage. Moreover, peak rate K_{pr} of cutting forces [22] is taken into account as well. A value exceeding a preset threshold (i.e., $K_{pr} > T_{pr}$, where T_{pr} is the threshold) indicates tool breakage.

It is very difficult to specify T_r and T_{pr}; in particular, many trial tests may be required. In this chapter, because we monitor conditions pass by pass, we can have dynamic T_r and T_{pr} without any trial tests.

In the first pass, we assume that there is no breakage, such that the maximum amplitudes of residual error and peak rate can be obtained. These two values are taken as the thresholds for the beginning of the second pass. From the second pass onward, as long as no breakage has been detected by vision after the *k*-th pass, the maximum amplitudes of the two features in the *k*-th pass can be taken as the thresholds for the beginning of the $(k + 1)$-th pass. Within each pass, if breakage is detected by force features but not verified by vision, i.e., if breakage does not really occur, the thresholds are updated by increasing them a little, say, 5%.

26.6   Experimental Results and Discussion

Before machining commenced, images of the fresh inserts were captured with the spindle rotating at a speed of 20 rpm. The current position of the tool was stored in the CNC machine controller as the image capture point. Images were processed to first get the reference line and critical area. When machining, the traverse force was low-pass-filtered (1 kHz low-pass filter) and sampled at 2 kHz via the data

TABLE 26.1 Parameters in Dry Milling for Online TCM

Parameters	Test 1	Test 2	Test 3	Test 4	Test 5	Test 6
Inserts (ISO SDKN42MT)	AC325, coated			A30N,		
# of inserts	2			uncoated		
Workpiece	ASSAB718HH steel					
Length of the workpiece [mm]	205					
Diameter of cutter [mm]	50					
Spindle speed [rpm]	600	1000	1200	1000	800	1200
[m/min]	94.2	157	188.4	157	125.6	188.4
Feed rate [mm/min]	200	200	150	200	300	150
Feed per tooth [mm/tooth]	0.17	0.1	0.063	0.1	0.188	0.063
Depth of cut [mm]	1					
Time/pass [s]	54	54	72	54	36	72
Immersion rate	Full					
# of machining passes	104	27	30	27	20	13
Conditions observed	Gradual wear	Breakage				

Source: From Wang, W.H., Wong, Y.S., Hong, G.S., and Zhu, K.P., *International Journal of Production Res.*, 2007, in press. With permission.

acquisition card PCI-1200. Four features of the cutting force were extracted in-cycle; namely, two features for breakage detection (residual error and peak rate) and two features based on average force and standard deviation. The first two features were checked immediately after extraction with a preset threshold to detect breakage. As the experiment aimed to look at the force trend after breakage, machining was continued with the detected broken tool insert till the end of that ongoing pass.

After one machining pass, the tool was programmed to park at the image capture point, and images of the inserts were captured while the spindle was rotating at a speed of 20 rpm. The wear value was then obtained by the successive image analysis. The chipped-away material along the cutting edge was also quantified to verify any breakage. If breakage was detected, machining was terminated. Otherwise, the SOM was trained in-cycle by using the wear increment value (measured in-cycle) and the extracted features (extracted in-process). After the training, a new machining pass was started. During this machining pass, the SOM used the two features extracted in-process to estimate the wear increment and, hence, the flank wear in the current pass. The experiment was repeated in this manner until the insert was seriously broken or worn out. The parameters for the tests are shown in Table 26.1. The experimental results for the six tests are shown in Figures 26.9 to 26.14 [10].

No breakage was observed or detected in Test 1. So, in Figure 26.9, only the wear estimation result is shown. For the other tests, breakage was detected in some pass, and the image of the insert was inspected by vision. Therefore, each of Figures 26.9 to 26.14: (1) shows the wear estimation result throughout the entire test (until the end of the pass listed in Table 26.1); (2) shows the breakage result inspected by vision, with gray-level image of the insert in left part, binary image in right-upper part, and breakage in right-bottom part; (3) shows the force and force features, i.e., residual error and peak rate against rotations, when breakage was detected in some pass by these two features (except Figure 26.10 for Test 2, in which no breakage was detected).

The following key findings may be observed:

1. The flank wear estimation result is good in association with the in-cycle measurement by vision, especially at the linear wear stage, although there is great deviation at the initial wear stage.
2. Breakage is successfully detected by vision.
3. Breakage can be detected by force features using dynamic thresholding. The residual error and peak rate are the features used in combination to detect breakage. Breakage is assumed to occur when detected by either feature, or simultaneously by both features.

Before the first machining pass, we assume no *a priori* knowledge of the data to train the SOM network, and set the wear estimation values for this pass to be zero. At the initial wear stage, the wear grows rapidly,

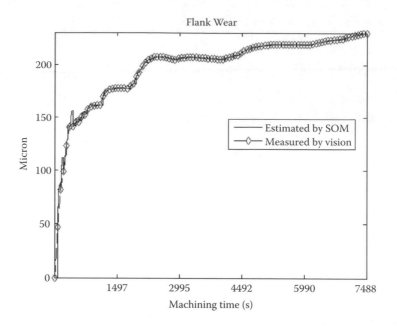

FIGURE 26.9   Online TCM result for Test 1. (Adapted from Wang, W.H., Wong, Y.S., Hong, G.S., and Zhu, K.P., *International Journal of Production Res.*, 2007, in press.)

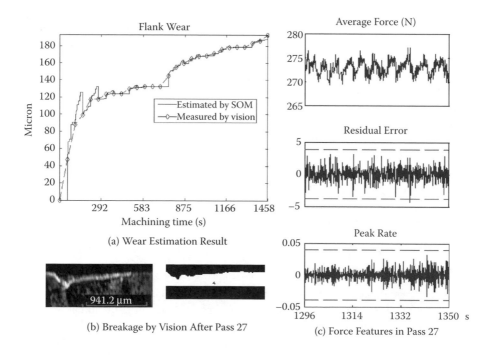

FIGURE 26.10   Online TCM result for Test 2. (Adapted from Wang, W.H., Wong, Y.S., Hong, G.S., and Zhu, K.P., *International Journal of Production Res.*, 2007, in press.)

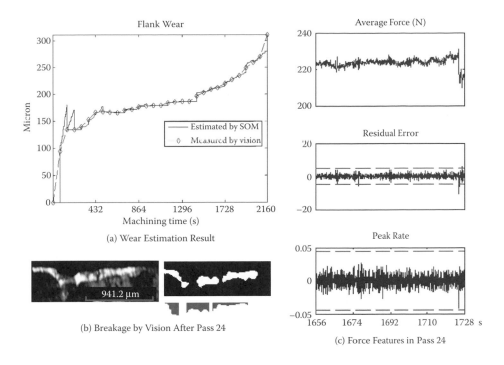

FIGURE 26.11 Online TCM result for Test 3. (Adapted from Wang, W.H, Wong, Y.S., Hong, G.S., and Zhu, K.P., *International Journal of Production Res.*, 2007, in press.)

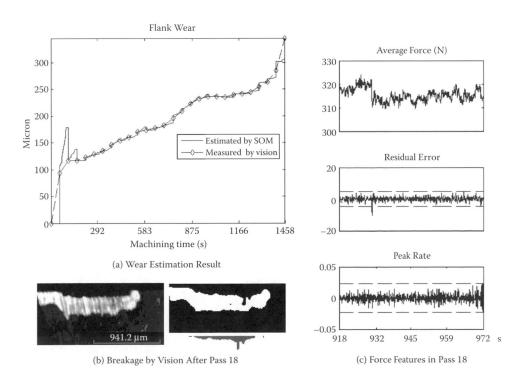

FIGURE 26.12 Online TCM result for Test 4. (Adapted from Wang, W.H., Wong, Y.S., Hong, G.S., and Zhu, K.P., *International Journal of Production Res.*, 2007, in press.)

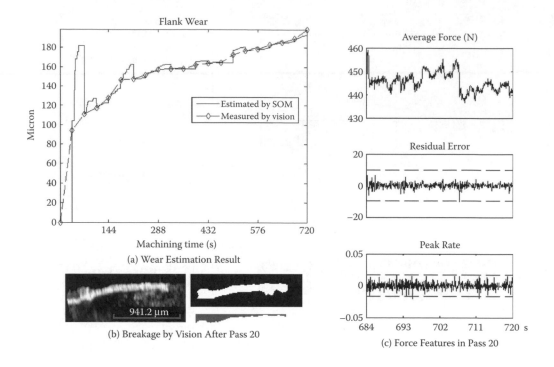

FIGURE 26.13 Online TCM result for Test 5. (Adapted from Wang, W.H., Wong, Y.S., Hong, G.S., and Zhu, K.P., *International Journal of Production Res.*, 2007, in press.)

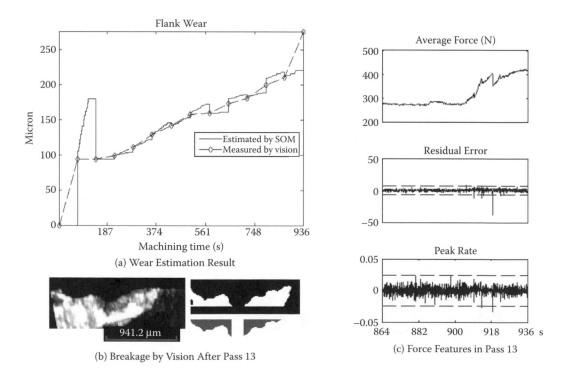

FIGURE 26.14 Online TCM result for Test 6. (Adapted from Wang, W.H., Wong, Y.S., Hong, G.S., and Zhu, K.P., *International Journal of Production Res.*, 2007, in press.)

and then the wear increases at a much slower rate at the linear wear stage. During this wear rate transition period, the SOM gives an estimate with greater deviation from that by the vision measurement. The reason for this deviation is that the SOM applied to a machining pass with a slow wear increment is trained with data from the previous machining pass with a more rapid wear rate. However, this deviation is tolerable because the tool is hardly worn out in this short period of time at the initial wear stage.

At the linear wear stage, the SOM tracks the wear value well. However, when the wear rate changes between two successive passes, i.e., wear in the current pass is slower or faster than the previous pass, there is some resultant deviation. This deviation is small, about 15 μm, which is tolerable, considering that change of a worn tool is usually recommended at the linear wear stage [23].

Too sensitive thresholds will lead to many cases of false breakage detection, i.e., many cycles of vision verification are required and thus much interruption of machining process. However, the proposed dynamic thresholding, determined by using the correlation between two successive passes, is not overly sensitive. In tests conducted, it has been observed that only after several points are the thresholds exceeded. Accordingly, the thresholds are reasonable.

Generally, the reliability of the system will be improved by virtue of the sensor integration with direct and indirect sensors. During machining, the tool is in close contact with the workpiece and the direct (vision) subsystem is totally blind to the region of interest. The only reliable signal source comes from the indirect sensor (force sensor). But the indirect signal alone is likely to give an estimation with a larger deviation in the long term, for example, after some passes. However, with the direct (vision) subsystem, the indirect signal can be calibrated well before its deviation becomes too large to accommodate. Although either a single direct or indirect sensor imposes a limitation on the reliability of the online system, the integration of SOM and vision, as presented in this chapter, makes use of the full benefit of the reliability of each subsystem. This works well in a local sense and thus achieves good overall reliability for the system in a global sense.

A comparison of the proposed sensor integration method and the methods mentioned in Section 26.1 is given in Table 26.2.

TABLE 26.2 Comparison of TCM Systems Using Indirect Sensors Plus Vision Scheme

Methods	Process/Condition	Indirect Signal and Processing	Image Processing	Fusion Method
1 [6,7]	Turning/wear	Force, an adaptive observer on wear–force relationship	Segmentation by thresholding	Vision system provides parameters for the observer and calibrates the observer
2 [8]	Turning/wear and breakage	Vibration, moving average of the root mean square (MARMS)	Segmentation by SOM	Vision system measures wear between cuts, vibration system gives breakage or other unforeseen situation. Wear value is not given online
3 [9]	Turning/wear classification	Sound, energy of two wavelet coefficients	Column projection and run length	Both features from sound and vision are fed to an RBF neural network to classify the degree of tool wear
Proposed	Milling/wear and breakage	Force, processed average force and standard deviation for wear estimation, residual error and peak rate for breakage	Moment-based edge detection, threshold-independent, and spindle rotates	Vision system provides wear value for the SOM network to estimate wear online with force features. SOM is updated with the wear value repeatedly. Breakage is detected via force features and verified by vision

26.7   Conclusion

Sensors are indispensable in the condition monitoring of mechatronic systems. A milling machine is a mechatronic system. In this chapter an integrated method of online flank wear monitoring was presented, which used both vision and force measurements, and an SOM estimator. The features of this method are:

- Sensor-fusion-integrating direct and indirect sensors are used.
- Spindle rotates during in-cycle vision measurement.
- In-cycle vision measurement based on successive image analysis is sufficiently accurate, robust, and fast, providing necessary reliable knowledge for the SOM estimator.
- Breakage can be successfully detected by vision.
- Breakage can be detected by force features in the time domain and verified by vision.
- The SOM network is used to estimate flank wear in-process, with the following advantages:
  1. Only two features of the cutting force are required and thus implementation is easy and fast, which is especially desirable in online implementations.
  2. Unsupervised batch training is used; hence, easier data collection and less computation time.
  3. The estimation is cutting-condition-independent. The SOM works in a repetitively updating mode: trained locally and applied immediately. It is thus applicable for various cutting conditions, as there is no need to train by collecting large amounts of force samples under different cutting conditions, as traditional neural networks do.

To conclude, this combination of vision and force overcomes the disadvantages inherent in single-sensor tool-monitoring procedures and can be implemented online easily and effectively.

References

1. Huang, P.T., Chen, J.C., and Chou, C.Y., A statistical approach in detecting tool breakage in end milling operations, *Journal of Industrial Technology*, Vol. 15, No. 3, 1–7, 1999.
2. Lin, S.C. and Lin, R.J., Tool wear monitoring in face milling using force signals, *Wear*, Vol. 198, 136–142, 1996.
3. Donnell, G.O., Young, P., Kelly, K., and Byrne, G., Towards the improvement of tool condition monitoring systems in the manufacturing environment, *Journal of Materials Processing Technology*, Vol. 119, 133–139, 2001.
4. Byrne, G., Dornfeld, D., Inasaki, I., Ketteler, G., Konig, W., and Teti, R., Tool condition monitoring (TCM)—the status of research and industrial application, *Annals of the CIRP*, Vol. 44, No. 2, 541–567, 1995.
5. Prickett, P.W. and Johns, C., An overview of approaches to end milling tool monitoring, *International Journal of Machine Tools and Manufacture*, Vol. 39, 105–122, 1999.
6. Park, J.J. and Ulsoy, A.G., On-line flank wear estimation using an adaptive observer and computer vision, part 1: theory, *ASME Journal of Engineering for Industry*, Vol. 115, 30–36, 1993.
7. Park, J.J. and Ulsoy, A.G., On-line flank wear estimation using an adaptive observer and computer vision, part 2: experiment, *ASME Journal of Engineering for Industry*, Vol. 115, 37–43, 1993.
8. Bahr, B., Motavalli, S., and Arfi, T., Sensor fusion for monitoring machine tool conditions, *International Journal of Computer Integrated Manufacturing*, Vol. 10, No. 5, 314–323, 1997.
9. Mannan, M.A., Kassim, Ashraf A., and Ma, J., Application of image and sound analysis techniques to monitor the condition of cutting tools, *Pattern Recognition Letters*, Vol. 21, 969–979, 2000.
10. Wang, W.H., Wong, Y.S., Hong, G.S., and Zhu, K.P., Sensor fusion for on-line tool condition monitoring in milling, *International Journal of Production Res.*, 2007 (In press).
11. Wang, W.H., Hong, G.S., and Wong, Y.S., Flank wear measurement by a threshold independent method with sub-pixel accuracy, *International Journal of Machine Tools Manufacturing*, Vol. 46, No. 2, 199–207, 2006.

12. Gonzalez, R.C. and Woods, R.E., *Digital Image Processing*, 2nd ed., Prentice Hall, Upper Saddle River, NJ, 2002, pp. 76–137.

13. Castleman, R.K., *Digital Image Processing*, Prentice Hall, Upper Saddle River, NJ, 1996, pp. 480–481.

14. Klette, R. and Zamperoni, P., *Handbook of Image Processing Operators*, John Wiley & Sons, New York, 1996, pp. 255–257.

15. Otsu, N., A threshold selection method from gray-level histogram, *IEEE Transactions on Systems, Man, and Cybernetics*, Vol. 9, No. 1, 62–69, 1979.

16. Kim, T.H., Moon, Y.S., and Han, C.S., An efficient method of estimating edge locations with subpixel accuracy in noisy images, *Proceedings of the IEEE Region 10 Conference*, TENCON 99, Cheju Island, South Korea, pp. 589–592.

17. Wang, W.H., Wong, Y.S., and Hong, G.S., Flank wear measurement by successive image analysis, *Computers in Industry*, Vol. 56(8–9), 816–830, 2005.

18. Kohonen, T., *Self-Organizing Maps*, 3rd ed., Springer, New York, 2001.

19. Elbestawi, M.A., Marks, J., and Papazafiriou, T.A., Process monitoring in milling by pattern recognition, *Mechanical Systems and Signal Processing*, Vol. 3, No. 3, 305–315, 1989.

20. Leem, C.S. and Dornfeld, D.A., A customized neural network for sensor fusion in on-line monitoring of cutting tool wear, *ASME Journal of Engineering for Industry*, Vol. 117, 152–159, 1995.

21. Altintas, Y., In-process detection of tool breakage using time series monitoring of cutting forces, *International Journal of Machine Tools and Manufacturing*, Vol. 28, No. 2, 157–172, 1988.

22. Zhang, D.Y., Han, Y.T., and Chen, D.C., On-line detection of tool breakages using telemetering of cutting force in milling, *International Journal of Machine Tools and Manufacturing*, Vol. 35, No. 1, 19–27, 1995.

23. Zhou, Q., Hong, G.S., and Rahman, M., A new tool life criterion for tool condition monitoring using a neural network, *Engineering Applications of Artificial Intelligence*, Vol. 8, No. 5, 579–588, 1995.

27

Modeling of Cutting Forces During Machining

Z.G. Wang

M. Rahman

Y.S. Wong

K.S Neo

J. Sun

H. Onozuka

Summary

Machine tools are mechatronic systems. In a machining operation, the cutting forces determine the proper performance of the machine and the quality of the product. It is costly and time consuming to investigate the machining mechanism by carrying out experiments, especially for machining processes of costly work materials and advanced cutting tools. Modeling and simulation of a machining process can help predict process variables, which can lead to better control and improved performance, productivity, and product quality. Oxley's comprehensive predictive machining theory has been used to predict the cutting forces, flow stress, and cutting temperature. His model assumes that the tool is perfectly sharp and the normal stress distributes uniformly at the tool—chip interface. These assumptions are not practical, particularly for high-speed machining. In this chapter, a new model is developed to predict cutting forces based on Oxley's theory, but without such assumptions. In the proposed model, the machining process is simulated with the finite element method (FEM) and the constitutive flow stress model. From the simulation results, the shear stress, the shear angle, and the strain rate constant can be estimated, which are then used to predict the cutting forces. Experiments have shown that the model can predict the cutting forces with good accuracy.

27.1 Introduction

Machine tools are mechatronic systems. In a machining operation, the cutting forces determine the proper performance of the machine and the quality of the product. Modeling of cutting forces during machining is very important for thermal analysis, tool life estimation, chatter prediction, and tool condition monitoring. Modeling and simulation of a machining process can help predict process variables, which can lead to better control and improved performance, productivity, and product quality. Many efforts have been made to develop cutting force models in metal cutting. According to the methods used, these models can be divided into three main categories: mechanistic, analytical, and numerical.

Mechanistic models are based on the relationship between the cutting forces and the undeformed area of cut, cutting tool geometry, cutting conditions, and workpiece geometry. The cutting force is usually assumed to be proportional to the undeformed area of cut. This mechanistic approach works without knowing parameters of the cutting force mechanics, such as shear angle, shear stress, and friction angle, and it is widely used. However, these models are commonly computer based and depend heavily on empirical cutting data for their modeling capacity [1]. Strictly speaking then, the mechanistic model is not a completely analytical model, because of its reliance on empirical cutting data.

Analytical models. The shear plane model was developed based on Merchant's shear plane theory [2]. This type of model is based on the assumption of continuous chip formation in a narrow zone, which is idealized as a plane with uniformly distributed shear stress. This method has been used quite successfully in the prediction of forces in several practical machining operations [3,4].

For the aforementioned shear plane model, a major problem is the uncertain magnitude of tool–chip friction and the shear stress at the shear plane [1]. In addition, this model is based on the assumption that the workpiece material deforms at constant flow stress. Actually, the flow stress of metal varies with strain, strain rate, and temperature. Oxley [5] developed a more effective model, which considered the variation of flow stress properties in terms of the strain, strain rate, and temperature. This model assumes a thin shear zone, chip equilibrium, and uniform shear stress in the secondary deformation zone at the tool–chip interface. Oxley's predictive machining theory is widely used to predict a comprehensive range of machining characteristic factors, such as the shear angle, cutting forces, flow stress, and so on. However, a major assumption in Oxley's predictive machining theory is that the tool is perfectly sharp. In practice, it is impossible for the cutting tools to be perfectly sharp. In addition, in Oxley's method, the normal stress at the tool–chip interface is assumed to be distributed uniformly, which is impractical for metal cutting. Furthermore, his work was mainly focused on the carbon steel work material. Thus, his model needs to be revised for the machining processes of other types of materials, such as titanium alloy Ti6Al4V and stainless steels.

Numerical models. In numerical modeling, techniques based on the finite element method (FEM) were found to be dominant [1]. In this approach, the solution region is first divided into many smaller elements so that various tool geometries, cutting conditions, and more sophisticated material and friction models can be incorporated [6]. Thus, this approach is able to predict chip flow, cutting forces, and especially the distribution of tool temperatures and stresses for various conditions. Ozel and Altan [7, 8] have done definitive work in this field. They developed a predictive model for high-speed milling based on FEM simulations. Using their model, primarily the resultant cutting forces, tool stresses, and temperatures in turning and flat end milling were predicted. More importantly, with fewer numbers of experiments, this method is able to estimate the variations of flow stress and friction conditions of high-speed machining. In their model, the tooth-pass was assumed to be circular; however, this approximation will cause some error for high-speed milling.

In this chapter, a predictive force model is presented. In the proposed model, Oxley's cutting force model is used with two modifications: (1) The Johnson–Cook (JC) strength model is used to describe the deformation behavior of target materials, and (2) the value of strain rate constant in Oxley's theory, C', is determined based on FEM simulation results; thus, the assumption about the uniform normal stress at the tool–chip interface is not needed.

27.2 Orthogonal Cutting Theory

The widely used shear plane model was first proposed by Merchant [2], based on the assumption of continuous chip formation in a narrow zone that is idealized as a plane with uniformly distributed shear stress. The forces in chip formation in metal cutting are shown in Figure 27.1. Here, t_1 is the undeformed chip thickness, t_2 is the chip thickness, α is the rake angle, ϕ is the shear angle, λ is the friction angle at the tool–chip interface, F_C is the force component in the cutting direction, F_T is the force component vertical to the cutting direction, F_f and F_n are the frictional force and the normal force at the tool–chip interface, F_N is the force normal to the shear plane, F_S is the shear force on the shear plane, F_r and F_t are force components in radial and tangential directions, and F_R is the resultant cutting force.

The area between the boundaries *CD* and *EF* is the chip formation zone or the shear zone. Based on the assumption that the boundaries *CD* and *EF* are parallel and equidistant from *AB*, Oxley [5] developed a parallel-sided shear zone theory to predict the cutting forces. He assumed that the shear flow stress k_{AB} along the shear plane was constant. Then the shear may be written as

$$F_S = k_{AB}wl = \frac{k_{AB}t_1w}{\sin\phi} \tag{27.1}$$

where w is the width of cut, l is the length of the shear plane, t_1 is the undeformed chip thickness, and ϕ is the shear angle.

According to the geometric relations shown in Figure 27.1, the following equations can be obtained:

$$
\begin{aligned}
V_{chip} &= \frac{V_C\sin\phi}{\cos(\phi-\alpha)} \\[4pt]
V_S &= \frac{V_C\cos\alpha}{\cos(\phi-\alpha)} \\[4pt]
t_2 &= t_1\cos(\phi-\alpha)/\sin\phi \\[4pt]
\theta &= \phi+\lambda-\alpha \\[4pt]
F_R &= \frac{F_S}{\cos\theta} \\[4pt]
F_C &= F_R\cos(\lambda-\alpha) = \frac{k_{AB}t_1w\cos(\lambda-\alpha)}{\sin\phi\cos\theta} \\[4pt]
F_T &= F_R\sin(\lambda-\alpha) = \frac{k_{AB}t_1w\sin(\lambda-\alpha)}{\sin\phi\cos\theta}
\end{aligned}
\tag{27.2}
$$

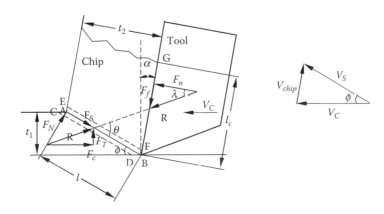

FIGURE 27.1 Cutting force diagram based on the shear plane model.

From Equation (27.2), the cutting forces can be predicted if the shear stress k_{AB}, shear angle ϕ, and the friction angle λ are known. Oxley [5] proposed a theory to estimate k_{AB}, ϕ, and λ. In this chapter, a new cutting force predictive model is proposed based on this method, and a brief overview of it is given next.

In the shear plane zone k_{AB} is calculated according to the following equation:

$$k_{AB} = \frac{\sigma_1 \varepsilon_{AB}{}^n}{\sqrt{3}} \tag{27.3}$$

where σ_1 and n are constants, and n is also called the strain-hardening index. In reality, σ_1 and n vary with strain rate and temperature [9]. For a combination of specific strain rate and temperature, σ_1 and n give a good fit in the following empirical stress–strain relation:

$$\sigma = \sigma_1 \varepsilon^n \tag{27.4}$$

where σ and ε are the effective flow stress and strain. In order to find the values of σ_1 and n, it is necessary to know the value of the velocity-modified temperature T_{mod}, which is defined as a function of strain rate ε and temperature T_{AB} as

$$T_{mod} = T_{AB}[1 - v \lg(\varepsilon/\varepsilon_0)] \tag{27.5}$$

The constants v and ε_0 are taken as 0.09 and $1/\text{s}^{-1}$ [5], respectively.

From Equations (27.3) through (27.5), in order to determine k_{AB}, three parameters are needed, which are the temperature at AB T_{AB}, together with the strain ε_{AB} and the strain rate ε_{AB} at AB. In Reference [5], the strain and strain rate are calculated as

$$\varepsilon_{AB} = \gamma_{AB}/\sqrt{3} \tag{27.6}$$

$$\varepsilon_{AB} = \gamma_{AB}/\sqrt{3} \tag{27.7}$$

where γ_{AB} and γ_{AB} are maximum strain and strain rate at AB, and they are assumed to be

$$\gamma_{AB} = \frac{1}{2} \frac{\cos\alpha}{\sin\phi\cos(\phi - \alpha)} \tag{27.8}$$

$$\gamma_{AB} = C' \frac{V_S}{l} \tag{27.9}$$

where C is the strain rate constant.

The temperature at the shear plane T_{AB} is calculated from the shear forces, shear velocities, and thermal properties (thermal conductivity and specific heat) of tool and workpiece materials. Because the cutting forces, thermal conductivity, and specific heat of workpiece material are temperature dependent, an initial temperature is assumed, to start the calculation. For the particular assumed temperature, the cutting forces and thermal properties of the workpiece material are determined, which can be used to calculate the temperature at AB. Then, the temperature at AB is updated with the calculated value. A new calculation begins with the replacement of a calculated temperature as the starting temperature. This process is repeated until the difference between the starting temperature and the calculated one is less than a given value, say, 0.1° K.

For the calculation of the chip temperatures, a similar iterative procedure is necessary. At first, the mean chip temperature is estimated for the initial calculation. Then, it is replaced with the newly calculated value. This process continues until the difference between the estimated and the calculated values is less than a specified tolerance. Then, Equation (27.5) is used to calculate the value of T_{mod} at the tool–chip interface with the temperature T_{int} and the average strain rate ε_{int} at the tool–chip interface. Specifically, ε_{int} is calculated as

$$\varepsilon_{int} = \gamma_{int}/\sqrt{3} = \frac{V_{chip}}{\sqrt{3}\delta t_2} \tag{27.10}$$

Finally, the shear flow stress in the chip at the tool–chip interface k_{chip} is given by

$$k_{chip} = \frac{\sigma_1}{\sqrt{3}} \tag{27.11}$$

where σ_1 is determined by the value of T_{mod} at the interface. In addition, the resolved shear stress at the tool–chip interface is found from the following equation:

$$\tau_{int} = \frac{F_f}{l_c w} \tag{27.12}$$

where l_c is the tool–chip contact length, given as

$$l_c = \frac{t_1 \sin\theta}{\cos\lambda \sin\phi}\left\{1 + \frac{C'n}{3[1 + 2(\pi/4 - \phi) - C'n]}\right\} \tag{27.13}$$

which is derived by considering moments about B of the normal stresses on AB to find the position of F_R. Assuming that the normal stress on the tool–chip interface is distributed uniformly, the resultant force F_R intersects the tool–chip interface at a distance $l_c/2$ from B. The angle θ is given by

$$\tan\theta = 1 + 2(\pi/4 - \phi) - C'n \tag{27.14}$$

By applying the appropriate stress equation along AB, it can be shown that ϕ should be in the range of $(0, \pi/4)$.

Based on the foregoing description, for given values of tool rake angle α, the cutting speed V_C, the thickness t_1 and the width of cut w of the undeformed chip, together with the thermal and flow stress properties of the workpiece material and the initial temperature of the work T_w (say, 20°C in all calculations), the following procedure is used. For a given δ and C', the equilibrium (when τ_{int} is equal to the value of k_{chip}) values of ϕ are found. If there is more than one shear angle ϕ that satisfies the equilibrium condition, the higher or the highest value of ϕ is chosen as the solution. Then, the required value of C' is determined from the stress boundary condition at B of Figure 27.1. According to the stress equilibrium equation [5] and the theory in which the tool–chip interface is assumed to be a direction of maximum shear stress, the normal stress σ_N' at B is given by

$$\frac{\sigma_N'}{k_{AB}} = 1 + \frac{\pi}{2} - 2\alpha - 2C'n \tag{27.15}$$

Based on the assumption of uniform normal stress along the tool–chip interface, the normal stress σ_N at B is also given by

$$\sigma_N = \frac{F_N}{wl_c} \qquad (27.16)$$

Then, C' can be found by fulfilling the condition $\sigma_N' = \sigma_N$.

The foregoing procedure is iterated for a given range of δ and C' until all the equilibrium conditions are satisfied. Three parameters are very important for the accuracy of Oxley's machining method: shear angle, strain rate constant, and the ratio of tool–chip interface plastic zone thickness to chip thickness. For practical computation, if the actual values of C' and δ are not given in the possible range, it is very difficult to use this method to find the accurate shear angle ϕ. Furthermore, this method is based on assumptions such as the uniform distribution of normal stress at the chip–tool interface; resultant force intersecting the tool face at a distance of $\overline{BG}/2$ from B. These assumptions will reduce the accuracy of the final value of ϕ to some extent.

In this chapter, an improved method based on Oxley's theory is proposed. In the developed model, the shear angle and the strain rate constant are estimated using FEM simulation. With this method, more accurate estimation of these two parameters can be ensured.

27.3 Modeling of Shear Stress Properties

The yield stress of metal under uniaxial conditions is defined as the flow stress or the effective stress, which depends on the strain and strain rate, material properties, and cutting temperature. The metals start deforming plastically when the applied stress reaches the values of the yield stress or flow stress [10].

The flow stress is mostly influenced by temperature, strain, strain rate, and material properties. Accurate and reliable flow stress models are very important for describing the deformation behavior of the workpiece material during practical machining processes. Many researchers have developed techniques to determine the flow stress of metals.

The widely used constitutive model of flow stress is the JC strength model, which was proposed by Johnson and Cook [11]. The JC model represents the flow stress $\bar{\sigma}$ of a material as the product of strain, strain rate, and temperature, as given by

$$\bar{\sigma} = [A + B(\bar{\varepsilon})^n]\left[1 + C\ln\left(\frac{\dot{\bar{\varepsilon}}}{\dot{\bar{\varepsilon}}_0}\right)\right]\left[1 - \left(\frac{T - T_r}{T_m - T_r}\right)^m\right] \qquad (27.17)$$

The parameter A is the initial yield strength of the material at room temperature and a strain rate of $1\ \text{s}^{-1}$; $\bar{\varepsilon}$ is the equivalent plastic strain; and $\dot{\bar{\varepsilon}}$ is the strain rate normalized by a reference strain rate $\dot{\bar{\varepsilon}}_0$. The temperature term is valid within the range from room temperature (T_r) to melting temperature of the workpiece material (T_m). The parameters B, C, m, and n are fitted to the experimental results obtained from the corresponding compression and tension tests. This model sacrifices the potential coupling of the effects of temperature on strain rate hardening, and so on, but it can be calibrated more easily. Therefore, some researchers chose the JC model as the constitutive equation for the deformation behavior of metals at higher strain rate and high temperature. Lee and Lin [12] investigated the deformation behavior of Ti6Al4V using the split Hopkinson bar (SHPB). They fitted the SHPB test results into the JC model at the strain-rate of $2 \times 10^3\ \text{s}^{-1}$. Meyer Jr. and Kleponis [13] also studied high strain rate behavior of Ti6Al4V and low-cost titanium.

Based on the published data listed in References [12,14,15], the Gauss–Newton algorithm with Levenberg–Marquardt modifications for global convergence was used to find the parameter estimates for the JC model. The estimated parameters are listed in Table 27.1, and those parameters for the JC model found by Lee and Lin [12] and Meyer Jr. and Kleponis [13] are also given in Table 27.1 for reference.

TABLE 27.1 Parameters of JC Constitutive Model for Ti-6Al-4V

A	B	n	C	m	Test	Reference
782.7	498.4	0.28	0.028	1.0	SHPB	Lee and Lin [12]
862.5	331.2	0.34	0.0120	0.8	SHPB	Meyer and Kleponis [13]
1165.5	236.6	0.29	0.0355	0.42	—	In this study

27.4 Modeling of Cutting Forces

In this section, a model for generating the cutting forces in milling is developed by modifying Oxley's model. Subsequently, simulation of the cutting forces using FEM is presented.

27.4.1 Modification of Oxley's Machining Theory

In Oxley's model [5], flow stress in the shear plane zone k_{AB} can be calculated according to equation (27.3) ($k_{AB} = \sigma_1 \varepsilon_{AB}{}^n/\sqrt{3}$). This is replaced by $k_{AB} = \overline{\sigma}/\sqrt{3}$, where $\overline{\sigma}$ is the effective flow stress along AB, which can be calculated using the constitutive Equation (27.17).

Based on the assumption of uniform normal stress distribution at the tool–chip interface, the value of C' is calculated from the equilibrium condition given in Equations (27.15) and (27.16), in Oxley's theory. In the present study, according to the chip formation model from Reference [5] and the constitutive equation of flow stress for Ti6Al4V, the value of C' is derived as described next.

The change rate of flow stress (dk/ds_2) normal to AB can be assumed to be only related to the actual strain rate. Therefore, dk/ds_2 can be derived using Reference [5]:

$$\frac{dk}{ds_2} = \frac{dk}{d\gamma} \frac{d\gamma}{ds_2} = \frac{dk}{d\gamma} \frac{d\gamma}{dt} \frac{dt}{ds_2} \tag{27.18}$$

where t denotes time. From Equation (27.17), the following equation can be obtained at AB:

$$\frac{d\overline{\sigma}}{d\varepsilon} = nB(\overline{\varepsilon}_{AB})^{n-1} \left[1 + C\ln\left(\frac{\overline{\varepsilon}}{\varepsilon_0}\right)\right]\left[1 - \left(\frac{T - T_r}{T_m - T_r}\right)^m\right] = \frac{nB(\overline{\varepsilon}_{AB})^{n-1}}{[A + B(\overline{\varepsilon}_{AB})^n]}\overline{\sigma}_{AB} \tag{27.19}$$

Then, the first term on the right-hand side of Equation (27.18) can be obtained as

$$\frac{dk}{d\gamma} = \frac{d\overline{\sigma}/\sqrt{3}}{\sqrt{3}d\varepsilon} = \frac{nB(\overline{\varepsilon}_{AB})^{n-1}}{\sqrt{3}[A + B(\overline{\varepsilon}_{AB})^n]}k_{AB} \tag{27.20}$$

The second term on the right-hand side of Equation (27.18) is the strain rate, which is given in Equation (27.9). The last term is the reciprocal of the cutting speed normal to AB, which can be presented as

$$dt/ds_2 = 1/(V_C \sin(\phi)) \tag{27.21}$$

By substituting for Equation (27.18) with Equations (27.9), (27.20), and (27.21), the following relation is obtained, and a similar derivation procedure is given in Reference [16]:

$$\frac{dk}{ds_2} = \frac{k_{AB}nB(\overline{\varepsilon}_{AB})^{n-1}}{\sqrt{3}[A + B(\overline{\varepsilon}_{AB})^n]} \frac{C'V_C \cos\alpha}{\cos(\phi - \alpha)l} \frac{1}{V_C \sin\phi} = \frac{2k_{AB}C'nB(\overline{\varepsilon}_{AB})^n}{l[A + B(\overline{\varepsilon}_{AB})^n]} \tag{27.22}$$

According to the stress equilibrium equation along AB from Reference [5], the following relation exists:

$$dp = \frac{dk}{ds_2} ds_1$$

(27.23)

By applying this equation along AB and substituting for dk/ds_2 from equation (27.22), one obtains

$$p_A - p_B = \frac{2k_{AB}C'nB(\bar{\varepsilon}_{AB})^n}{[A + B(\bar{\varepsilon}_{AB})^n]}$$

(27.24)

where p_A and p_B are the hydrostatic stresses at points A and B, respectively. Finally the unknown parameter C' is given by

$$C' = \frac{(p_A - p_B)[A + B(\bar{\varepsilon}_{AB})^n]}{2k_{AB}nB(\bar{\varepsilon}_{AB})^n}$$

(27.25)

In Oxley's theory, the angle θ made by the resultant force R with AB is expressed as

$$\tan\theta = 1 + 2\left(\frac{\pi}{4} - \phi\right) - \frac{\Delta k}{2k_{AB}} \frac{l}{\Delta s_2}$$

(27.26)

With the empirical stress-strain relation $\sigma = \sigma_1\varepsilon^n$, Equation (27.26) is represented as Equation (27.14), whereas in the present study, by substituting Equation (27.22) into Equation (27.26), the following equation is obtained:

$$\tan\theta = 1 + 2\left(\frac{\pi}{4} - \phi\right) - C'n\frac{B\bar{\varepsilon}_{AB}^n}{A + B\bar{\varepsilon}_{AB}^n}$$

(27.27)

Based on the model presented earlier, for given values of the tool rake angle α, cutting speed V_C, thickness t_1, and the width of cut w of the undeformed chip, together with the thermal and flow stress properties of the workpiece material and the initial temperature of the work T_w (say, 20°C in all calculations), a simulation using finite elements may be employed to predict the cutting forces.

27.4.2   FEM Simulation of Machining Process

Recently, the application of FEM has achieved significant progress in view of dramatic advances in both computer hardware and software. Many researchers have developed user-oriented general-purpose programs to facilitate FEM analyses. In the present study, FEM is used to simulate the cutting process in machining, which is an extension of FEM application in the analysis of metal forming. Heat is generated during practical machining operations. At high cutting temperatures, the workpiece material properties will vary considerably. Thus, the consideration of temperature effects in the analysis of plastic deformation during a machining process is very important. In a cutting process, workpiece materials deform viscoplastically under the cutting load. Although there still exists elastic deformation, viscoplastic strains outweigh elastic strains. Therefore, it is reasonable to assume a cutting tool and workpiece combination with a rigid-viscoplastic material behavior.

Kobayashi et al. [17] found that the governing equations for viscoplastic deformation are formally identical to those of plastic deformation, except that the effective stress is a function of strain, strain rate, and temperature. For the rigid-plastic materials, the deformation process is a boundary-value problem. Solutions to the boundary-value problem are the velocity distribution that satisfies the governing equations and the boundary conditions. The boundary conditions are such that the velocity vector u is prescribed on a part of surface S_u together with traction F on the remainder of the surface S_F. The governing equations can be expressed mathematically in the form of the following two equations [17]:

$$\pi = \int_V \bar{\sigma}\dot{\bar{\varepsilon}}dV - \int_{S_F} F_i u_i dS \tag{27.28}$$

$$\delta\pi = \int_V \bar{\sigma}\delta\dot{\bar{\varepsilon}}dV - \int_{S_F} F_i \delta u_i dS + K \int_V \dot{\bar{\varepsilon}}_V \delta\dot{\bar{\varepsilon}}_V dV \tag{27.29}$$

where $\bar{\sigma}$ is the effective stress, $\bar{\varepsilon}$ is the effective strain rate, and F_i represents the surface tractions. The stresses exist in the region V, and traction F_i is applied over the surface S. Also, u_i is the admissible velocity, $\bar{\varepsilon}_V$ is the volumetric strain rate, and K is a penalty constant, which is a very large positive constant.

In heat transfer analysis of deforming material, the energy balance equation can be expressed by Reference [17]

$$\int_V k_1 T_{,ii}\delta T dV - \int_V \rho c \dot{T}\delta T dV + \int_V k\sigma_{ij}\dot{\varepsilon}_{ij}\delta T dV = 0 \tag{27.30}$$

where k_1 is the thermal conductivity; $T_{,ii}$ is used for $T_{,i,i}$ with the comma denoting differentiation and repeated subscript denoting summation (Laplace differential operator applied to temperature T); ρ is the density; c is the specific heat; and k is the heat generation efficiency.

Equations (27.28) through (27.30) are the basic equations for FEM discretization. FEM is used to obtain the closed-form solution to the velocity distribution. Once the solution for the velocity field that satisfies the basic equation is obtained, the corresponding stress can be calculated using the flow rule and the known mean stress distribution.

The velocity distribution can be calculated with the minimum work rate principle. According to this principle, the material should always flow in the path of least resistance. Because of the effects of cutting heat, the equations for flow analysis and temperature calculation are strongly coupled. The following iterative process is used to determine the velocity distribution together with the cutting temperature:

Step 1: Assume an initial temperature field.
Step 2: Calculate the initial velocity field at the initial temperature.
Step 3: Calculate the initial temperature-rate field and the quantity.
Step 4: Update the nodal point positions and effective strain for the next step.
Step 5: Use the velocity field at the previous step to calculate the temperature.
Step 6: Calculate a new velocity field with the solution from Step 5.
Step 7: Use the new velocity to calculate the second temperature field.
Step 8: Repeat steps 6 and 7 until both converge.
Step 9: Calculate the new temperature rate field.
Step 10: Repeat steps 4 to 9 until the desired deformation state is reached.

The main problem in FEM simulation is determination of the boundary conditions. The boundary conditions along the cutting tool–chip interface are very complicated. It is extremely difficult to determine the frictional stress at the tool–workpiece interface. The frictional stress has been influenced by factors such as cutting speed, feed rate, and rake angle. Much work has been done to investigate the friction mechanism at the tool–chip interface. Zorev [18] considered that at the tool–chip interface there are two regions on the tool rake face: sticking region and sliding region, as shown in Figure 27.2. In the sticking region whose length is l_p, frictional stress is known to be equal to the local shear stress (k_{chip}). In the sliding region $\overline{BC}$, shear stress decreases on the rake face, and Coulomb's friction law can be applied to

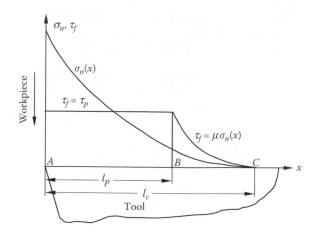

FIGURE 27.2 Curves representing normal stress (σ_n) and frictional stress (σ_f) distributions on the tool rake face proposed by Zorev. (From Zorev, N.N., *Proceedings of the Conference on International Research in Production Engineering*, ASME, New York, 1963, pp. 42–49. With permission.)

calculate the frictional stress. Ozel [7] found that variable friction coefficients as a function of normal pressure at the tool rake surface can ensure more accurate results in FEM simulations. In the present study, the same friction model is used. For this friction model, the frictional stress τ_f at the tool–chip interface is defined according to

$$\tau_f = \begin{cases} k_{chip}, & 0 \leq x \leq l_p \\ \mu_p \sigma_n(x), & l_p < x \leq l_c \end{cases} \tag{27.31}$$

where μ_p is a constant friction coefficient.

From Equation (27.31), it is seen that there are two major unknown parameters: shear flow stress k_{chip} of the chip and the constant friction coefficient μ_p. According to Oxley's theory [5], the shear flow stress in the chip at the tool–chip interface can be calculated with Equation (27.11). In the practical simulation, an initial value of μ_p is assumed to start the simulations. In this study, the initial value of μ_p is chosen at 1.5 according to the estimated value of the friction coefficient from experimental results.

In FEM simulation at different instantaneous chip thicknesses, the values of shear angle ϕ are found by fitting the maximum strain rate along the thin shear zone. Simultaneously, the position of shear plane AB can be determined. Then, the shear stress k_{AB} and the hydrostatic stresses at points A and B are found at different instantaneous chip thickness values. The required value of C' is determined from Equation (27.25), and the value of angle θ is obtained from Equation (27.27). Finally, the cutting forces are found using Equation (27.2). With this procedure, the cutting forces can be calculated accurately and realistically. So the assumption of uniform normal stress at the tool–chip interface used in Oxley's method is not needed.

27.5   Verification of the Cutting Force Model

For the verification, a new tool material called binderless CBN is used to mill Ti6Al4V at high cutting speed. The details of the tool material, cutting conditions, and experimental setup are given in Reference [19].

27.5.1   FEM Simulation of High-Speed Milling of Ti6Al4V

A typical FEM simulation of high-speed milling of Ti6Al4V is shown Figure 27.3. As described in Figure 27.1, there exists a deformation zone.

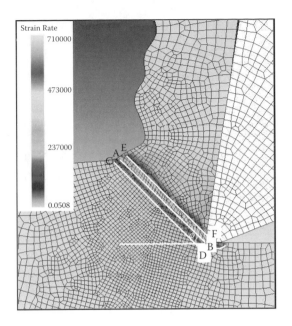

FIGURE 27.3 Deformation zones of FEM simulation in machining of Ti6Al4V.

27.5.2 Prediction of Cutting Forces

As described before, the cutting forces can be predicted using the modified Oxley's model. In particular, tangential $(F_t(\varphi))$ and radial $(F_r(\varphi))$ cutting forces acting on the cutter can be derived. Normally, the radial component F_r of the cutting force is assumed to be ξF_t, where ξ is a constant. Finally, the resultant cutting forces F_r in the z direction can be obtained. The horizontal (x direction) and the normal (y direction) components (F_x and F_y) of the cutting forces acting on the cutter are derived from the equilibrium diagram, shown in Figure 27.4, as

$$F_x = -F_t \sin\varphi + F_r'\cos\varphi = -F_t \sin\varphi + F_r \cos\varphi \sin\left(\theta_0 + \frac{\theta_e}{2}\right) \tag{27.32}$$

$$F_y = -F_t \cos\varphi - F_r'\sin\varphi = -F_t \cos\varphi - F_r \sin\varphi \sin\left(\theta_0 + \frac{\theta_e}{2}\right) \tag{27.33}$$

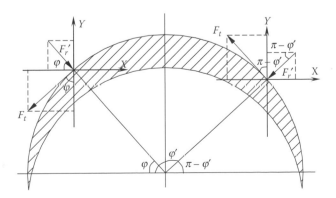

FIGURE 27.4 Mechanics of the milling process.

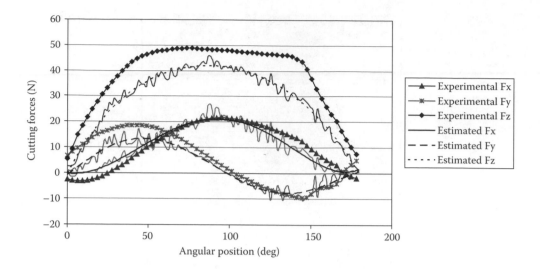

FIGURE 27.5 Experimental and estimated cutting forces at $a = 0.075$ mm, $f = 0.075$ mm/r, and $v = 300$ m/min. (From Wang, Z.G., Rahman, M., Wong, Y.S., and Li, X.P., *CIRP Annals*, Vol. 54, No. 1, pp. 71–74, 2005. With permission.)

It is assumed that $\tan\varphi_0 = \xi\sin(\theta_0 + \frac{\theta_e}{2})$. Then $F_r' = \tan\varphi_0 \cdot F_t$.

$$F_x = -F_t\sin\varphi + F_t\tan\varphi_0\cos\varphi = -\frac{F_t}{\cos\varphi_0}\sin(\varphi - \varphi_0) \qquad (27.34)$$

$$F_y = -F_t\cos\varphi - F_t\tan\varphi_0\sin\varphi = -\frac{F_t}{\cos\varphi_0}\cos(\varphi - \varphi_0) \qquad (27.35)$$

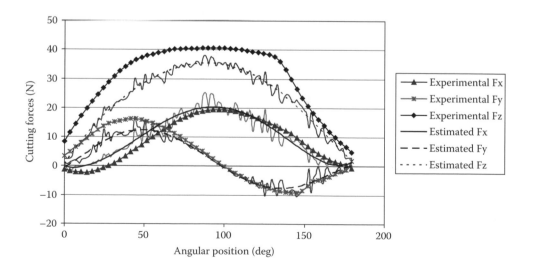

FIGURE 27.6 Experimental and estimated cutting forces at $a = 0.075$ mm, $f = 0.075$ mm/r, and $v = 350$ m/min. (From Wang, Z.G., Rahman, M., Wong, Y.S., and Li, X.P., *CIRP Annals*, Vol. 54, No. 1, pp. 71–74, 2005. With permission.)

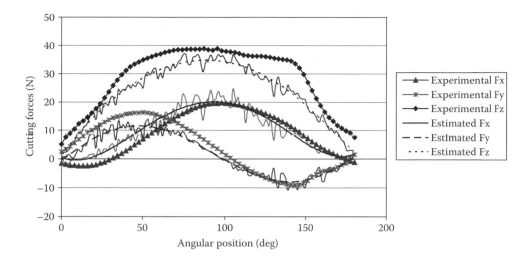

FIGURE 27.7 Experimental and estimated cutting forces at $a = 0.075$ mm, $f = 0.075$ mm/r, and $v = 400$ m/min. (From Wang, Z.G., Rahman, M., Wong, Y.S., and Li, X.P., *CIRP Annals*, Vol. 54, No. 1, pp. 71–74, 2005. With permission.)

Because of the effect of φ_0, F_x and F_y cannot reach their maximum values when F_r and F_t nearly reach their maximum at $\varphi \approx \pi/2$. The detailed information about the prediction of cutting forces is given in References [19,20]. The varying values of F_x and F_y are simulated in the next section.

27.5.3 Comparison of Predicted and Experimental Cutting Forces

Figures 27.5 to 27.7 show the predicted or estimated cutting forces from FEM simulation and the corresponding experimental cutting forces under three different cutting conditions, where a is the axial depth of cut, f is the feed rate per tooth revolution, and v is the cutting speed. Unlike the measured cutting forces, there is much fluctuation in the estimated cutting force. The possible reason for this is the frequent remeshing of the cutting area and the discretization of the cutting area. Similar observations have also been made by Ozel [7]. After fitting the trend of fluctuating estimated cutting forces, the mean values of the estimated cutting forces are obtained (the solid line and the dashed line in Figures 27.5 to 27.7). The estimated cutting force values of F_x and F_y are similar to those of the experimental results. This indicates that the cutting forces can be predicted with reasonable accuracy for these two directions. But for all three cases, the estimated cutting forces of F_z are slightly smaller than the experimental values. However, the difference is not so large; so, it means that the cutting forces can be predicted with reasonable accuracy for all three directions.

27.6 Conclusion

Machine tools are mechatronic systems. In a machining operation, the cutting forces determine the proper performance of the machine and the quality of the product. It is costly and time consuming to investigate the machining mechanism by carrying out experiments, especially for machining processes of costly work materials and advanced cutting tools. Modeling and simulation of a machining process can help predict process variables, which can lead to better control and improved performance, productivity, and product quality. In this chapter, a brief overview of Oxley's predictive machining theory was given firstly. Then, based on Oxley's theory and FEM simulation, a new predictive force model was established. In this model, Oxley's cutting force model is used with two modifications: (1) the JC model is used to describe the deformation behavior of Ti6Al4V, which has been fitted based on Gauss–Newton algorithm with Levenberg–Marquardt modifications for global convergence; and (2) the value of C is determined based on FEM simulation results, so that the assumption about the uniform normal stress at the tool–chip interface is not needed. Finally, milling experiments were conducted to verify the developed force model. It was found that the force model could predict the cutting forces with reasonable accuracy.

References

1. Van Luttervelt, C.A., Childs, T.H.C., Jawahir, I.S., Klocke, F., and Venuvinod, P.K., Present situation and future trends in modelling of machining operations. Progress report of the CIRP working group modelling of machining operations, *CIRP Annals*, Vol. 47, No. 2, 587–626, 1998.
2. Merchant, M.E., Basic mechanics of the metal-cutting process, *ASME Journal of Applied Mechanics*, Vol. 11, A168–A175, 1944.
3. Armarego, E.J.A., Wang, J.P., and Deshpande, N.P., Computer-aided predictive cutting model for forces in face milling allowing for tooth run-out, *CIRP Annals*, Vol. 44, No. 1, 43–48, 1995.
4. Altintas, Y. and Budak, E., Analytical prediction of stability lobes in milling, *CIRP Annals*, Vol. 44, No. 1, 357–362, 1995.
5. Oxley, P.L.B., *The Mechanics of Machining: An Analytical Approach to Assessing Machinability*, E. Horwood, Chichester, England, 1989, pp. 23–135.
6. Altintas, Y., Modeling approaches and software for predicting the performance of milling operations, *Machining Science and Technology*, Vol. 4, No. 3, 445–478, 2000.
7. Ozel, T., Investigation of High Speed Flat End Milling Process: Prediction of Chip Formation, Cutting Forces, Tool Stresses and Temperatures, Ph.D. thesis, Ohio State University, Columbus, OH, 1998.
8. Ozel, T. and Altan, T., Process simulation using finite element method—prediction of cutting forces, tool stresses and temperatures in high-speed flat end milling, *International Journal of Machine Tools and Manufacture*, Vol. 40, No. 5, 713–738, 2000.
9. Usui, E. and Shirakashi, T., Mechanics of machining—from descriptive to predictive theory, *American Society of Mechanical Engineers, Manufacturing Engineering Division*, Vol. MED 7, pp. 13–35, 1982.
10. Altan, T., Oh, S.I., and Gegel, H.L., *Metal Forming: Fundamentals and Applications*, American Society for Metals, Metals Park, OH, 1983, pp. 45–46.
11. Johnson, G.R. and Cook, W.H., A constitutive model and data for metals subjected to large strains, high strain rates and high temperatures, *Proceedings of the 7th International Symposium on Ballistics*, The Hague, Netherlands, 1983, pp. 541–547.
12. Lee, W.S. and Lin, C.F., High-temperature deformation behavior of Ti6Al4V alloy evaluated by high strain-rate compression tests, *Journal of Materials Processing Technology*, Vol. 75, No. 1–3, 127–136, 1998.
13. Meyer, H.W., Jr. and Kleponis, D.S., Modeling the high strain rate behavior of titanium undergoing ballistic impact and penetration, *International Journal of Impact Engineering*, Vol. 26, No. 1–10, 509–521, 2001.
14. Lee, W.S. and Lin, C.F., Plastic deformation and fracture behavior of Ti-6Al-4V alloy loaded with high strain rate under various temperatures, *Materials Science and Engineering A*, Vol. 241, No. 1–2, 48–59, 1998.
15. Majorell, A., Srivatsa, S., and Picu, R.C., Mechanical behavior of Ti-6Al-4V at high and moderate temperatures—Part I: Experimental results, *Materials Science and Engineering A*, Vol. 326, No. 2, 297–305, 2002.
16. Huang, Y. and Liang, S.Y., Cutting forces modeling considering the effect of tool thermal property—application to CBN hard turning, *International Journal of Machine Tools and Manufacture*, Vol. 43, 307–315, 2003.

17. Kobayashi, S., Oh, S.I., and Altan, T., *Metal Forming and the Finite-Element Method,* Oxford University Press, New York, 1989, pp. 1–110.

18. Zorev, N.N., Interrelationship between shear processes occurring along tool face and on shear plane in metal cutting, *Proceedings of the Conference on International Research in Production Engineering,* ASME, New York, 1963, pp. 42–49.

19. Wang, Z.G., High-Speed Milling of Titanium Alloys: Modeling and Optimization, Ph.D. thesis, Department of Mechanical Engineering, National University of Singapore, 2005.

20. Wang, Z.G., Rahman, M., Wong, Y.S., and Li, X.P., A hybrid cutting force model for high-speed milling of titanium alloys, *CIRP Annals,* Vol. 54, No. 1, pp. 71–74, 2005.

28

Mechatronics in Landmine Detection and Removal

T. Nanayakkara

L. Piyathilaka

A. Subasinghe

Summary

Robots are mechatronic systems. This chapter focuses on robotic applications developed for humanitarian landmine detection. An introduction to some popular available technologies that are used in the field is given. Details of a legged robot developed for landmine detection in a vegetated environment is presented. This laboratory-developed mobile robot is a fully embedded platform with simple sensors such as bumper switches and a sonar sensor. In its operation, a fuzzy neural network generalizes the statistical information of the environment and maps it to robotic behaviors. The algorithm has been implemented using commercially available PIC18F452 microcontrollers. Experiments have been carried out in different settings to cover the diversity of the tropical environment in Sri Lanka. The experimental results show that the robot suits the minefields in a tropical country such as Sri Lanka.

28.1 Introduction

Robots are popular mechatronic devices. Their applications are representative and important in the field of mechatronics. This chapter concerns one such application. Landmines remain a significant barrier to economic and social development in more than sixty countries. For most of these developing

FIGURE 28.1   Metal detector used by a manual deminer.

countries, landmine removal is a prerequisite to economic development. Once a conflict comes to an end, the areas where landmines have been laid have to be cleared for human resettlement. Demining is an operation accompanied by great risk to human deminers because a minor mistake can cause death and destruction. Manual demining, as illustrated in Figure 28.1, is boring and repetitive. Analysis of the associated actions shows that some of them may be more easily and safely performed by robotic systems. From this perspective, these robotic systems appear to have an important role to play in finding and removing millions of landmines from around the world.

28.1.1   Conventional Methods in Landmine Detection

First, let us briefly look at the conventional methods used by the manual deminers.

Manual demining using metal detectors. In this method, mines are located with the help of a metal detector. The signal from the metal detector may be from a mine or from a metal fragment. If it is a landmine, it is taken out and subsequently destroyed. This method is useful for the clearance of areas contaminated with antipersonnel landmines. Despite being a slow process that is associated with safety concerns for the person involved, this method has proved to be practical and reliable.

Manual demining using trained dogs. There is a trend in the use of animals with good olfactory capability. Among them, dogs have been very popular because they can be tamed and trained easily. When a dog gives an indication of the presence of a landmine, as shown in Figure 28.2, it is rechecked with a highly sensitive metal detector and then investigated manually. Deminers then dispose of any fragments or the mines that are found. This method is particularly appropriate and useful in the clearance of areas with a low intensity of mines and areas contaminated with minimum metal mines. This method has proved to be relatively fast, but safety concerns remain.

Demining using earth-processing machines. Although large earth-processing machines, as shown in Figure 28.3, seem to be very efficient for removing or detonating landmines, they cost more than the manual mine clearance and are not very reliable. In addition, the maintenance of such machines is extremely difficult in mine-affected countries. For instance, in Sri Lanka, the agricultural industry in the North and the East of the country depends on a thin layer of fertile soil. These earth-processing machines have to remove the top soil layer in order to remove landmines. Therefore, this method is not conducive to achieving the final goal of supporting the farming community to resettle in the affected areas.

FIGURE 28.2 Trained dogs used to sniff for landmines.

28.1.2 Mechatronic Applications

Mechatronic applications, which use an optimal integrated system of mechanical engineering, control engineering, computer science, and electronics engineering, have attracted considerable attention in the recent past because of the opportunities for automation, learning and adaptation, machine intelligence, and so on, which can significantly reduce the risk to humans. Now, we will briefly discuss some of the robotic solutions developed thus far [13].

STAR robot. Figure 28.4 shows the Spiral Track Autonomous Robot (STAR), which was developed at the Lawrence Livermore National Laboratory (LLNL). It can be fitted with multiple sensor packages to complete a variety of desired missions. The STAR is compact, measuring 38 in.^{2} and 30 in. high, and has a low center-of-gravity, allowing the system to climb steep terrain. The electronic control system,

FIGURE 28.3 Flails used to remove the top soil layer.

FIGURE 28.4   The STAR robot.

communications hardware, and software provide the STAR with sufficient intelligence and capabilities to operate remotely or autonomously. A computer system gives the system enough intelligence to autonomously negotiate environments.

MR-1 robot. The MR-1 shown in Figure 28.5 was developed by Engineering Services, Inc. Canada. It is based on the latest technologies for hazardous environmental operations. This multipurpose robot has become globally recognized for its precision, robustness, and reliability. The MR-1 is a remote-controlled mobile robot, consisting of a robust chassis and a reconfigurable robotic arm. The manipulator arm, with its special bulkhead, allows the operator to instantly configure the robot with modules appropriate for the specific mission. These modules include extension links, grippers, cameras, disrupter mounts, and various other equipment.

MR-2 robot. MR-2 shown in Figure 28.5 is an off-road mobile robot developed by the Engineering Services, Inc. Canada is designed to detect landmines, including those with minimal metal content, and UXO. Equipped with MR-1 Arm and vision system, MR-2 performs neutralization of landmines under the supervision of a remotely located operator. MR-2 uses only one metal detector and adaptively follows the terrain by avoiding the obstacles.

DERVISH robot. The Dervish shown in Figure 28.6 is a three-wheeled vehicle with wheel axles pointing to the center of a triangle. If all wheels were driven at the same speed, then it would merely rotate about this center and make no forward progress. In the normal mine-detonating mode, the Dervish advances at about 1 m a minute, a rate set by the requirement that there should be no mine-sized gaps between its wheel tracks. The Dervish can carry a metal detector such as the Schiebel AN 19/2 in a thorn-resistant protective shroud with the sensor head just inboard of the wheel radius at 60 degrees from a wheel.

FIGURE 28.5   Left: The MR-1 robot; right: the MR-2 robot.

FIGURE 28.6   The DERVISH robot designed by Prof. Salter, Department of Mechanical Engineering, University of Edinburgh, Scotland.

28.2   Robots for Tropical Mine Clearance

In this section, specific issues and requirements related to robotic mine clearance are given and their implications are indicated.

28.2.1   Issues with Current Robotic Solutions

Unfortunately, almost all robots developed for landmine detection and removal have been based on the assumption that the robot will not be deployed in highly vegetated environments. For example, all the pictures shown above are for desertlike environments. These robots may work well in countries in the Middle East, but may fail in countries such as Vietnam and Sri Lanka. Another practical issue that keeps robots away from real-life humanitarian demining is the cost of the existing complex robots. The robots should be lightweight and inexpensive to be useful in a real-world humanitarian demining projects. This will also minimize the damage caused in the case of an accidental explosion of a landmine.

In this chapter we focus on low-cost robotic platforms that could be deployed in tropical countries. The fact that they should be low cost implies that they should perform well with a minimum number of inexpensive sensors, use commercially available inexpensive microcontrollers, and the path planning and navigation algorithms should be sufficiently simple and robust to tolerate the non-uniformity of the environment.

28.2.2   Implications of a Tropical Environment for the Design

In any robotic application, the physical morphology of the robot and the algorithms that run in the embedded microprocessors should match the environment with which they are intended to interact [1,2]. In the present case, the robot has to move in a vegetated environment with rough terrain conditions because the land is a former battlefield. The knowledge the user should have to operate the robot should be minimal because a deminer is in a highly stressful situation within a dangerous environment and wearing excessive body armor. Moreover, he or she has to follow strict procedures even to handle simple tools such as rakes. Clearly, a robotic platform will eliminate the danger to humans and accelerate the demining process. Yet, there are key issues to which one must pay attention when designing a robot. One is the maximum allowable weight of the robot. Most antipersonnel landmines explode under any load beyond 7 kg. Therefore, the robot should be as lightweight as possible in order to avoid detonating landmines during navigation. Although a very lightweight robot is desired, requirements such as the robot's ability to carry commercially available landmine detectors impose a lower limit to the size and weight of the robot. The most cost-effective detectors are the commercially available metal detectors. There is a minimum required size for a metal detector for it to achieve an acceptable penetration into the ground, usually a dial of about 15 cm diameter and weighing approximately 1 kg. This size and weight of the commercially available metal detectors limit how small the robot can be in order to move in a tropical minefield while carrying a commercially available metal detector. Another key issue is the cost-effectiveness of the robot. Sophisticated sensors and actuators sometimes cannot justify their cost even for a very hazardous task such as landmine detection. Therefore, the sensing and navigation mechanism should be as simple as possible. The challenge the associated simple algorithms have to face in this application is navigation in forests. The robot should not treat each tree as an obstacle. If the robot can go over some plants such as weeds, it should do so in order to cover as much area as possible. Even if the robot encounters a large tree, it has to go as close to the tree as possible and skirt around the tree to maximize the area covered by the mine detectors. It follows that the navigation algorithms and the mechanical structure of the robot have to be unique to the particular application.

28.3   The Legged Robot Murali (Moratuwa University Robot for Anti-Landmine Intelligence)

After a number of experiments with different robotic platforms, we designed a robot that looks and walks like an iguana, which is frequently seen in the tropical forests of Sri Lanka. Figure 28.7 shows the laboratory prototype of the robot, which weighs 4 kg. The legged robot consists of two independent units, each having the shape of an iguana. The two units are kept together using two rods hinged at the front and the rear ends of each unit. The two units are driven by two geared DC motors, which can be controlled independently. Consequently, turning of the robot involves one unit moving faster than the other or one unit reversing while the other unit moves forward.

Robotic perception. The developed robot is equipped with sensors to carry out proactive behavior planning and implementation as well as reactive navigation. Proactive behavior planning and implementation is done by recognizing the environment in the vicinity of the robot. These calculations are done while moving in an uncharted environment. This is achieved by sweeping a sonar proximity sensor back and forth within a 60° angle around the center axis parallel to the two units of the robot as shown in Figure 28.8. In one sweep, ten proximity readings are taken to calculate the nature of the vegetation in front of the robot. Details of how these statistics are calculated and then used to plan the behaviors are given in Section 28.3.4. Although the robot performs proactive planning by estimating the nature of the vegetation just in its vicinity, it is required to use tactile-sensor-based reactive navigation around obstacles such as large trees to make sure that as much land as possible is scanned for landmines. Simple bumper switches attached to the front of the robot provides this additional tactile sensory feedback for the robot to follow the surface of large obstacles. In this method, the robot navigates in forests by executing local proactive plans that are derived based on rough estimates of the nature of the vegetation in its vicinity. These planned behaviors are further tuned by reactive behaviors stimulated by tactile sensory feedback.

FIGURE 28.7   The legged mobile robot MURALI (Moratuwa University Robot for Anti-Landmine Intelligence).

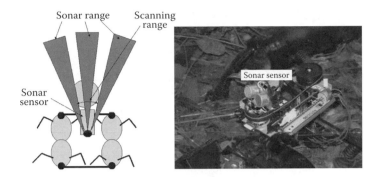

FIGURE 28.8 The scanning action of the sonar sensor and the platform that carries it.

28.3.1 Designing MURALI

The details of the mobile robot named MURALI, which has been designed and developed in our laboratory, are given in this section.

28.3.1.1 Sonar Sensor Interface

Ultrasonic sonar sensors are used extensively in robot navigation, distance measurement, and obstacle avoidance because infrared object detection alone is not adequate in some applications and environments. In the present application, we have used the Devantech SRF04 ultrasonic range finder because it provides a precise, noncontact option for distance measurement in the range of approximately 1.2 in. to 3.3 yd. The SRF04 operates by emitting a 40 kHz ultrasonic burst consisting of 8 cycles, and receiving its echo.

Once the Trigger Pulse Input pin is given a 5 V pulse, the SRF04 emits a 40 kHz ultrasonic burst and sets the Echo Pulse Output pin to HIGH (5 V). The 40 kHz burst travels through air at the speed of sound (~1.1 ft/ms or 1087 ft/s). If the SRF04 receives a reflected wave, it sets the Echo Pulse Output pin to LOW (0 V). By setting interrupts to capture Echo Pulse Output HIGH and Echo Pulse Output LOW, we can measure the time taken between these two events and, consequently, the time of flight of the ultrasonic burst. If no echo is detected, then it will automatically time out after about 30 ms.

28.3.1.2 PicBasic Pro Commands

Trigger pulse. The PULSOUT command of the PicBasic Pro compiler may be used to generate an output trigger pulse. For instance, to generate a pulse on *Pin* of specified *Period*, one may use the command:
PULSOUT *Pin, Period*

The pulse is generated by toggling the pin twice; thus, the initial state of the pin determines the polarity of the pulse. *Pin* is automatically made to be on output mode. *Pin* may be a constant in the range 0 to 15, or a variable that contains a number in the range 0 to 15 (e.g., B0), or a pin name (e.g., PORTA.0).

The resolution of **PULSOUT** depends on the oscillator frequency. If a 4 MHz oscillator is used, the *Period* of the generated pulse will be in 10 µs increments. If a 20 MHz oscillator is used, *Period* will have 2 µs resolution. Defining an OSC value has no effect on **PULSOUT**. The resolution always changes with the actual oscillator speed.

Example

Sends a pulse 1 ms long (at 4 MHz) to Pin 5
PULSOUT PORTB.5, 100

The echo pulse. The PicBasic Pro PULSIN command is used to measure the width of the pulse output on the SRF04 echo pin. Resolution for PULSIN is the same as with the PULSOUT command, i.e., 10 µs at 4 MHz and 2 µs at 20 MHz.

The SRF04 outputs a pulse on the Echo pin after each ultrasonic measurement. Using the PicBasic Pro pulsin command, one may easily read and record this pulse.

Example

Measures high pulse on Pin 4 and store it in W3
PULSIN PORTB.4, 1, W3
Distance calibration. This 40 kHz ultrasonic burst travels through air at the speed of sound (3.3×10^2 cm in 1 μs). The **pulsin** command has a 10 μs resolution when used with a 4 MHz oscillator. The factor for use in calibration can be calculated assuming a 4 MHz crystal.

If the **pulsin** command returns w as the length of the echo pulse, then

$$\text{Echo pulse width} = w \times 10 \ \mu s$$

$$\text{Actual distance traveled by ultrasonic burst} = 3.3 \times 10^{-2} \times w \times 10 \text{ cm}$$

$$= w \times 3.3 \times 10^{-1} \text{ cm}$$

The actual distance to the obstacle is half of this since the burst signal is reflected by the obstacle.

$$\text{Actual distance to the obstacle is} = w \times 3.3 \times 10^{-1}/2 \text{ cm}$$

$$= w/6 \text{ cm}$$

It follows that the distance to the obstacle may be calculated in centimeters by dividing the return value of the **pulsin** command by 6.

28.3.1.3 Body Design to Suit the Intended Behavior

In the present application, the robot body systematically evolved over a number of experiments to suit the target environment. In this context we were inspired by a reptile called the iguana that is frequently found in the tropical forests of Sri Lanka. Unlike other animals, the iguana's legs are very short compared to the size of its body. It achieves the motion partly by swinging its body in a wavy motion. Figure 28.9 shows the overall kinematic structure of the robot and how the crawling motion has been achieved using only two motors.

In Figure 28.9d, let r be the radius of the sphere containing the legs of the robot, l_1 be that portion of the leg above the knee, l_2 be that portion of the leg below the knee, AA be the axis around which the sphere rotates, β be the angle between the axis that goes through the center of the sphere parallel to that portion of the leg above the knee and the axis AA, θ be the angle that the sphere rotates from the vertical

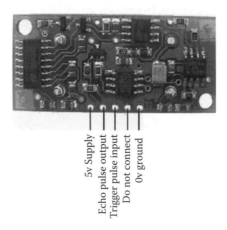

FIGURE 28.9 The SRF04 sensor module.

FIGURE 28.9(A) The overall kinematic structure.

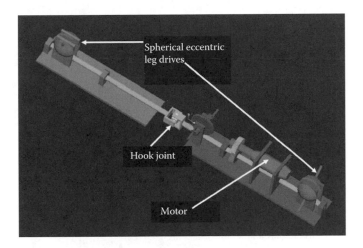

FIGURE 28.9(B) The driving mechanism in a single module.

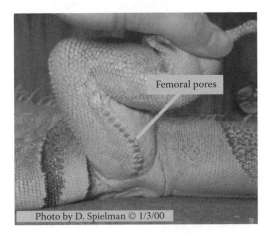

FIGURE 28.9(C) The anatomy of the leg of a green iguana.

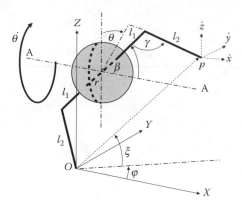

FIGURE 28.9(D)   The kinematical structure of a pair of legs.

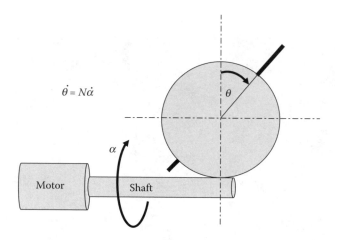

FIGURE 28.9(E)   How the motor shaft and a pair of legs are coupled.

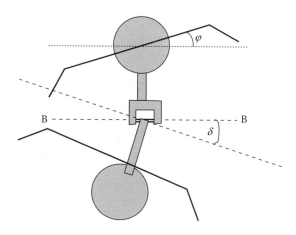

FIGURE 28.9(F)   How the two sections are coupled.

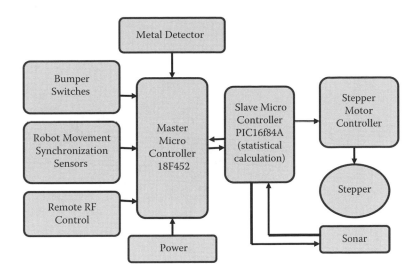

FIGURE 28.10 Hardware architecture of the robot motherboard.

axis, γ be the angle between the upper portion and the lower portion of a leg at the knee, α be the angle of rotation of the motor shaft, N be the gear ratio at the point connecting the motor shaft and the sphere, and δ be the twist of one section of one module of the robot relative to the other section when a pair of legs are rotated by φ in the horizontal plane. Then,

$$\text{Angular speed of one section relative to the other around axis } BB = \left(\frac{\varphi}{\pi}\right) N\alpha$$

$$\text{Distance between } O \text{ and } p = 2[l_2 \cos\gamma + (l_1 + r)]$$

28.3.1.4 Designing the Brain

The thinking is done within the on-board fully embedded microcontrollers. Because the robot is required to explore for landmines in harsh outdoor environments, a PIC16F877A microcontroller was used to interface the sensors and actuators. Figure 28.10 shows how an embedded microcontroller is interfaced to the sensors and actuators.

28.3.2 DC Motor Control for Robotic Applications

Permanent magnet DC internally geared motors were used to generate motion of the robot. A worm gear set is used as well, together with the motor. Worm gears are threaded and matched to a lead screw attached to the shaft of the motor. In this manner the output motion is turned by a right angle from the shaft. Motors were chosen by considering the cost, operating voltage, torque, weight, and speed.

28.3.2.1 DC Interfacing Motors

A microprocessor cannot drive a motor directly, because it cannot supply enough current. In addition, there must be some interfacing circuitry so that the motor power is supplied from another power source, and only the control signals are derived from the microprocessor. This interfacing circuitry can be implemented using a variety of technologies such as relays, bipolar transistors, power MOSFETs, and motor drive integrated circuits. In all technologies however, the basic topology of the interface circuitry is usually the same. This circuitry is known as an H-bridge.

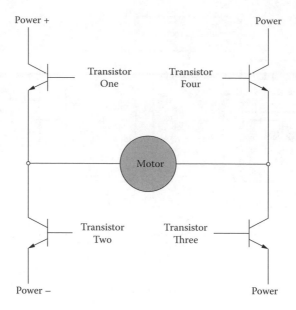

FIGURE 28.11　The H-bridge circuit.

28.3.2.1.1　*The H-Bridge Circuit*

A circuit known as the *H-bridge* (named for its topological similarity to the letter H) is commonly used to drive motors. In the circuit shown in Figure 28.11, two of the four transistors are selectively enabled to control the current flow through a motor at a given time.

As shown in Figure 28.12, an opposite pair of transistors (Transistor One and Transistor Three) is enabled, allowing current to flow through the motor. The other pair is disabled, and can be thought of as not in the circuit.

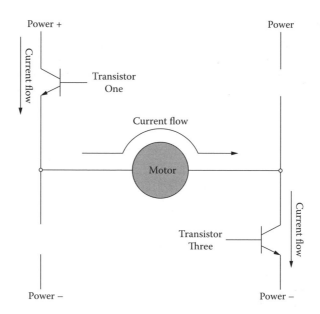

FIGURE 28.12　H-bridge with left-to-right current flow.

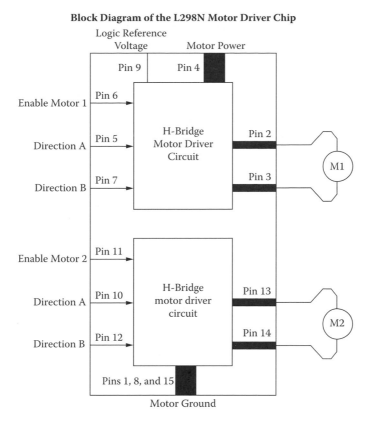

Block Diagram of the L298N Motor Driver Chip

FIGURE 28.13 Block diagram of the L298N motor driver chip.

By determining which pair of transistors is enabled, current can be made to flow in either of the two directions through the motor. Because permanent-magnet motors reverse their direction of turn when the current flow is reversed, this circuit allows bidirectional control of the motor.

28.3.2.1.2 L298N Motor Driver Chip

L298N motor control chip incorporates two H-bridge motor-driving circuits into a single 15-pin package. Figure 28.13 shows a block diagram of this useful integrated circuit. The motor control chip is shown in Figure 28.14.

Control inputs and the corresponding output motor function are given in Table 28.1. The schematic diagram of the motor circuit shows how the L298N chips control the robot. Specifically, their bits are used to control a motor. Two of the bits (**INPUTA, INPUTB**) determine the direction of the motors and one bit (**ENABLE MOTOR**) determines whether the motor is on or off. The speed of a motor may be controlled by pulsing the enable bit of its associated controller chip on and off. This technique is called *pulse width modulation* (PWM).

The L298 is an integrated monolithic circuit in a 15-lead multiwatt package. It is a high voltage, high-current dual full-bridge driver designed to accept standard TTL logic levels and drive inductive loads such as relays, solenoids, and DC and stepping motors. Two enable inputs are provided to enable or disable the device (on or off the motor) independently of the input signals. The emitters of the lower transistors of each bridge are connected together, and the corresponding external terminal can be used for the connection of an external sensing resistor. An additional supply input is provided so that the logic works at a lower voltage.

The Enable pin is used to enable or disable the motor. One may use input1 and input2 for selecting the direction of the corresponding motor.

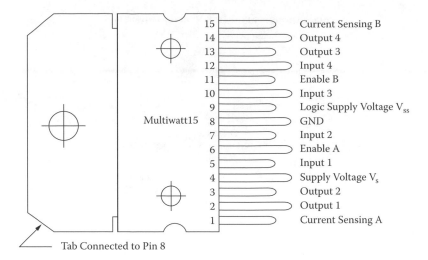

FIGURE 28.14   L298 motor control chip.

28.3.2.2 Speed Control of DC Motors by PWM

Speed of motors can be controlled by varying the current to the motor. Current can be controlled by varying the applied voltage to the motor terminals. The common technique used for this purpose is called PWM.

The basic idea is shown in Figures 28.15a and 28.15b. The input voltage to the motor is switched on and off depending on the voltage to be applied to the motor. The PWM signal has two important properties. One is Duty Cycle, and the other is PWM Frequency. PWM period (inverse of frequency) of the signal should be far below the robot's dynamic time constant and also should satisfy the switching frequency of the motor driver chip. Duty Cycle controls the average voltage output of the signal as shown in Figures 28.15a and 28.15b. For example, if the duty cycle is 25%, then (Figure 28.15b):

$$\text{Output voltage} = \text{supply voltage} \times 0.25$$

The market leaders in the microcontroller industry now include a hardware-implemented PWM signal generator (HPWM) in their microcontrollers. Such a signal generator consumes very little CPU time compared to a software-implemented PWM signal generator. The programmer has to deal with only very few registers to activate and run the PWM signal generator. The PWM signal is supplied to the enable pin of the L298 motor driver chip, which turns the motor supply voltage on and off. The configuration of this motor driver is shown in Figure 28.16.

28.3.2.3 Software for Driving Motors

Microchips PIC 16F 877 A MCU together with PicBasicPro programming language is used to write firmware for the motor control of the robot. Two pins are assigned for direction control, i.e., PortB.1 and Port B.2, and the hardware PWM pin of the MCU (PortC.2) for speed control.

TABLE 28.1   Motor Control Logic

	Inputs	Function
	DIRA=H DIRB=L	Turn Right
ENABLE MOTOR=H	DIRA=L DIRB=H	Turn Left
	DIRA=DIRB	Fast Motor Stop
ENABLE MOTOR=L	DIRA=X DIRB=X	Motor Stop

Note: H = Logic High, L = Logic Low, X = Don't Care.

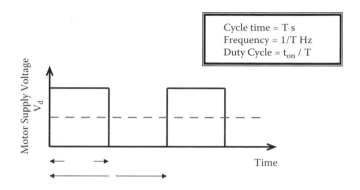

FIGURE 28.15 (a) Pulse width modulation at 50% duty.

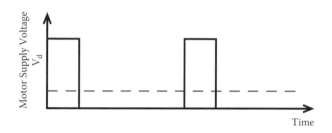

FIGURE 28.15 (b) Pulse width modulation at 25% duty.

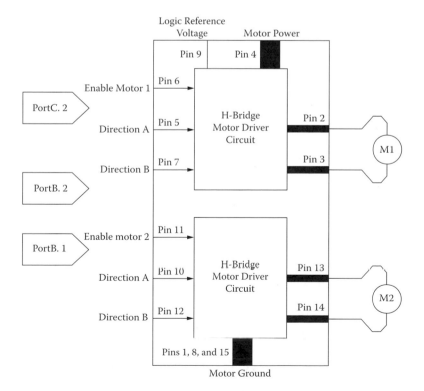

FIGURE 28.16 L298 motor driver configuration.

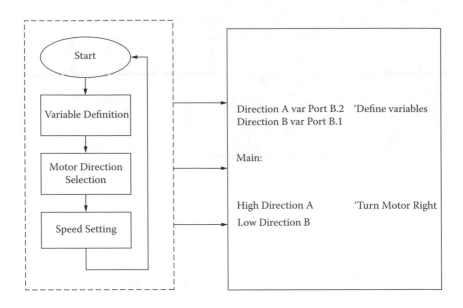

FIGURE 28.17 A full-scale illustration of the motor driver.

Example

High Portb.1	'set the pins output voltage to 5 V. I.e. logic high
Low Portb.1	'set the pins output voltage to oV. I. e logic zero

The speed of the motors can be controlled by using HPWM command of PicBasicPro.

28.3.2.3.1 *HPWM Channel, Duty Cycle, and Frequency*

The HPWM outputs a pulse width modulated pulse train using PWM hardware available on the pic PICmicro MCUs. It can run continuously in the background while the program is executing other instructions. *Channel* specifies which hardware PWM channel to use. Some devices have one, two, or three PWM channels. In devices with two channels, the *Frequency* must be the same for both channels. *Dutycycle* specifies the on/off (high/low) ratio of the signal. It ranges from 0 to 255, where 0 is off (low all the time) and 255 is on (high) all the time. A value of 127 gives a 50% duty cycle (square wave). *Frequency* is the desired frequency of the PWM signal. Not all frequencies are available at all oscillator settings. The highest frequency at any oscillator speed is 32,767 Hz. The operation of the motor drive system is illustrated in Figure 28.17.

Example

HPWM 1,127,1000   'Send a 50% duty cycle PWM signal at 1KHz
HPWM 1,64,2000 'Send a 25 % duty cycle PWM signal at 2kHz

28.3.3 Behavior Planning in a Vegetated Environment

Most conventional obstacle avoidance algorithms (see Reference [8] for example) treat each tree as an obstacle. This does not suit the present application, because the objective of the whole operation is not

FIGURE 28.18 Situation 1: Large tree surrounded by few small plants spread over a mean distance of 2.3 m, with variance/mean ratio 4.3.

collision avoidance but minimizing the land area explored by the detectors even if the whole area is a forest. Rather than opting for algorithms that require sophisticated vision systems and high-end computers to process the sensory information, we were required to design a method that could be implemented using inexpensive sensors and embedded microcontrollers commonly available in developing countries. Therefore, generalizing the situation in front of the robot was more appropriate than zooming into individual obstacles before planning the behaviors. Once a general behavior is chosen, skirting around a particular tree was effected using simple bumper switches.

28.3.4 A Statistical Approach to Situation Recognition

The experiments were carried out in a Sri Lankan forest environment. Based on the biodiversity of the area, we identified five sites that represented the diversity of the forest. These environments can be seen in Figures 28.18 to 28.22.

Table 28.2 indicates how each environment is interpreted based on visual inspection. A human is able to judge how the robot should move based on the terrain condition, the density of the trees, the size of the robot, and the power of the motors given the type of trees around it, and so on. In this manner, we have been able to reduce the mathematical complexity of generalizing a complex vegetated environment by depending on human linguistic interpretations.

The next task is to characterize each environment using a simple set of parameters that adequately represent the criteria that a human observer considers before coming to a conclusion as to what behavior should be activated. In order to achieve this, we adopted a very simple statistical method that could be implemented on board the robot using commercially available inexpensive microcontrollers.

FIGURE 28.19 Situation 2: Open area between robot and a cluster of densely located trees spread over a mean distance of 3.4 m, with variance/mean ratio 2.8.

FIGURE 28.20 Situation 3: Large tree compared to the size of the robot, with a mean distance of 0.53 m and variance/mean ratio 4.3.

FIGURE 28.21 Situation 4: Open space with trees located sufficiently far away from the robot (>4 m). Variance reset to zero.

FIGURE 28.22 Situation 5: Cluster of sparsely distributed trees located with a mean distance of 1.9 m and variance/mean ratio 5.5.

TABLE 28.2 Behavior Recommendations Based on the Interpretations of the Environments

Situation	Interpretation and the Behavior Recommendation
Situation 1	There is a large tree at a medium distance from the robot. The rest of the environment is fairly clear except for a few plants over which the robot can go. Therefore, it is recommended to navigate with a reduced speed directly toward the tree with the hope of skirting around it.
Situation 2	There is a fairly comfortable space before the robot will encounter a dense cluster of trees. Therefore, it is recommended to navigate with a reduced speed while scanning the environment to locate a gap between the trees.
Situation 3	The robot has come very close to a large obstacle. It is recommended to start a skirting behavior with the help of tactile sensors.
Situation 4	There is an open field. Navigate with normal speed.
Situation 5	There is a sparsely distributed cluster of trees. Navigate with the normal speed unless it comes closer to a tree.

The robot simply looked at the mean and the variance of the distances to the trees in front, by swinging a sonar sensor back and forth within a 60° angle as shown in Figure 28.8. In each scan, 10 readings were taken to calculate the mean (μ) and the variance (σ^2) of the sonar readings. The purpose of taking an array of sonar readings was not only to assess the complexity of the vegetated environment, but also to encapsulate an assessment of the terrain complexity [9,10]. Here, we satisfy both requirements because the readings were taken while the robot was moving. Therefore, even if the robot walks toward a wall that would ideally give almost zero variance, the terrain conditions will add a variance component because of the jolt of the plane in which the sonar sensor sweeps back and forth.

The mean distance (μ) to trees and the ratio between the variance to mean ($V = \sigma^2/\mu$) were calculated in these five environments. Based on the average of five trials in each environment, we obtained the values given in Table 28.3. Now we have a mechanism to tap the human interpretations of the environment relative to the capabilities of the robot by only looking at some simple parameters of the environment. For a new environment, we work in the reverse order, where μ and V are calculated first. The fuzzy sets are then used to classify the new environment into one of the five situations we had identified earlier. Thereafter, we use the recommendations given by the human consultant to perform behavior planning. This method can be extended to any number of preidentified situations depending on the biodiversity of the forest in which the robot will work

The linguistic labels (fuzzy sets) for the mean distance and the ratio between the variance to mean are shown in Figure 28.23. The shape and the location of the labels were derived from the experimental data given in Table 28.3.

The fuzzy sets are given by

$$\mu_S = \begin{cases} 1 & : \mu \leq 1 \\ -0.67\mu + 1.67 & : 1 \leq \mu \leq 2.5 \end{cases}$$

$$\mu_M = \begin{cases} 0.67\mu + 0.67 & : 1 \leq \mu \leq 2.5 \\ 0.67\mu + 2.68 & : 2.5 \leq \mu \leq 4 \end{cases}$$

$$\mu_L = \begin{cases} 0.67\mu - 1.67 & : 2.5 \leq \mu \leq 4 \\ 1 & : 4 \leq \mu \end{cases}$$

and

$$V_S = \begin{cases} 1 & : \mu \leq 2 \\ -\mu + 3 & : 3 \leq \mu \leq 4 \end{cases}$$

$$V_M = \begin{cases} \mu - 2 & : 3 \leq \mu \leq 4 \\ \mu + 4 & : 4 \leq \mu \leq 5 \end{cases}$$

$$V_L = \begin{cases} \mu - 3 & : 4 \leq \mu \leq 5 \\ 1 & : 5 \leq \mu \end{cases}$$

The final step is behavior arbitration. Behavior arbitration in this case is carried out by a winner-take-all approach. A simple fuzzy neural network may be used to associate the statistical properties of the distance from the robot to the trees in the respective environments when scanned using a sonar sensor, with a set of behaviors. Figure 28.24 shows how this is done using a simple Takagi–Sugeno-type fuzzy neural network [11], [12] in which the mean distance and the ratio between the variance and the mean in each scan are mapped to a set of (behavior, confidence) pairs given by (B_i, β_i), $i = 1, 2, \quad ,5$. The behavior with the maximum confidence is taken for implementation.

TABLE 28.3 Statistical Properties of the Distance from the Robot to the Trees in the Environments Scanned, Using a Sonar Sensor

	Statistics	
Situation	Mean (Meters)	Variance/Mean
Situation 1	2.3	4.3
Situation 2	3.4	2.8
Situation 3	0.53	4.3
Situation 4	>4	0
Situation 5	1.9	5.5

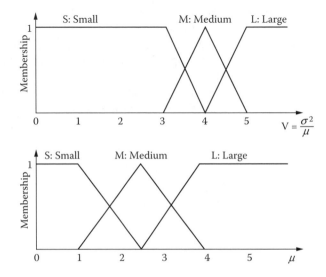

FIGURE 28.23 The designed fuzzy rules.

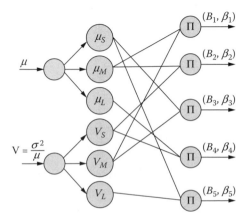

FIGURE 28.24 The designed neural network.

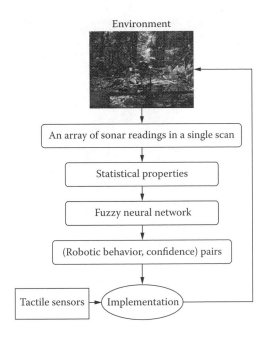

FIGURE 28.25 The overall relationship between the environment and the behavior implementations.

When a particular behavior is implemented, the tactile sensors attached to the front of the robot feeds additional information to the behaviors for reactive navigation around trees or over small plants. Figure 28.25 summarizes the overall relationship between the environment, how it is generalized from far, how behaviors are planned, and how they are finally implemented using tactile sensors.

28.4 Experimental Results in a Vegetated Environment

Figure 28.26 shows how the robot goes very close to a large tree and then tries to circumnavigate it. This was enabled by the statistical approach we adopted. Within a few scans, the robot was able to distinguish between the two situations using the FNN shown in Figure 28.24.

Figure 28.27 shows how the navigation algorithm arrives at a decision to neglect a small plant that the robot could force to bend. We wish to highlight the fact that a conventional navigation algorithm would treat the plant as an obstacle and try to avoid it, and would have left a considerable area unexplored for landmines. Therefore, the strong feature in the developed algorithm is that it strikes a meaningful balance between the primary purpose of scanning the ground for landmines and the less important purpose of avoiding obstacles.

28.5 Conclusion

A low-cost and robust robotic solution to landmine detection in vegetated environments was developed in this chapter. The design placed much emphasis on reducing the number of sensors and the need for sophisticated vision systems in order to make it as reliable as possible. Consequently, only a single sonar proximity sensor and two bumper switches attached to the front of the robot were needed. Because there was no vision involved, the robot could work round the clock. Moreover, the light weight (4 kg) ensures that the robot will not detonate landmines in case it steps on one (in case of failure to detect it). All calculations were performed using commercially available PIC18F452 microcontrollers. The navigation algorithm was a simple one to suit the limited memory and the processing capacity of the embedded processors, where the mean and the variance of the proximity readings in each scan were calculated first.

FIGURE 28.26 The experimental results in a different environment.

These statistics were then used to select a behavior out of many possible ones for different environments. To the best of the authors' knowledge, this is the first time a low-cost robot and a simple navigation algorithm that can be supported by a single sonar sensor and two bumper switches interfaced to very inexpensive microcontrollers were tested in a real-world vegetated environment in a tropical country infested with landmines. [13] However, further improvements are needed in the learning of behaviors to suit different environments. Embedded circuits have to be improved to support such learning algorithms, or the learning algorithms themselves need to be simplified to suit simple commercially available microcontrollers. Robot localization is still an issue to be solved. At present, if the robot locates a landmine, the only option it has is to drop a mark at that place and move forward. On-board energy is still a grave

FIGURE 28.27 The robot goes very close to a small plant that it could concur.

issue to be resolved. At present, the autonomous robots have to carry the batteries with them. Using one metal detector to cover a sufficiently large area remains a problem to be solved. One solution would be to deploy a colony of simple robots well coordinated to achieve a synergetic effect.

Acknowledgment

This work was funded by the National Science Foundation (NSF), Sri Lanka, under the contract number: RG/2004/E/02

References

1. Steels, L. and Brooks, R.A. (Eds.), *The Artificial Life Route to Artificial Intelligence: Building Embodied Situated Agents*, Lawrence Erlbaum Associates, Hillsdale, NJ, 1995.
2. Brooks, R.A., Intelligence without representation, *Artificial Intelligence Journal*, Vol. 47, 139–159, 1991.
3. Stentz, A., Optimal and efficient path planning for partially-known environments, *Proceedings of the IEEE International Conference on Robotics and Automation*, Vol. 4, May 1994, San Diego, CA, pp. 3310–3317.
4. Nicoud, J.D., Light weight demining robots, *Proceedings of the Symposium on Autonomous Vehicles in Mine Countermeasures*, April 1995, Monterey, CA, pp. 8.81–8.85.
5. Wedeward, K., Bruder, S., Yodaiken, V., and Guilberto, J., Low-cost outdoor mobile robot: a platform for landmine detection, *Proceedings of the Midwest Symposium on Circuits and Systems*, Vol. 42; No. 1, 2000, Las Cruces, NM, pp. 131–134.
6. Kato, K. and Hirose, S., Development of the quadruped walking robot, TITAN-IX—mechanical design concept and application for the humanitarian demining robot, *Advanced Robotics*, Vol. 15, No. 2, 191–204, June 2001.
7. Borenstein, J. and Koren, Y., Error eliminating rapid ultrasonic firing for mobile robot obstacle avoidance, *IEEE Transactions on Robotics and Automation*, Vol. 11 , No. 1, 132–138, 1995.
8. Ram, A., Arkin, R.C., Moorman, K., and Clark, R.J., Case-based reactive navigation: a method for on-line selection and adaptation of reactive robotic control parameters, *IEEE Transactions on Systems, Man, and Cybernetics* — Part B, Vol. 27, No. 1, 376–393, June 1997.
9. Lumelsky, V.J., Mukhopadhyay, S., and Sun, K., Dynamic path planning in sensor-based terrain acquisition, *IEEE Transactions on Robotics and Automation*, Vol. 6, No. 4, 462–472, 1990.
10. Hodgins, J.K. and Raibert, M.N., Adjusting step length for rough terrain locomotion, *IEEE Transactions on Robotics and Automation*, Vol. 7, No. 3, 289–298, 1991.
11. Takagi, T. and Sugeno, M., Fuzzy identification of systems and its applications to modeling and control, *IEEE Transactions on Systems, Man, and Cybernetics*, Vol. SMC-15, No. 1, January–February 1985.
12. Kiguchi, K. and Fukuda, T., Fuzzy selection of fuzzy-neuro robust force controllers in an unknown environment, *Proceedings of the IEEE ICRA-1999*, Michigan, pp. 1182–1187.
13. Web site: http://diwww.epfl.ch/w3lami/detec/minelinks.html, last viewed on May 2, 2006.

29

Online Monitoring and Fault Diagnosis of Ship Propulsion Systems

Z. Feng

Z. Chen

Y. Wang

Summary

A ship propulsion system is a mechatronic system. Online monitoring and fault diagnosis are crucial to achieving high performance levels from such a system. In this chapter, an online monitoring and fault diagnosis system for torsional vibration in the shafting of a ship propulsion system is presented. First, the typical fault types of ship shafting are outlined. Second, the fault diagnosis principles based on the back-propagation neural network and the Kohonen self-organization neural network are presented. Third, a simulated test bed is developed, and its architecture and functions are introduced. In conclusion, experimental results are presented and discussed to validate the developed approaches.

29.1 Introduction

Health monitoring, fault detection, and fault diagnosis are important for uninterrupted and high-quality performance of a mechatronic system. Mechanical vibration is a popular signature that is used for this purpose [1]. The ship propulsion system is a mechatronic system. As an important part of the ship power system, the diesel propulsion system will inevitably generate torsional vibrations under the effects of the periodic driving torque of the diesel engine and the propeller. With time, torsional vibration will cause fatigue failure of the shafting, thereby reducing its lifetime and increasing the potential for accidents, and even the danger of the ship sinking. Although the shafting torsional vibration of the propulsion system is tested carefully to meet the design requirements before it comes intouse, torsional vibration accidents still happen from time to time because of the complexity of the operating conditions. An effective method for reducing and avoiding accidents is to use various monitoring and diagnosis technologies to carry out online monitoring of the propulsion system.

Fault monitoring of ship propulsion systems through the detection and analysis of vibration signals has been researched by many. Through online monitoring of the shafting vibrations and analyzing for possible faults, potential accidents and their consequences can be forecasted in advance so that one could avoid underlying dangers and economic losses. However, it is usually very difficult to monitor and analyze faults in a real ship and, consequently, it is not easy to predict various ship navigation states and other factors such as material performances, manufacture quality, assembly approaches, and measurement methods. Therefore, it is necessary to improve the capability of identifying vibration fault types and the reliability of the monitoring systems.

Usually, there exists a fuzzy correspondence in the relationship between a fault symptom and its reason in a propulsion system. Because the neural network technology has unique advantages in the field of fuzzy mapping, it is advantageous to bring the neural network technology into the field of fault diagnosis of shafting vibration in the form of an intelligent diagnosis technology [2,3].

Fault diagnosis of a mechanical equipment includes status monitoring, fault detection, and diagnosis. The first task focuses on signal acquiring and processing, the second task involves signal analysis and the detection of the presence of a fault, and the third task focuses on further analysis to diagnose the detected faults. At present, status-monitoring instruments from inertial vibration meters to noncontact vibration analyzers are developed, which implement automatic sampling, processing, and recording of signals thanks to the advances of microprocessor technology. This chapter proposes an intelligent fault diagnosis and analysis system for torsional vibration of ship shafting using neural networks. In addition, simulation and experimental results are presented to validate its effectiveness and feasibility. The experimental results indicate that it is a good diagnosis system for shafting faults.

29.2 Typical Faults of Ship Shafting

The rotating machinery that is widely employed today forms an important mechanical device in industries such as electric power, ship, and aviation. In ship-propulsion systems, the torsional vibration is usually caused by an imbalanced rotor, misalignment of the shafting, a rotor crack, and wearing out of the bearings or the rotor because of rubbing [4,5]. Most of the faults in rotating machinery can be detected by analyzing its vibration signals.

29.2.1 Imbalanced Rotor

Rotor imbalance is a common fault in rotating machines. When rotor imbalance exists, because the periodic centrifugal force of inertia has an excitation effect on the rotor, it will cause a forced oscillation, thereby affecting the rotorís normal rotation and even damaging the rotor [6,7,8].

Reasons that cause rotor imbalance include unsatisfactory structural design, manufacturing and assembly errors, nonuniform rotor material, nonuniform thermal effects, corrosion and degradation of the rotor and shafting, loose fitting of rotor parts, and so on [11].

Rotor imbalance possibly causes serious problems, including rotor fatigue failure resulting from alternating bending and internal stresses, making the machine very noisy and producing excessive vibration, and resulting in reduced machine life and efficiency. In addition, rotor vibrations can affect the foundation through the bearings, the machinery base, engine mounts, and so on, which make working conditions worse.

If a rotor has an unbalanced mass, its vibration signal usually has the following characteristics [11]:

1. The primitive waveform of the vibration signal is a sine wave.
2. In the frequency spectrograph of the vibration signal, its base-frequency component represents a larger proportion, and the multiplication-frequency component represents a relatively small proportion.
3. In the acceleration/deceleration processes, when the rotational speed is smaller than the critical speed, the oscillation amplitude increases with the rotational speed, and the resulting stress directions of the two bearings are almost the same. However, when the rotational speed is greater than the critical speed, the oscillation amplitude decreases with the rotational speed and gradually tends to a small constant value.

29.2.2 Out of Alignment of the Shafting

Shafting may be out of alignment because the coupling part is not installed properly, the middle line of the bearing is out of alignment, or the rotor is deformed; this may cause vibrations and even failures. The main characteristics of vibration signals of a rotor with these type of faults are as follows [9]:

1. The primitive waveform of the vibration signal is an abnormal sine wave.
2. The more serious the fault is, the greater the proportion of the multiplication-frequency component.
3. In the frequency spectrum of the longitudinal vibration, the largest magnitude appears at frequency doubling.

29.2.3 Rotor Crack

If the rotor of a rotating machine is not designed correctly, or the manufacturing method is improper, stress concentrations will appear, which will result in cracks on the rotor. Rotor cracks may have three types of shapes: closed cracks, open cracks, and open and closed cracks. Although rotor cracks can affect the vibration characteristics of the rotor, most of them are not very sensitive. At present, the effective method to detect this fault is to analyze the amplitude rate in the starting/stopping processes of the machine.

The main characteristics of the vibration signal corresponding to rotor cracks are as follows [10]:

1. Critical rotational speeds of different orders are smaller than their normal counterparts, especially when the crack tends to be serious.
2. Because the crack causes uneven rigidity in the rotor, there exist multiple resonant speeds in the rotor.
3. At constant rotational speed, the magnitude and phase of the frequency-multiplication components, especially the frequency-doubling component, are unstable.

29.2.4 Rotor Wear

Because of mass imbalance, deformation, deflection, and so on, there will be a frictional force between the rotor and the stator that further intensifies hot bending of the rotor and aggravates the friction.

The main characteristics of this fault are as follows:

1. Before the rotor jitters, the component of the frequency spectrum is rich and the axle center assumes an irregular trajectory.
2. After the rotor jitters, the waveform assumes a serious distortion and the trajectory of the axle center diverges.
3. When the rotor bumps slightly, the magnitude at the same frequency fluctuates.
4. When the rotor has a serious friction, the amplitudes in various frequency components increase rapidly.
5. The rigidity of the system increases, the area of the critical speed stretches, and the phase of various orders changes.

29.3   Fault Diagnosis Using Neural Networks

Neural networks have been studied for many years as a type of intelligent diagnosis approach, leading to a number of important and several practical applications. It is expected that the neural network technology will become a trend in the field of fault diagnosis. The neural network approach, by its characteristic of nonlinear mapping, can lead to an effective diagnosis approach for complex engineering systems. Moreover, as a powerful nonlinear tool, it has been used extensively in such fields as signal processing, pattern recognition, knowledge engineering, and automatic control. In particular, the back-propagation (BP) network and the self-organization network are widely used and have many successful applications.

29.3.1   Diagnosis Model and Algorithm for the BP Network

The BP network is one type of supervised reversal network that requires a large amount of samples and complex training [12]. Its original samples have a significant influence on the diagnosis results. The forward-feed structure of the BP network with three layers is shown in Figure 29.1, which includes the input layer, the hidden layer, and the output layer. In each layer, it has nodes that represent neurons.

Assume that there are n inputs in the BP network, as given by

$$X = (x_1, x_2, \quad, x_n)^T \tag{29.1}$$

According to the network principle, the input weighted sum of j-th node of the hidden layer is s_j as given by

$$s_j = \sum_{i=1}^n w_{ij} \cdot y_i - \theta_j = \sum_{i=0}^n w_{ij} \cdot y_i \tag{29.2}$$

In equation (29.2), n is the node number of the input layer; $i = 12n$; w_{ij} are the connection weights from the i-th node of the input layer to the j-th node of the hidden layer; y_i is the output of the i-th node of the input layer; and $_j$ is the threshold value of the j-th node of the hidden layer according to $w_{0j} = -\theta_j, y_0 = 1$.

The output of the j-th node of the hidden layer is

$$y_j = f(s_j) \tag{29.3}$$

In equation (29.3), $f(x)$ is the transfer function of the hidden layer, and usually the Sigmoidal function is used, as given by

$$f(x) = \frac{1}{1 + e^{-x}} \tag{29.4}$$

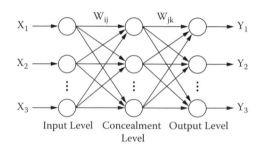

FIGURE 29.1   Three-layer forward-feed BP neural network.

The input weighted sum of k-th node of the output layer is

$$s_k = \sum_{j=1}^{m} w_{jk} y_j - \theta_k = \sum_{j=0}^{m} w_{jk} y_j \qquad (29.5)$$

In equation (29.5), m is the node number of the hidden layer; $j = 12m$; w_{jk} are the connection weights from the j-th node of the hidden layer to the k-th node of the output layer; y_j is the output of the j-th node of the hidden layer; and $_j$ is the threshold value of the k-th node, according to $w_{0k} = -\theta_k, y_0 = 1$.

The output of the k-th node of the output layer is

$$y_k = F(s_k) \qquad (29.6)$$

In equation (29.6), $F(s)$ is the transfer function of the output layer, expressed using the Sigmoid function.

The objective function for the BP network is given by

$$E = \frac{1}{2} \sum_{k=0}^{l} (T_k - y_k)^2 = \frac{1}{2} \sum_{k=0}^{l} \left\{ T_k - F\left[\sum_{j=0}^{m} w_{jk} \cdot f\left(\sum_{i=0}^{n} w_{ij} \cdot y_i \right) \right] \right\}^2 \qquad (29.7)$$

In equation (29.7), T_k is the expected output of the k-th node of the output layer, $k = 1, 2l$.

The connection weight vector is given by

$$W(t+1) = W(t) + \eta(t) \frac{\partial E}{\partial W}\bigg|_{W=W(t)} \qquad (29.8)$$

29.3.2   The Diagnosis Model and Algorithm of the Kohonen Network

In 1987, one type of competition network—Kohonen neural network which simulates the self-organization mapping characteristic of the human brain, was developed by Kohonen. It solves the problems that exist in the diagnosis of the BP neural network. In the Kohonen network, the standard samples and the candidate modes can be simultaneously input into the network, and in the competition layer, the classification result is shown in the form of a two-dimensional matrix structure. The Kohonen network outperforms the BP neural network in dealing with sick samples and in the aspect of computational load. It has advantages such as simpler algorithm, nonsupervised self-organization learning, and automatic pattern recognition, which enable this network to possess a very good application prospect in the field of fault diagnosis. The Kohonen network architecture is shown in Figure 29.2. [12]

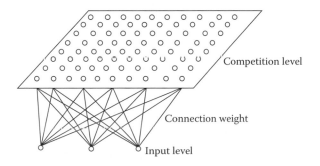

FIGURE 29.2   The Kohonen neural network.

There are n neurons in the input layer, and $m \times m$ neurons in the competition layer, which constitutes a two-dimensional array. The neurons of the input layer and the competition layer are connected with each other. Assume that the network has q input patterns, $P_k = (p_1^k, p_2^k, \ldots, p_n^k), k = 1, 2, \ldots, q$, and the connection weights vector from neuron j in the competition layer to the neurons in the input layer are $W_j = (w_{j1}, w_{j2}, \ldots, w_{jn}), j = 1, 2, \ldots, m^2$. The learning algorithm of the network is as follows:

1. Assign the network connection weights w_{ij} as random values within 0 to 1. Initialize the learning rate $\eta(t)$ and the neighborhood $N_g(t)$ (commonly $0 < \eta(0) < 1$), as well as the number T of the learning steps.
2. In the q input patterns, randomly select a pattern of P_k. Supply it to the network's input layer and carry out the normalization operation:

$$\overline{P_k} = \frac{P_k}{\|P_k\|} = \frac{\left(p_1^k, p_2^k, \ldots, p_n^k\right)}{\left[\left(p_1^k\right)^2 + \left(p_2^k\right)^2 + \ldots + \left(p_n^k\right)^2\right]^{1/2}} \tag{29.9}$$

3. Normalize the connection weight vector $W_j = (w_{j1}, w_{j2}, \ldots, w_{jn})$, and calculate the Euclidean distance between $\overline{W_j}$ and $\overline{P_k}$.

$$\overline{W_j} = \frac{W_j}{\|W_j\|} = \frac{(w_{j1}, w_{j2}, \ldots, w_{jn})}{[(w_{j1})^2 + (w_{j2})^2 + \ldots + (w_{jn})^2]^{1/2}} \tag{29.10}$$

$$d_j = \left[\sum_i^n \left(\overline{p_i^k} - \overline{w}_{ji}\right)^2\right]^{1/2}, \quad j = 1, 2, \ldots, m^2 \tag{29.11}$$

4. Find the minimum distance d_g, and make certain the winner neuron g is given by

$$d_g = min[d_j], \quad j = 1, 2, \ldots, m^2 \tag{29.12}$$

5. Adjust the neighborhood connection weights according to

$$w_{ji}(t+1) = w_{ji}(t) + \alpha(d_j)\eta(t)\left[p_i^k - w_{ji}(t)\right], \quad j \in N_g(t) \tag{29.13}$$

In the formula, $\alpha(d_j)$ is a function of the distance d_j, as given by $\alpha(d_j) = e^{-\lambda\left(\frac{d_j}{m}\right)^2}$, $\lambda = 0.3 \sim 10$
6. Select the next learning pattern, return to step 2, and repeat the computations.
7. Update the study rate $\eta(t)$ and the neighborhood $N_g(t)$ as follows:

$$\eta(t) = \eta(0)\left(1 - \frac{t}{T}\right) \tag{29.14}$$

$$N_g(t) = INT\left[N_g(0)\left(1 - \frac{t}{T}\right)\right] \tag{29.15}$$

In these formulas, t is the number of learning steps, and $0 < t < T$.
8. $t = t + 1$. Return to step 2 until $t = T$.

If the candidate pattern is input into the network, after carrying out the computations according to the preceding learning algorithm, the winner neuron in the competition layer will be excited. Therefore, it can recognize the corresponding category of input pattern, i.e., the fault diagnosis process is implemented.

29.4   Measurement Circuit and Principles

The torsional vibration signal of a ship propulsion system, as measured by an electro-optical sensor, needs some preprocessing operations such as enlargement and reshaping before it can be sent to the input end of the neural network. The measurement principle of the torsional vibration signal is shown in Figure 29.3.

In the figure, the torsional vibration signal provided by the noncontact electro-optical sensor is a series of rectangular pulses after being amplified and reshaped, which enters an AND gate. Simultaneously, a time-base signal generated by a crystal oscillator is converted into a low-frequency pulse signal by the frequency-dividing circuit, which also enters the AND gate. The output of the AND gate is transmitted to the data acquisition board in a computer, and the number of time-base pulses within one optical-electricity pulse is calculated. Because the frequency of the time-base pulse can reach as high as 2 MHz, the measurement accuracy is quite high.

When the shaft rotates at a constant speed, the frequency of the optical-electricity pulse is also constant, but when the propulsion system runs in an unstable manner because of the torsional vibration, the frequency of the optical-electricity pulse will change along with it.

Suppose that the torsional vibration of the shaft is a simple harmonic oscillation. Its vibration angular displacement and angular speed, respectively, are

$$\varphi = A\text{Sin}(\omega t - \psi) \tag{29.16}$$

and

$$\frac{d\varphi}{dt} = A\omega\text{Cos}(\omega t - \psi) \tag{29.17}$$

In the preceding formulas, A is the oscillation amplitude, ω is the frequency of the torsional vibration, and ψ is the initial phase. Then, the fluctuation of the rotational speed caused by the torsional vibration is

$$\Delta n = \frac{1}{2\pi}\frac{d\varphi}{dt} = \frac{A\omega}{2\pi}\text{Cos}(\omega t - \psi) \tag{29.18}$$

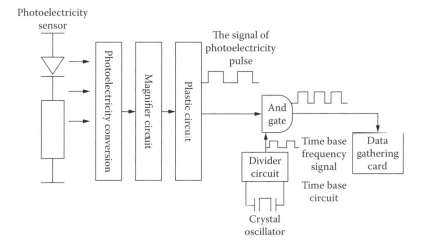

FIGURE 29.3 The measurement circuit of the torsional vibration signal.

The frequency of the optical-electricity signal is

$$F = \frac{Z(n_0 + \Delta n)}{60} = F_0 + \Delta F \tag{29.19}$$

In this formula, Z is the number of teeth in the tooth plate of the optical-electricity sensor, n_0 is the average rotational speed of the shaft, F_0 is the constant component in the optical-electricity pulse frequency, and Δn and ΔF, respectively, are the components in the rotational speed and the optical-electricity pulse frequency, which are caused by the torsional vibration.

29.5 System Software and Functions

29.5.1 Standard Samples

Standard samples have to be established for training the neural network before the fault diagnosis process is carried on. In the diagnosis using a three-layer BP network, the node number of the input layer is equal to the number of system patterns, the node number of the hidden layer is equal to the node number of the input layer, and the fault number is equal to the node number of the output layer. Such a design guarantees the convergence rate of the network and identification of the distribution characteristics of the frequency spectrum. In the diagnosis using the Kohonen network, the spectrograph is divided equally into eight sections along the frequency axis, and the amplitudes of each frequency value in every section are added together. In this manner, a vector containing eight elements is obtained, which represents the distribution characteristic of the frequency spectrum of the vibration signal.

In the Kohonen network, there are 8 neurons in the input layer, and $6 \times 6 = 36$ neurons in the competition layer. To obtain the standard training samples, vibration signals are acquired under different working conditions of the test bed, and the distribution characteristics of their frequency spectra are extracted, which are regarded as the standard samples for training the neural network.

29.5.2 Diagnosis Software Design

The developed intelligent fault diagnosis system using neural networks runs on a computer with Microsoft Windows platform. The system integrates online monitoring with the offline analysis and implements continuous and preventive monitoring and diagnosis. It is composed of thirteen function modules including online monitoring, offline analysis, parameter estimation, alarm system, and so on. The software adopts the mixed programming technology of Visual Basic (VB) and MATLAB®. The software architecture is shown in Figure 29.4.

29.5.3 Mixed Programming Technology

The modules of user interface, data acquisition, vibration computation, and database operation are developed using Microsoft VB, whereas modules of fast Fourier transform (FFT) and neural network are developed using MATLAB. This type of mixed programming has fully utilized and combined the respective advantages of the individual development kits, enabling greater overall performance of the system. Because VB and MATLAB are two independent development kits, data cannot be transferred directly between them. In the present work, this problem has been solved using the ActiveX technology. The version of MATLAB 5.0 for windows supports the ActiveX Automation Server Protocol, in which a VB program (control end) is permitted to control a MATLAB program (server end). Based on this protocol, the VB program can exchange data/commands easily with the MATLAB program.

29.5.4 Diagnosis Flow

First, the software starts the data acquisition board to acquire data from the three channels simultaneously. After 1000 data samples are acquired, a profile of the rotational speed is calculated, and the FFT transformation

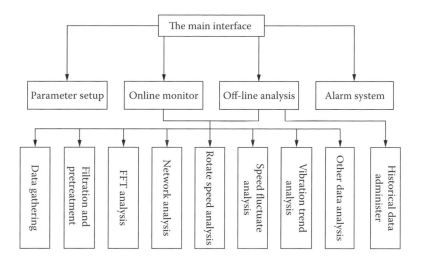

FIGURE 29.4 The software modules.

to the data is carried out to obtain its frequency spectrum. Then, the rotational speed wave profile and the vibration frequency spectrum of the shafting are displayed on the computer screen. Meanwhile, the feature vector of the frequency spectrum is extracted and is sent to the neural network to forecast and diagnose the fault of the shafting. Once a fault is detected, the software will start the alarm circuit immediately to give an acousto-optic alarm. The vibration data acquired by the software system is online, preserved in an SQL server database in the local hard disk for use in offline analysis.

29.5.5 User Interface

The user interface is shown in Figure 29.5. The curve at the top left shows the instant rotational speed and the torsional angle, whereas the curve at the top right shows its frequency spectrum. In addition,

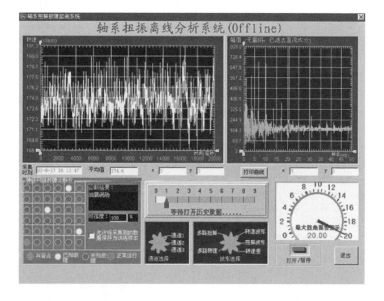

FIGURE 29.5 The software interface.

the maximum torsional angle is shown at the bottom right. The two green knobs in the bottom middle are used to select the sensor channel and the curve category. At the bottom left there exists an indicator of the self-organization Kohonen neural network in which 36 competition-layer neurons constitute a 6 × 6 two-dimensional table. In this table, the yellow dot represents the known fault, the green dot represents the unknown fault, the red dot represents the current excitation neuron in the competition layer, and the amaranth dot represents the status spot when the shafting works normally. By choosing the channel and the curve type, the rotational speeds and the torsional vibrations of the propulsion system at various measurement points can be directly monitored, and the fault diagnosis results of the neural network system can be displayed.

29.6   Simulated Test Bed

To verify the approaches presented in this chapter, a test bed, which is shown in Figure 29.6, is developed in-house to simulate the operating conditions of diesel engines and the shafting system of a real ship. The test bed includes three subsystems: the velocity modulation subsystem, the excitation subsystem of torsional vibration, and the simulated shafting subsystem.

The velocity modulation subsystem is composed of SGV-T110 AC motor and the belt transmission system. The rated power of the motor is 13 kW, and the adjustable range of its frequency is 0.5–400 Hz. Its rotational speed can be continuously adjusted within the range of 0–2000 rpm.

The excitation subsystem of torsional vibration is simulated by an oil injection pump of the 6135III diesel engine. The normal working sequence of its six cylinders is 1—5—3—6—2—4, and the angle between two adjacent cylinders is 60°. Through changing the oil pressure, the fuel feed, the feed order, and the excitation torque of torsional vibration are simulated.

The simulated shafting subsystem is composed of the belt pulley, the shaft coupling, the flywheel, and three sections of shaft, which constitute a torsional vibration system with three masses. The parameters of the test bed are given in Table 29.1.

The architecture of the fault diagnosis system is presented in Figure 29.7.

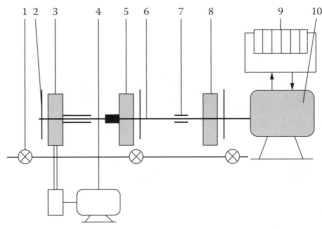

1. Sensor	6. Middle wheelset
2. Speed measurement gear	7. Middle bearing
3. Pulley	8. Flywheel
4. Frequency modulation electromotor	9. Injector
5. Coupling	10. Fuel injection pump

FIGURE 29.6　The simulated test bed.

TABLE 29.1　The Main Parameters of the Test Bed

Item	Model Specification	Manufacturer
Oil injection pump	Model number: 6135-3	The diesel engine plant of Shanghai, China
Signal preprocessor	TVFDS-1	Developed in-house
Crankshaft	$\Phi25 \times 65$ mm	
Middle shaft	$\Phi18 \times 50$ mm	
Screw shaft	$\Phi22 \times 50$ mm	
Flywheel	$\Phi180 \times 40$ mm	
Pulley	$\Phi160 \times 60$ mm	
Coupling	$\Phi180 \times 40$ mm	
AC motor	JZTW52-4	

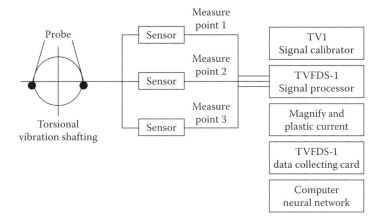

FIGURE 29.7　The architecture of the fault diagnosis system.

29.7　Experimental Results

29.7.1　Experimental Conditions

In the following experiments, the rotational speed is controlled within 30 to 500 rpm. The shafting load is changed by adjusting the fuel feed of the oil injection pump. There are four load conditions, which are idling (0.1 L/min), light load (0.2 L/min), medium load (0.3 L/min), and heavy load (0.42 L/min). In addition, there are three fuel feed states of the oil pump: normal feed, breaking the fuel feed every other cylinder, and breaking the fuel feed cylinder by cylinder.

29.7.2　Testing Methods

The fault of rotor imbalance is simulated by adding a mass at the flywheel edge. The fault of the bearing wear is simulated by adjusting the clearance of the bearing. The fault of out-of-alignment is simulated by adjusting the alignment of the crank and the intermediate shaft. The effect of the excitation force is tested by changing the fuel feed rate and the oil pressure.

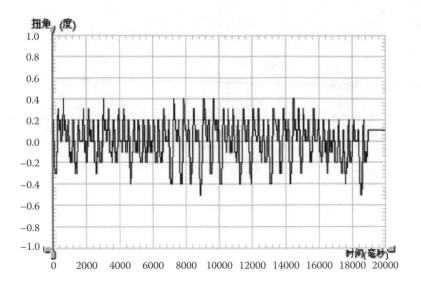

FIGURE 29.8   Waveform of torsional vibration under normal working conditions.

29.7.3   Experimental Results

29.7.3.1   Effect of the Imbalanced Rotor

Figures 29.8 and 29.9 present the torsional vibration waveform in channel #1 and its FFT frequency spectrum under stable idle conditions with a speed of 200 rpm and normal fuel feed. Here, the maximum torsional angle is 0.4°. Figures 29.10 and 29.11 present the torsional vibration waveform in the same channel and its corresponding FFT frequency spectrum under stable idle conditions with a speed of 200 rpm, normal fuel feed, and an imbalanced rotor simulated by attaching a mass of 2.5 kg to the flywheel edge. Here, the maximum torsional angle is 0.42°.

From Figures 29.8 through 29.11, it is clear that the imbalanced mass does not have a significant influence over the torsional vibration signal except for its amplitude in the low-frequency range ($f < 3.3$ Hz). This result agrees with standard conclusions. The difference in the low-frequency section is not caused by torsional vibration but by transverse vibration because of the imbalanced mass.

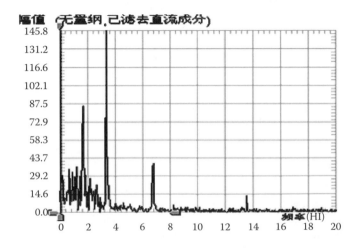

FIGURE 29.9   The FFT spectrum of torsional vibration under normal working conditions.

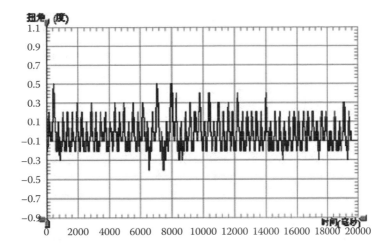

FIGURE 29.10 Waveform of torsional vibration with imbalanced rotor.

In Figure 29.12, the relationship between the torsional angle and the rotational speed with and without the imbalanced rotor are compared. From Figure 29.12, it is clear that the imbalanced mass does not play a major role in this type of situation.

29.7.3.2 Effect of the Excitation Force

The excitation force can be changed by adjusting the fuel feed rate of the oil pump, the duration angle of fuel injection, and the maximum injection pressure. Under working conditions, at a rotational speed of 250 rpm, idle load, and normal fuel feed, the waveform of the shafting torsional vibration and its FFT spectrum are measured. Its maximum torsional angle is 0.32°. Figures 29.13 and 29.14 show, respectively, the waveform of the shafting torsional vibration and its FFT spectrum at a rotational speed of 250 rpm, with a heavy load. In this case, the maximum torsion angle is 0.85°, which means that the vibration has increased nearly two times.

From Figures 29.13 and 29.14, it is clear that, because of higher injection pressure and heavier load, the shafting excitation force has increased and that it causes a much greater torsional vibration. It follows that the excitation force is the main cause of torsional vibration of the shafting.

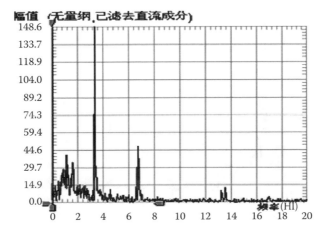

FIGURE 29.11 The FFT spectrum of torsional vibration with imbalanced rotor.

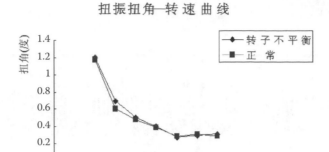

FIGURE 29.12   Relationship between torsional angle and rotational speed in the presence of an imbalanced rotor.

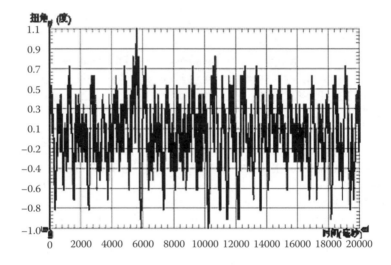

FIGURE 29.13   Waveform of torsional vibration with a heavy load.

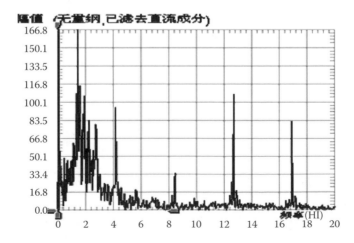

FIGURE 29.14   The FFT spectrum of torsional with a heavy load.

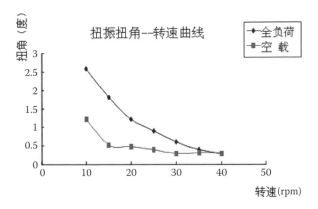

FIGURE 29.15 The relationship between torsional angle and rotational speed in the presence of excitation force.

How the excitation force affects the torsional vibration under different rotational speeds is shown in Figure 29.15.

29.7.3.3 The Influence of Oil Break in the Cylinders

Figure 29.16 shows how the oil break in every other cylinder affects the frequency spectrum of the shafting torsional vibration under working conditions at a rotational speed of 300 rpm, whereas Figure 29.17 shows how the oil break, cylinder by cylinder, affects the frequency spectrum of the shafting torsional vibration under the same operating conditions.

Table 29.2 presents the relationships between the rotational speed and the frequency spectrum obtained when the fuel injection pump of #1 cylinder breaks the oil. It can be concluded that the closer the rotational speed to the critical speed area, the greater the amplitude of torsional vibration.

When the fuel injection pump works at full load, normal fuel feed, and oil break in a single cylinder, the relationship between the amplitude of torsional vibration and the rotational speed is shown in Figure 29.18. When the rotational speed is close to the critical speed, the frequency spectrum of the torsional vibration is significantly influenced by the oil break.

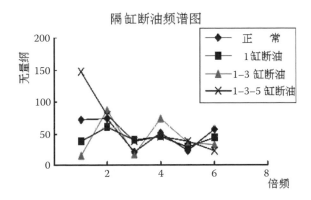

FIGURE 29.16 The FFT spectrum of torsional vibration due to oil break in every other cylinder.

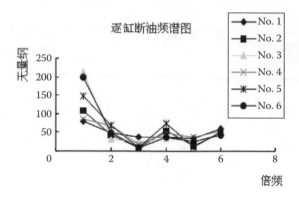

FIGURE 29.17   The FFT spectrum of torsional vibration due to oil break, cylinder by cylinder.

TABLE 29.2   Relationship between the Rotational Speed and the Frequency Spectrum
When the Oil in the #1 Cylinder is Cut Off

| Rotational Speed (rpm) | Magnitude of torsional vibration | | | | | |
| | dimensionless | | | | | |
	Base-Frequency f (Hz)	$2 \times f$ (Hz)	$3 \times f$ (Hz)	$4 \times f$ (Hz)	$5 \times f$ (Hz)	$6 \times f$ (Hz)
100	480	187	141	88.9	28.9	13.2
150	415	148.7	11.7	23.2	28.4	38.4
200	95.2	220.7	60.3	27.1	18.3	90.7
250	79.4	73.2	12.5	16.8	24.2	121.5
300	120.5	80.2	26.4	47.2	22.8	48.4
350	52.6	56.8	34.9	31.8	12.5	20.6
400	85.7	43.2	63.2	63.0	19.3	12.8

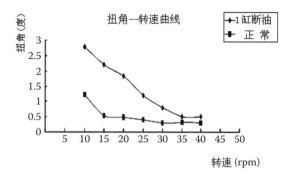

FIGURE 29.18   Relationship between the torsional angle and the rotational speed in the presence of an oil break in #1 cylinder.

29.8 Conclusion

Health monitoring, fault detection, and fault diagnosis are important for uninterrupted and high-quality performance of a mechatronic system. Mechanical vibration is a popular signature that is used for this purpose. The ship propulsion system is a mechatronic system. In this chapter, an online monitoring and fault diagnosis system for ship propulsion systems was presented. A neural network was employed to identify the features of torsional vibration of the shafting. A simulated test bed was developed in-house, which implemented stepless speed regulation, adjusting the excitation force continuously and simulating four typical faults including rotor imbalance and misalignment of the shafting. Experimental results were presented to validate the effectiveness of the developed approaches. Faults such as rotor imbalance, shaft deflection and bearing wear usually cause transverse vibrations in the shafting but have little influence on the torsional vibration. The excitation force is the primary cause of torsional vibration. The experimental results indicated that the characteristics of the simulated test bed agreed with those in a real ship, and they were consistent with the calculated results from theoretical analysis. This shows that the developed system is effective and feasible.

References

1. Chen, J., *Vibration Detection and Fault Diagnosis of Mechanical Equipment*, Shanghai Jiao Tong Press, Shanghai, China, 1999.
2. Feng, Z., Wang, Y., and Hu, Z., Kohonen network based fault diagnosis and its experiments, *Journal of Agricultural Machinery*, Vol. 33, No. 6, pp. 103–106, 2002.
3. Kuo, H., Wu, L., and Chen, J., Neural-fuzzy fault diagnosis in a marine propulsion shaft system, *Journal of Materials Processing Technology*, Vol. 122, 12–22, 2002.
4. Ciringione, J.L., Apparatus for Testing Torsional Vibration Dampers, U.S. Patent 3,054,284, 1962.
5. Zobrist, G.S., Torsional Exciter for a Rotating Structure, U.S. Patent 4,283,957, 1981.
6. Hao, Z., Li, J., and Xue, Y., Research and application of a simulated test-bed of torsional vibration of the diesel engine shafting, *Diesel Engine Engineering*, Vol. 12, No. 1, pp. 74–80, 1991.
7. Xu, M., Luo, Z., and Yan, J., *Vibration, Shock and Measurement of Ship Power Machinery*, Press of National Defense Industry, Beijing, China, 1981.
8. Wang, Q., *Torsional Vibration of the Diesel Engine Shafting*, Dalian Polytechnic Unversity Press, Dalian, China, 1991.
9. Liu, Z., Removal of rolling vibration in the diesel engine shafting, *Wuhan Shipbuilding*, No. 4, pp. 22–25, 1995.
10. Chen, X., Calculation and testing of torsional vibration of ship shafting, *Ship Engineering*, No. 1, pp. 22–26, 2002.
11. De Silva, C.W., *Vibration—Fundamentals and Practice*, 2nd ed., Taylor & Francis, CRC Press, Boca Raton, FL, 2006.
12. Karray, F. and de Silva, C.W., *Soft Computing and Intelligent Systems Design*, Addison-Wesley, New York, 2004.

Index